Functional Linear Algebra

Textbooks in Mathematics

Series editors:
Al Boggess, Kenneth H. Rosen

https://www.routledge.com/Textbooks-in-Mathematics/book-series/
CANDHTEXBOOMTH

Functional Linear Algebra

Hannah Robbins
Roanoke College

CRC Press
Taylor & Francis Group
Boca Raton London New York

CRC Press is an imprint of the
Taylor & Francis Group, an **informa** business
A CHAPMAN & HALL BOOK

First edition published 2021
by CRC Press
6000 Broken Sound Parkway NW, Suite 300, Boca Raton, FL 33487-2742

and by CRC Press
2 Park Square, Milton Park, Abingdon, Oxon, OX14 4RN

Library of Congress Cataloging-in-Publication Data

Names: Robbins, Hannah, 1980- author.
Title: Functional linear algebra / Hannah Robbins.
Description: First editon. | Boca Raton, FL : CRC Press, 2021. | Includes bibliographical references and index.
Identifiers: LCCN 2020046750 (print) | LCCN 2020046751 (ebook) | ISBN 9780367486877 (hardback) | ISBN 9781003042280 (ebook)
Subjects: LCSH: Algebras, Linear. | Functional analysis.
Classification: LCC QA184.2 .R6215 2021 (print) | LCC QA184.2 (ebook) | DDC 512/.5--dc23
LC record available at https://lccn.loc.gov/2020046750
LC ebook record available at https://lccn.loc.gov/2020046751

ISBN: 978-0-367-48687-7 (hbk)
ISBN: 978-1-003-04228-0 (ebk)

To all my linear algebra students: past, present, and future

and

To my linear algebra teacher: David Perkinson

Contents

Introduction for Students

Linear algebra occupies an important place in the world of math and science because it is an extremely versatile and useful subject. It rewards those of us who study it with powerful computational tools, lessons about how mathematical theory is built, examples for later study in other classes, and much more. Even if you think you know why you are studying linear algebra, I encourage you to learn about and appreciate those aspects of the course which don't seem immediately related to your original motivation. You may find that they enhance your general mathematical understanding, and you may need them for unexpected future applications. I initially loved linear algebra's lessons about how to generalize outward from familiar mathematical environments, but I have recently used tools from linear algebra to help a biology colleague with an application to her research.

As you work your way through this book, it is important to make sure you understand the basic ideas, definitions, and computational skills introduced. The best way to do this is to work through enough examples and problems to make sure you have thoroughly grasped the material. How many problems constitute "enough" will vary from person to person, but you'll know you're there when a type of problem or idea elicits boredom instead of confusion or stress. If you work your way through all of the exercises on a particular idea and still need more, I encourage you to look up that topic in other linear algebra books or online to find more problems.

The answers to the odd problems are in Appendix A.3. If your answers don't match up, I encourage you to seek help quickly so you can get straightened out before you move on. Math is not a subject which rewards an "I'll come back to that later" mentality! Realize that help is available in many places including your teacher, classmates, any tutoring resources available at your school, online tutorials, etc. Take the time to figure out which type of help works best for you – not everyone's brain responds the same way and it is much easier to work with your learning style rather than fight it.

Most of the computational techniques used in this book can be done either by hand or using technology. I encourage you to do enough work with each one by hand to understand its properties, and also to learn how to do it quickly using a calculator or computer software. This book specifically addresses how to use Mathematica, but feel free to use whichever technological tool best suits your needs.

Finally, welcome to linear algebra! I hope you find this book helpful and interesting.

Introduction for Instructors

A first course in linear algebra is fairly unique in that it may appear in different places within the mathematics curriculum depending on the department's planned progression of classes. This results in a wide array of first linear algebra courses, each of which caters to a different set of student circumstances. This book is (entirely selfishly) written for needs of the linear algebra class as it is taught at Roanoke College, where we teach a one-semester linear algebra course whose only prerequisite is first-year calculus. In particular, our students may not have taken an introduction to proofs class. With that student background in mind, I've made the following choices:

- For the vast majority of the book, I stick to vector spaces over $\mathbb{R}$. Complex vector spaces are briefly discussed in Appendix A.2.

- I freely use the label "Theorem," but proofs are called explanations or justifications to avoid causing students anxiety.

- More emphasis is placed on the idea of a linear function, which is used to motivate the study of matrices and their operations. This should seem natural to students after the central role of functions in calculus.

- Row reduction is moved further back in the semester and vector spaces are moved earlier to avoid an artificial feeling of separation between the computational and theoretical aspects of the course.

- Applications from a wide range of other subjects are introduced in Chapter 0 as motivation for students not intending to be math majors.

The chapters and sections are designed to be done in the order they appear, however, there are a few places where some judicious editing is possible.

- Section 2.9 on large matrix calculations can be skipped, except for the introduction of elementary matrices, which could be introduced where they next appear in Section 4.2.

- Sections 3.2 and 3.3 could be skipped, since students already have $m \times n$ matrices as an example of a vector space besides $\mathbb{R}^n$. However, this will require you to skip many later problems which include polynomial vector spaces and $\mathbb{C}$.

- Sections 5.1–5.3 depend only on the material from Chapter 1.

- Section 5.4 depends only on the material from Chapter 1 and the idea of a basis in $\mathbb{R}^n$, which could be taught there rather than in Section 3.1 if you wanted to move Chapter 5 earlier in the course.

I hope you find, as I do, that this book fills a gap in the literature and so is helpful to you and your students.

Acknowledgments

While I have wanted to write a linear algebra book for quite some time, I couldn't have completed this project without help from many people who definitely deserve to have their contributions recognized.

First of all, David Taylor deserves many, many thanks for not only supporting my sabbatical proposal, which gave me the time off to write, but also providing writing advice, formatting help, and reassurance when LaTeX was being uncooperative. Thanks also to Karin Saoub for all the lessons I learned watching her write her first book as well as her support when I was feeling stressed by the writing process. Roland Minton provided good feedback on my first draft. Steve Kennedy helped deepen the book's motivation of definitions and encouraged me to provide a more geometric approach. Maggie Rahmoeller taught out of a draft version of the book and provided invaluable feedback as well as reassurance that another instructor could appreciate my book's approach.

As I moved outside my area of expertise, many people helped inspire and check my motivating examples: Rachel Collins introduced me to the use of matrices in population modeling in biology, Skip Brenzovich and Gary Hollis explained how matrices are used to model molecular structures in chemistry, Adam Childers helped me find applications of matrices in statistics, Chris Santacroce talked with me about coordinate systems in machining, and Alice Kassens and Edward Nik-khah discussed how matrices are used in economics modeling.

Thanks to Bob Ross for helping me think about how to position the book in the marketplace and shepherding me through the publication process.

Finally, a special thank you to my Fall 2018 linear algebra students, in particular Gabe Umland, who provided the first field test of this book, and found many typos and arithmetic errors.

0

Introduction and Motivation

Linear algebra is widely used across many areas of math and science. It simultaneously provides us with a great set of tools for solving many different types of problems and gives us a chance to practice the important art of mathematical generalization and theory building. In this initial section, we'll do both of these things. First we'll explore a variety of different problems which will all turn out to yield to linear algebra's methods. Next we'll pick out some of the common characteristics of these problems and use them to focus our mathematical inquiries in the next chapters. As we develop our linear algebra toolbox, we'll return to each of these applications to tackle these problems.

A common problem type in chemistry is balancing chemical reactions. In any given chemical reaction, an initial set of chemicals interact to produce another set of chemicals. During the reaction, the molecules of the input chemicals break down and their constituent atoms recombine to form the output chemicals. Since the atoms themselves are neither created nor destroyed during a reaction, we must have the same number of atoms present in the input and output chemicals. To balance a chemical reaction, we need to figure out how many molecules of each chemical (both input and output) are present so that the number of atoms balances, i.e., is the same in the inputs as in the outputs. This must be done for each type of atom present in the chemical reaction, and the complication is that the quantities of molecules we use must balance each type of atom simultaneously. To make this less abstract, let's look at an example reaction.

Example 1. When propane is burned, it combines with oxygen to form carbon dioxide and water. Propane molecules are composed of three carbon atoms and eight hydrogen atoms, so they are written as C_3H_8. Oxygen molecules each contain two atoms of oxygen, so they are written as O_2. Carbon dioxide molecules each contain one carbon atom and two oxygen atoms, so are written CO_2. Water molecules are made up of two hydrogen atoms and one oxygen atom (which is why some aspiring chemistry comedians call water dihydrogen monoxide), so are written H_2O. Using this notation, our chemical reaction can be written $C_3H_8 + O_2 \rightarrow CO_2 + H_2O$.

There are three different types of atoms involved in this reaction: carbon, hydrogen, and oxygen. For each molecule involved, we need to keep track of how many carbon, hydrogen, and oxygen atoms it has. It is important that we keep these quantities separate from each other since they record totally different properties of a molecule. For example, each carbon dioxide molecule contains 1 carbon atom, 0 hydrogen atoms, and 2 oxygen atoms.

If we have multiple copies of a molecule, we can calculate the number of atoms of each type by multiplying each number of atoms by the number of molecules present. For example, if our reaction produces 15 molecules of carbon dioxide, we know this can also be viewed as producing with 15 carbon atoms, 0 hydrogen atoms, and 30 oxygen atoms. However, we don't just produce carbon dioxide, we also produce water, each molecule of which contains 0 carbon atoms, 2 hydrogen atoms, and 1 oxygen atom. If our reaction produces 20 molecules of water, we've produced 0 carbon atoms, 40 hydrogen atoms, and 20 oxygen atoms.

However, our reaction will actually produce molecules of both carbon dioxide and water. If we produce 15 molecules of carbon dioxide and 20 molecules of water, we've produced 15 carbon atoms (15 from the carbon dioxide and 0 from the water), 40 hydrogen atoms (0 from the carbon dioxide and 40 from the water), and 50 oxygen atoms (30 from the carbon dioxide and 20 from the water).

To balance our reaction, we'd need to figure out quantities of each of our four different molecule types so that the number of each type of atom is the same before and after the reaction. In our example computation above, that would mean we'd need to find quantities of propane and oxygen molecules which together have 15 carbon atoms, 40 hydrogen atoms, and 50 oxygen atoms to use as our inputs since those are the quantities of atoms from our output molecules.

In physics, we often need to figure out the net effect of a group of forces acting on the same object. If we're thinking of our object as living in three dimensions (which is how most of us think about the world we live in), then each force can push on the object to varying degrees along each of our three axes. Let's agree to call these axes up/down, North/South, and East/West. Different forces can either reinforce each other if they are acting in the same direction along one or more of these axes, or they can cancel each other out if they are acting in opposite directions. The overall force acting on the object can be found by figuring out the combined effect of the separate forces along each of these axes. Again, let's look at an example.

Example 2. A person jumps upward and Northward against a wind which is pushing them Southward and Eastward while gravity pulls them downward. The force of their jump is 1000 Newtons up and 650 Newtons North, the wind's force is 200 Newtons South and 375 Newtons East, and the force of gravity is 735 Newtons down.

To figure out the overall force acting on this person, we need to combine the various forces' actions in each of our three directions. The first step is to realize that although we are describing the forces above as if North and South are separate, they are actually opposites. This means we need to treat each pair of directions as one directional axis and assign one of the two directions along that axis as positive and the other as negative. I'll stick with typical map conventions and assign North to be positive and South to be negative along the North/South axis, West as positive and East as negative along the East/West axis, and up as positive and down as negative along the up/down axis. With these conventions, we can restate the components of our three forces acting on this person. The person's jump has 1000 Newtons along the up/down axis, 650 Newtons along the North/South axis, and 0 Newtons along the East/West axis. The force of the wind has 0 Newtons along the up/down axis, −200 Newtons along the North/South axis, and −375 Newtons along the East/West axis. The force of gravity has 0 Newtons along the up/down axis, 0 Newtons along the North/South axis, and −735 Newtons along the East/West axis.

Now we can combine the various components of each force, but we must be careful to add them up separately for each directional axis. Thus the total force on the person is 265 Newtons along the up/down axis (1000 from the jump, 0 from the wind, and −735 from gravity), 450 Newtons along the North/South axis (650 from the jump, −200 from the wind, and 0 from gravity), and −375 Newtons along the East/West axis (0 from the jump, −375 from the wind, and 0 from gravity). Eventually we'll develop tools to do this not just along the standard axes, but along any set of axes at right angles to each other.

In economics, it is important to model the relationships between different types of resources in the production process. To manufacture or produce a given product requires different amounts of various resources, possibly even including some of the same product being produced. (Think of the seed corn required to grow the next corn crop.) In order to produce certain quantities of a variety of different products, we will need to calculate how much of each of the input resources are required for our production process. An example of this is given below.

Example 3. A tortilla factory manufactures both corn and flour tortillas. It takes 2 cups of cornmeal, 1.5 cups of water, and 0.5 teaspoons of salt to make 24 corn tortillas, and 2 cups of flour, 0.75 cups of water, 0.25 cups of oil, and 1 teaspoon of salt to make 20 flour tortillas.

If the factory wants to make 100 flour tortillas and 200 corn tortillas, they will need to calculate the amount of each ingredient to order. One way to do this is to divide the ingredient quantities in each recipe by the number of tortillas that recipe produces (24 for corn tortillas and 20 for flour tortillas) and then multiply by the number of tortillas to be produced (200 for corn tortillas and 100 for flour tortillas). This can also be done in one step using a fraction whose numerator is the desired number of tortillas and whose denominator is the number of tortillas produced by that recipe ($\frac{200}{24}$ for corn tortillas and $\frac{100}{20}$ for flour tortillas).

Multiplying the quantities in the corn tortilla recipe by $\frac{200}{24}$ shows us that the factory will need $16 \frac{2}{3}$ cups of cornmeal, 12.5 cups of water, and $4 \frac{1}{6}$ teaspoons of salt to make 200 corn tortillas. Multiplying the quantities in the flour tortilla recipe by $\frac{100}{20}$ shows that the factory will need 10 cups of flour, 3.75 cups of water, 1.25 cups of oil, and 5 teaspoons of salt to make 100 flour tortillas.

Combining the quantities of the ingredients needed for each type of tortilla, this means that to make 100 flour tortillas and 200 corn tortillas the factory will need $16 \frac{2}{3}$ cups of cornmeal, 10 cups of flour, 16.25 cups of water, 1.25 cups of oil, and $9 \frac{1}{6}$ teaspoons of salt.

In statistics, a tool called regression is often used to model the relationship between variables so that the value of one variable, y, (often called the response variable) can be predicted based on the values of a set of other variables $x_1, \ldots, x_p$. In multiple linear regression, we want to use a data set to find a set of constant coefficients $\beta_0, \beta_1, \ldots, \beta_p$ so that $y = \beta_0 + \beta_1 x_1 + \cdots \beta_p x_p$. The number β_i describes the effect of x_i on y if the other x variables are held constant. The constant term β_0 is the value of y when all the other variables equal zero. The goal of this model is to choose the values of the β's which give us the most accurate predictions for the value of y.

Example 4. A group of scientists wanted to predict people's VO$_2$max (a measure of peak oxygen consumption during exercise) based on their age, weight, heart rate (during low-intensity exercise), and gender. Since they wanted to predict VO$_2$max, that is their response variable y. The other variables are the x_i's, so we can assign x_1 to be age, x_2 to be weight, x_3 to be heart rate, and x_4 to be gender.

The values of all these variables were recorded for 100 people. The scientists found the following regression equation based on their data.

$$y = 87.83 - 0.165x_1 - 0.385x_2 - 0.118x_3 + 13.208x_4$$

Examining the first coefficient $\beta_1 = -0.165$, we can say that if a person's weight and heart rate remain the same but they are a year older, then their VO$_2$max decreases by 0.165.

In computer graphics, the world is currently restricted to our two-dimensional screen. The geometry of this 2D space is usually mapped out in terms of a point's horizontal and vertical position. In many computer games the screen gives our point of view as we move through the world. If we move straight ahead, this means the screen needs to give us the feeling of zooming into the center point of the screen to produce the impression that we're traveling toward that point. This can be done by moving the position of everything on the screen outward at the same rate. One way to do this is to multiply both the horizontal and vertical positions by the same number. One such example is explored below.

Example 5. Suppose we have an object on the computer screen whose horizontal position is -5 and whose vertical position is 12.

If we multiply both the horizontal and vertical positions by 2, we get a horizontal position of -10 and a vertical position of 24. To see how this has moved our point, let's look at a picture.

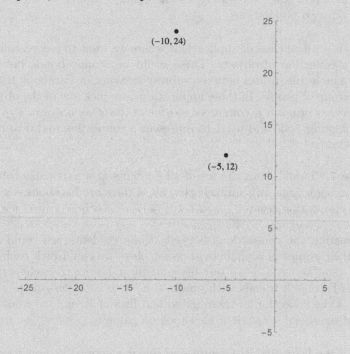

Both of these points are on the same line through the origin, but the point where we multiplied the coordinates by 2 is much farther out. You can imagine that if you did this to every point in the plane at once it would produce the feeling of moving toward the center.

In biology there is a class of problems where we want to model a population of organisms over time to see whether that population is growing or declining. This is usually done by dividing the life cycle of an organism up into different stages and observing both the number of organisms present in each stage and the probability that an organism moves from a given stage into each other stage. This data is then used to extrapolate the population over long stretches of time to see if it is sustainable and survives or is not sustainable and shrinks away to extinction.

Example 6. The smooth coneflower is a native Virginian plant. It's life cycle can be divided into five stages: seedling, small plant, medium plant, large plant, and flowering plant. Data was collected for two different geographically isolated populations of coneflowers.

For each of these two populations, there were different numbers of plants in each of the five stages of growth. For example, in one population there were 88 large plants and in the other population there were only 16. If we want to combine data about these two populations, we need to combine the numbers for each life cycle stage separately. In the example above we have 104 large plants in our two populations.

There is a whole class of applications where we want to record connections between a collection of objects. These could be atomic bonds between the atoms in a molecule, routes between subway stations, or Facebook friendships within a group of people. In these applications, we pick one of the objects and record whether or not it is connected to each of the other objects. Typically we mirror computer code and use 1 to represent a connection and 0 to represent no connection.

Example 7. Think about the set of all students at a particular college who have a Facebook page. It is unlikely that all of them are Facebook friends with everyone else in this group.

To describe the connections between these students, we could start by putting their names in alphabetical order. Next we could pick each of them one at a time and run down our list putting a 1 next to each person that student is Facebook friends with, and a 0 next to each person they are not. Looking at the collection of these individual lists of 1s and 0s, would give us a sense of the social network of Facebook on campus.

One of the least typical examples of linear algebra I've come across is a scene in Neal Stephenson's book *Cryptonomicon*. In it, several siblings gather after their mother's death to divide up their parents' belongings. One brother is a mathematician and he explains that each person will be allowed to rank each item based on two quantities: their perception of its monetary value and their emotional attachment to it. These are represented visually as a 2D grid. After each sibling has assigned their values to every object, the objects will be divided. Using their valuation, each sibling can now compute their perception of the total monetary and emotional value of their share versus the share of each other sibling. The goal is to divide up the objects so that each person thinks that their share is worth at least as much (both monetarily and emotionally) as everyone else's shares.

Example 8. Suppose one person receives a table, a car, and a set of dishes as their share. In monetary terms, they think the table is worth $450, the car is worth $2000, and the dishes are worth $25. In emotional terms, they've assigned the table a value of 60, the car a value of 5, and the dishes a value of 100.

This person has assigned each object a pair of numbers to represent its monetary and emotional value. The table is worth $450 monetarily and 60 emotionally, the car is worth $2000 monetarily and 5 emotionally, and the dishes are worth $25 monetarily and 100 emotionally. This person feels their share has a total monetary value of $450 + $2000 + $25 = $2475 and a total emotional value of 60 + 5 + 100 = 165.

To decide whether or not this person is satisfied with this division, we'd need to use their valuations of the objects that make up each other person's share to make sure that they don't believe anyone else's share was worth more than $2475 or 165 in emotional value.

Now that we've seen a variety of different examples, let's shift our focus to looking for common characteristics between these problems to help us decide where to start our exploration of linear algebra in Chapter 1.

Although the examples in this section may not seem closely related, they have certain underlying similarities. In each problem we had several different numbers associated to a given object. These numbers meant separate things or described separate qualities about that object, so needed to be kept separate from each other. For example, when we looked at forces acting on a person we kept the components of our forces in the up/down, North/South, and East/West directions separate from each other.

In several of these problems we needed to be able to multiply all numbers associated with a particular object by a constant. For example, if we used 15 water molecules (which are each made of hydrogen and oxygen) in a chemical reaction then we'd multiply both the number of hydrogens and the number of oxygens by 15.

We also needed to add the collections of numbers associated to different objects together, but in such a way that the result was a collection of numbers where we only combined numbers which meant the same thing. For example, when adding a sibling's valuation of several objects together we'd need to add up their monetary values and emotional values separately to get a monetary total and an emotional total.

Noticing this pattern is our first step in the process of mathematical generalization which was mentioned at the beginning of this section. It allows us to strip away the differences between these various example applications to focus on the underlying similarities we need to study and understand. In the next chapter we'll develop a way to write down these collections of associated numbers, figure out how to add two such collections, and how to multiply such a collection by a constant. Once we've solidified our understanding of these basic processes, we can tackle each of these example problems using the same basic mathematical tools.

1

Vectors

1.1 Vector Operations

As we saw in Chapter 0, we often have several separate numeric values associated with the same data point. Vectors provide us with a way to simultaneously record all of those quantities in a compact way. Whether we are looking at an object's emotional and monetary value or its position in different directions, a vector quickly sums things up by providing multiple descriptive numbers and using the position of those numbers within the vector to tell us what each one means. We will start with the most basic definition of a vector below.

Definition. A *vector* is an ordered column of real numbers written $\begin{bmatrix} v_1 \\ v_2 \\ \vdots \\ v_n \end{bmatrix}$.

To help distinguish which variables represent vectors, I'll write them with arrows over the top of the variable as in $\vec{v}$.

Notice that this definition doesn't tell us what the various positions inside the vector represent. This is deliberate, because it allows us the flexibility to assign the appropriate meanings for our current use. However, it does mean we'll need to get into the habit of clearly communicating the meaning of our vector's entries in each application we do.

Example 1. Write down the vector which records the number of carbon, hydrogen and oxygen atoms in a molecule of carbon dioxide (CO_2).

We saw in Example 1 from Chapter 0 that a carbon dioxide molecule contains 1 carbon atom, 0 hydrogen atoms, and 2 oxygen atoms. There are several possible ways to put these three numbers into a vector, depending on the order of the entries. However, if we standardize our notation so that the first entry counts carbon atoms, the second hydrogen, and the third oxygen, we get that carbon dioxide's atomic vector is

$$\vec{v} = \begin{bmatrix} 1 \\ 0 \\ 2 \end{bmatrix}.$$

Example 2. The vector $\vec{v} = \begin{bmatrix} 1 \\ 2 \\ 3 \end{bmatrix}$ can be thought of as describing a position in 3-space.

Points in 3-space are usually described using three axes (often denoted by x, y, and z). Our vector $\vec{v}$ describes the point we get to by starting at the origin where the three axes meet and moving 1 unit along the first axis (usually x), 2 units along the second axis (usually y), and 3 units along the third axis (usually z). Notice here that as in Example 1, both the entries of the vector and their positions within the vector are important. Here each entry tells us how many units to move, and its position in the vector tells us which axis to move along.

Geometrically, we will draw our vectors as arrows that start at the origin and end at the point in the plane or 3-space whose coordinates match the entries of the vector as in Figure 1.1. (In some other situations, vectors may instead be drawn between any two points in the plane or in 3-space.)

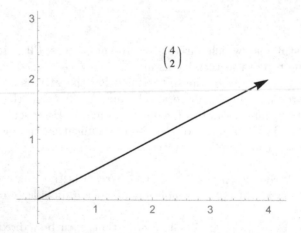

Figure 1.1: A geometric representation of a vector

This often gives rise to the geometric description of a vector as having both a length (or magnitude) and a direction. This interpretation of vectors also shows up in physics where they are used to model forces acting on objects. We can expand this geometric perspective from a single vector to a set of vectors.

Example 3. The set of all 2-vectors is the plane, $\mathbb{R}^2$, where $\mathbb{R}$ stands for the set of real numbers.

We can write this more explicitly as $\mathbb{R}^2 = \left\{ \begin{bmatrix} x_1 \\ x_2 \end{bmatrix} \right\}$. Thinking of these vectors geometrically as in Example 2, we can see that this set of vectors corresponds to the set of points in the plane. If we rewrite our 2-vectors as $\mathbb{R}^2 = \left\{ \begin{bmatrix} x \\ y \end{bmatrix} \right\}$, it becomes even clearer that this is the familiar 2D space we use to graph one variable functions.

Similarly, using the interpretation of the entries from Example 2, the set of all 3-vectors is $\mathbb{R}^3$. We can expand this to the set of all vectors of any fixed size.

Definition. The set of all vectors with n entries is $\mathbb{R}^n$.

I will often refer to a vector with n entries as an "n-vector". The dash makes it clear that we're talking about the number of entries in the vector rather than the number of vectors involved. For example, saying "suppose we have two 3-vectors" means that we have two vectors which each have three entries, i.e., two vectors from $\mathbb{R}^3$.

Because both the numbers in a vector and the positions of those numbers matter greatly for the vector's meaning, two vectors are considered equal only when each corresponding pair of entries is equal. Since vectors from $\mathbb{R}^n$ and $\mathbb{R}^m$ with $n \neq m$ don't have the same number of entries, they are automatically not equal.

Example 4. $\vec{v} = \begin{bmatrix} -1 \\ 0 \\ 3 \end{bmatrix}$ equals $\vec{w} = \begin{bmatrix} -1 \\ 0 \\ 3 \end{bmatrix}$.

Both $\vec{v}$ and $\vec{w}$ are in $\mathbb{R}^3$, so it is reasonable to ask whether they are equal. Starting at the top and moving down $\vec{v}$ and $\vec{w}$, we can see that each pair of their corresponding entries matches up. Therefore $\vec{v} = \vec{w}$.

Example 5. $\vec{v} = \begin{bmatrix} -1 \\ 0 \\ 3 \end{bmatrix}$ is not equal to $\vec{w} = \begin{bmatrix} -1 \\ 3 \\ 0 \end{bmatrix}$.

Again, these are both 3-vectors, so it's reasonable to ask if they are equal. Their first entries do match up, but their second and third entries do not. (Remember that it is important not only that the vectors contain the same numeric entries, but that they have those numbers in the same positions.)

Our motivating examples in Chapter 0 involved more than just vectors, since in Examples 1 and 5 we wanted to multiply a vector by some real number. To distinguish between vectors and real numbers, we make the following definition.

Definition.   A number r in $\mathbb{R}$ is also called a *scalar*.

Thus -5 is a scalar while $\begin{bmatrix} -5 \\ 0 \end{bmatrix}$ is a vector.

Scalars will be represented by variables without arrows over them, so a is a scalar while $\vec{a}$ is a vector.

In Chapter 0 we saw several problems where we were interested in adding vectors or multiplying vectors by scalars, so let's work out how this type of addition and multiplication should work. As we define these operations, we need to keep in mind the properties we want them to have. The biggest advantage of using vectors to express information is that different entries can describe different numeric properties. This means we want our vector operations to respect the position of a vector's entries. We also only want to combine like entries together, so that the resulting vector's information can be interpreted in the same way as the original vectors' entries.

Let's start with vector addition. Suppose we want to use vector addition to find the number of atoms of carbon, hydrogen, and oxygen contained in a molecule of carbon dioxide and a molecule of water. Let's use the same notation as in Example 1 so that our vector's entries will record the number of carbon, hydrogen, and oxygen atoms in that order. We saw in Example 1 that carbon dioxide's atomic vector was $\begin{bmatrix} 1 \\ 0 \\ 2 \end{bmatrix}$. Water is H_2O, so each water molecule contains 0 carbon atoms, 2 hydrogen atoms, and 1 oxygen atom giving us the atomic vector $\begin{bmatrix} 0 \\ 2 \\ 1 \end{bmatrix}$. Our notion of vector addition should add together the pair of numbers representing carbon and place that sum into the entry which represents carbon, and similarly for hydrogen and oxygen.

Applying this pattern to the general case of two n-vectors, we get the following definition.

Definition.   Let $\vec{v} = \begin{bmatrix} v_1 \\ v_2 \\ \vdots \\ v_n \end{bmatrix}$ and $\vec{w} = \begin{bmatrix} w_1 \\ w_2 \\ \vdots \\ w_n \end{bmatrix}$ be vectors in $\mathbb{R}^n$. We define

$$\vec{v} + \vec{w} = \begin{bmatrix} v_1 + w_1 \\ v_2 + w_2 \\ \vdots \\ v_n + w_n \end{bmatrix}.$$

We need to add the condition here that $\vec{v}$ and $\vec{w}$ are both in the same $\mathbb{R}^n$ to make sure that they have the same number of entries. Otherwise, it is impossible to add their entries pairwise since some of the entries in the longer vector won't have corresponding entries in the shorter vector.

Example 6. Use vector addition to find the atomic vector of a molecule of carbon dioxide and a molecule of water.

From the discussion before our definition of vector addition, we know carbon dioxide's atomic vector is $\begin{bmatrix} 1 \\ 0 \\ 2 \end{bmatrix}$ and water's atomic vector is $\begin{bmatrix} 0 \\ 2 \\ 1 \end{bmatrix}$, where our vector entries represent carbon, hydrogen, and oxygen (in that order).

The combined atomic vector of these two molecules can be found by adding their individual atomic vectors. This gives us

$$\begin{bmatrix} 1 \\ 0 \\ 2 \end{bmatrix} + \begin{bmatrix} 0 \\ 2 \\ 1 \end{bmatrix} = \begin{bmatrix} 1+0 \\ 0+2 \\ 2+1 \end{bmatrix} = \begin{bmatrix} 1 \\ 2 \\ 3 \end{bmatrix}$$

which means that together these two molecules contain 1 carbon atom, 2 hydrogen atoms, and 3 oxygen atoms.

Notice that our definition of addition did exactly what we asked: it combined only pairs of numbers recording the same type of atom and placed each sum into the appropriate spot in the vector for that atom.

Example 7. Explain why it is impossible to compute $\vec{v}+\vec{w}$ where $\vec{v} = \begin{bmatrix} 2 \\ -10 \\ 13 \end{bmatrix}$ and $\vec{w} = \begin{bmatrix} -5 \\ 8 \end{bmatrix}$.

Since $\vec{v}$ is a 3-vector and $\vec{w}$ is a 2-vector we can't do this addition, because we don't have $\vec{v}$ and $\vec{w}$ both in $\mathbb{R}^n$ for a single n. Another way to think about this is that the third entry of $\vec{v}$ has no corresponding entry to be added to in $\vec{w}$.

In $\mathbb{R}^2$, vector addition has a striking geometric pattern. If we write our 2-vectors in the form $\begin{bmatrix} x \\ y \end{bmatrix}$, we know that the x-coordinate of our sum is the sum of the x-coordinates of the vectors being added and similarly for the y-coordinates. However, if we draw a parallelogram from the two vectors we're adding, then their sum forms that parallelogram's diagonal as shown in Figure 1.2.

We can do something similar in $\mathbb{R}^3$, but it is too complicated to draw in a 2D book.

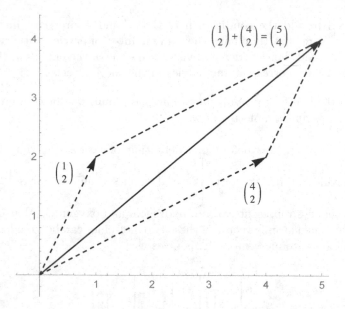

Figure 1.2: Adding vectors geometrically

Now that we've explored vector addition, let's turn our attention to multiplying a vector by a scalar. Suppose we want to count the numbers of the various types of atoms in 15 molecules of carbon dioxide. We can do this by multiplying the number of carbon, hydrogen, and oxygen atoms in one molecule by 15 and placing the results into the appropriate entries of our new vector.

This motivates the following general definition for multiplying an n-vector by a scalar.

Definition. If $\vec{v} = \begin{bmatrix} v_1 \\ v_2 \\ \vdots \\ v_n \end{bmatrix}$ is a vector in $\mathbb{R}^n$ and r is a scalar, then $r \cdot \vec{v} = \begin{bmatrix} rv_1 \\ rv_2 \\ \vdots \\ rv_n \end{bmatrix}$.

Example 8. Compute $15 \begin{bmatrix} 1 \\ 0 \\ 2 \end{bmatrix}$.

To do this, we multiply each entry of our vector by 15 which gives us

$$15 \begin{bmatrix} 1 \\ 0 \\ 2 \end{bmatrix} = \begin{bmatrix} 15(1) \\ 15(0) \\ 15(2) \end{bmatrix} = \begin{bmatrix} 15 \\ 0 \\ 30 \end{bmatrix}.$$

Note that our vector is the atomic vector for carbon dioxide from Example

1. This means our computation above gives us the atomic vector for 15 molecules of carbon dioxide.

As with vector addition, there is a nice geometric interpretation of scalar multiplication in $\mathbb{R}^2$. Since the pattern of scalar multiplication is slightly more complicated to describe than the pattern of addition, we'll start by exploring its pattern in the next two examples.

Example 9. Compute $4\vec{v}$ and explain geometrically what this scalar multiplication did to $\vec{v}$, where $\vec{v} = \begin{bmatrix} 3 \\ 1 \end{bmatrix}$.

Computing $4\vec{v}$ means multiplying each entry of $\vec{v}$ by 4. This gives us

$$4\vec{v} = \begin{bmatrix} 4(3) \\ 4(1) \end{bmatrix} = \begin{bmatrix} 12 \\ 4 \end{bmatrix}.$$

To see what this looks like geometrically, let's plot $\vec{v}$ and $4\vec{v}$ together.

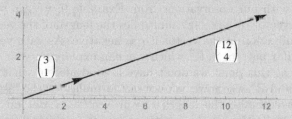

Here we can see that $4\vec{v}$ and $\vec{v}$ lie along the same line in $\mathbb{R}^2$, but $4\vec{v}$ is 4 times as long as $\vec{v}$.

As we saw in the previous example, multiplying a vector by a positive scalar doesn't change the direction it points, but does change its length. In fact I've often wondered if scalars got their name because multiplying a vector by a positive real number scales the length of the vector by that number. On the other hand if our scalar is negative, we get a slightly different picture.

Example 10. Compute $-\frac{1}{2}\vec{v}$ and explain geometrically what this scalar multiplication did to $\vec{v}$, where $\vec{v} = \begin{bmatrix} 3 \\ 1 \end{bmatrix}$.

As in Example 9, computing $-\frac{1}{2}\vec{v}$ means multiplying each entry of $\vec{v}$ by $-\frac{1}{2}$. This means

$$-\frac{1}{2}\vec{v} = \begin{bmatrix} -\frac{1}{2}(3) \\ -\frac{1}{2}(1) \end{bmatrix} = \begin{bmatrix} -\frac{3}{2} \\ -\frac{1}{2} \end{bmatrix}.$$

Again, let's plot $\vec{v}$ and $-\frac{1}{2}\vec{v}$ together to get a visual idea of what this looks like.

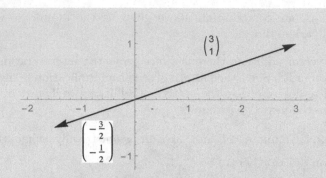

These two vectors still lie on the same line in $\mathbb{R}^2$, but they are pointing opposite directions along that line. In addition to pointing the opposite way, our new vector $-\frac{1}{2}\vec{v}$ is also half the length of $\vec{v}$.

From the picture in the example above, we can see that the only difference when our scalar is negative is that our vector's direction is switched. Putting this together with our observations from Example 9, we can say that in $\mathbb{R}^2$ multiplying a vector by a scalar r multiplies the length of the vector by $|r|$ and switches the direction of the vector if r is negative. As with vector addition, there is a similar pattern in $\mathbb{R}^3$ which you can explore on your own.

Note that at this point we don't have a good idea of what length means except in $\mathbb{R}^2$ and $\mathbb{R}^3$ where we can use our formulas for the distance between two points to compute the length of our vectors. Later on in Chapter 5 we'll create a definition of the length of a vector which doesn't depend on geometry and hence can be extended to higher dimensions. However, we can still compute the product of a vector and a scalar for any size of vector even if we can't visualize its geometric effect.

Now that we've defined these two vector operations, let's explore some of their nice properties. You'll notice that many of these properties are similar to what we have in $\mathbb{R}$, which means we can use our intuition about how addition and multiplication work in the real numbers. This is an important consideration, because whenever a newly defined operation shares its name with a familiar operation, it is tempting to assume they behave exactly the same. This is not always true, so it pays to check carefully! We'll start with properties of vector addition.

From the definition of vector addition, we know that if $\vec{v}$ and $\vec{w}$ are both in $\mathbb{R}^n$, their sum $\vec{v} + \vec{u}$ is also in the same $\mathbb{R}^n$. We will call this property closure of addition.

If we have $\vec{v}$ and $\vec{w}$ in $\mathbb{R}^n$, then $\vec{v} + \vec{w} = \vec{w} + \vec{v}$. To see this, we can compute

$$\vec{v} + \vec{w} = \begin{bmatrix} v_1 + w_1 \\ v_2 + w_2 \\ \vdots \\ v_n + w_n \end{bmatrix}.$$

Since addition of real numbers is commutative (i.e., order doesn't matter), we know $v_i + w_i = w_i + v_i$ for $i = 1, \ldots, n$. This means

$$\begin{bmatrix} v_1 + w_1 \\ v_2 + w_2 \\ \vdots \\ v_n + w_n \end{bmatrix} = \begin{bmatrix} w_1 + v_1 \\ w_2 + v_2 \\ \vdots \\ w_n + v_n \end{bmatrix}.$$

Since the right-hand side is $\vec{w} + \vec{v}$, we have $\vec{v} + \vec{w} = \vec{w} + \vec{v}$. This means vector addition is commutative.

If we have $\vec{v}$, $\vec{u}$, and $\vec{w}$ in $\mathbb{R}^n$, then $(\vec{v} + \vec{u}) + \vec{w} = \vec{v} + (\vec{u} + \vec{w})$. To see this, we compute

$$(\vec{v} + \vec{u}) + \vec{w} = \left(\begin{bmatrix} v_1 + u_1 \\ v_2 + u_2 \\ \vdots \\ v_n + u_n \end{bmatrix} \right) + \begin{bmatrix} w_1 \\ w_2 \\ \vdots \\ w_n \end{bmatrix} = \begin{bmatrix} (v_1 + u_1) + w_1 \\ (v_2 + u_2) + w_2 \\ \vdots \\ (v_n + u_n) + w_n \end{bmatrix}.$$

Since addition of real numbers is associative (i.e., where we put our parentheses in addition doesn't matter), we know $(v_i + u_i) + w_i = v_i + (u_i + w_i)$ for $i = 1, \ldots, n$. This means

$$\begin{bmatrix} (v_1 + u_1) + w_1 \\ (v_2 + u_2) + w_2 \\ \vdots \\ (v_n + u_n) + w_n \end{bmatrix} = \begin{bmatrix} v_1 + (u_1 + w_1) \\ v_2 + (u_2 + w_2) \\ \vdots \\ v_n + (u_n + w_n) \end{bmatrix}$$

and since

$$\begin{bmatrix} v_1 + (u_1 + w_1) \\ v_2 + (u_2 + w_2) \\ \vdots \\ v_n + (u_n + w_n) \end{bmatrix} = \begin{bmatrix} v_1 \\ v_2 \\ \vdots \\ v_n \end{bmatrix} + \left(\begin{bmatrix} u_1 + w_1 \\ u_2 + w_2 \\ \vdots \\ u_n + w_n \end{bmatrix} \right) = \vec{v} + (\vec{u} + \vec{w})$$

we get $(\vec{v} + \vec{u}) + \vec{w} = \vec{v} + (\vec{u} + \vec{w})$ as claimed. Therefore vector addition is associative.

In the real numbers, we have 0 as our additive identity, i.e., $0 + r = r$ for any real number r. In $\mathbb{R}^n$, our additive identity is the n-vector whose entries are all 0, which we'll write as $\vec{0}$. For any $\vec{v}$ in $\mathbb{R}^n$, we have

$$\vec{v} + \vec{0} = \begin{bmatrix} v_1 \\ v_2 \\ \vdots \\ v_n \end{bmatrix} + \begin{bmatrix} 0 \\ 0 \\ \vdots \\ 0 \end{bmatrix} = \begin{bmatrix} v_1 + 0 \\ v_2 + 0 \\ \vdots \\ v_n + 0 \end{bmatrix} = \begin{bmatrix} v_1 \\ v_2 \\ \vdots \\ v_n \end{bmatrix} = \vec{v}.$$

This shows $\vec{0}$ is the additive identity of vector addition.

In the real numbers, every number r has an additive inverse, $-r$, so that $-r + r = 0$. For every $\vec{v}$ in $\mathbb{R}^n$ we have the additive inverse $-\vec{v} = (-1) \cdot \vec{v}$ so that $-\vec{v} + \vec{v} = \vec{0}$. We can check this by computing

$$-\vec{v} + \vec{v} = \begin{bmatrix} -v_1 \\ -v_2 \\ \vdots \\ -v_n \end{bmatrix} + \begin{bmatrix} v_1 \\ v_2 \\ \vdots \\ v_n \end{bmatrix} = \begin{bmatrix} -v_1 + v_1 \\ -v_2 + v_2 \\ \vdots \\ -v_n + v_n \end{bmatrix} = \begin{bmatrix} 0 \\ 0 \\ \vdots \\ 0 \end{bmatrix} = \vec{0}.$$

Thus vector addition has additive inverses. (Notice that we set $-\vec{v} + \vec{v}$ equal to $\vec{0}$, because $\vec{0}$ is the additive identity of $\mathbb{R}^n$. This mirrors how $-r + r = 0$ in $\mathbb{R}$, because 0 is the additive identity of $\mathbb{R}$.)

Now that we've explored vector addition, let's turn our attention to scalar multiplication of vectors.

From our definition of scalar multiplication, we know that if $\vec{v}$ is in $\mathbb{R}^n$ then $r \cdot \vec{v}$ is also in $\mathbb{R}^n$. We call this property closure of scalar multiplication.

If we have $\vec{v}$ in $\mathbb{R}^n$ and two scalars r and s, then $r \cdot (s \cdot \vec{v}) = (rs) \cdot \vec{v}$. To see this, we compute

$$r \cdot (s \cdot \vec{v}) = r \left(\begin{bmatrix} sv_1 \\ sv_2 \\ \vdots \\ sv_n \end{bmatrix} \right) = \begin{bmatrix} rsv_1 \\ rsv_2 \\ \vdots \\ rsv_n \end{bmatrix}.$$

Pulling a factor of rs out of each entry on the right-hand side gives us

$$\begin{bmatrix} rsv_1 \\ rsv_2 \\ \vdots \\ rsv_n \end{bmatrix} = (rs) \begin{bmatrix} v_1 \\ v_2 \\ \vdots \\ v_n \end{bmatrix} = (rs) \cdot \vec{v}.$$

This means scalar multiplication of vectors is associative.

In the real numbers, our multiplicative identity is 1, because for any real number r we have $1 \cdot r = r$. This is also true for multiplication of vectors by scalars, since for any $\vec{v}$ in $\mathbb{R}^n$ and scalar r we have

$$1 \cdot \vec{v} = \begin{bmatrix} 1 \cdot v_1 \\ 1 \cdot v_2 \\ \vdots \\ 1 \cdot v_n \end{bmatrix} = \begin{bmatrix} v_1 \\ v_2 \\ \vdots \\ v_n \end{bmatrix} = \vec{v}.$$

Thus 1 is the identity for scalar multiplication of vectors.

To finish up our exploration of the properties of addition and scalar multiplication of vectors, let's see how these two operations interact with each other.

If $\vec{v}$ is in $\mathbb{R}^n$ and r and s are scalars, then $(r+s) \cdot \vec{v} = r \cdot \vec{v} + s \cdot \vec{v}$. To see this we compute

$$(r+s) \cdot \vec{v} = \begin{bmatrix} (r+s)v_1 \\ (r+s)v_2 \\ \vdots \\ (r+s)v_n \end{bmatrix} = \begin{bmatrix} rv_1 + sv_1 \\ rv_2 + sv_2 \\ \vdots \\ rv_n + sv_n \end{bmatrix} = \begin{bmatrix} rv_1 \\ rv_2 \\ \vdots \\ rv_n \end{bmatrix} + \begin{bmatrix} sv_1 \\ sv_2 \\ \vdots \\ sv_n \end{bmatrix}.$$

Factoring r out of each entry of the first vector and s out of each entry of the second vector gives us

$$\begin{bmatrix} rv_1 \\ rv_2 \\ \vdots \\ rv_n \end{bmatrix} + \begin{bmatrix} sv_1 \\ sv_2 \\ \vdots \\ sv_n \end{bmatrix} = r \begin{bmatrix} v_1 \\ v_2 \\ \vdots \\ v_n \end{bmatrix} + s \begin{bmatrix} v_1 \\ v_2 \\ \vdots \\ v_n \end{bmatrix} = r \cdot \vec{v} + s \cdot \vec{v}.$$

Therefore scalar multiplication distributes over addition of scalars.

If $\vec{v}$ and $\vec{u}$ are in $\mathbb{R}^n$ and r is a scalar, then $r \cdot (\vec{v} + \vec{u}) = r \cdot \vec{v} + r \cdot \vec{u}$. To see this we compute

$$r \cdot (\vec{v} + \vec{u}) = r \left(\begin{bmatrix} v_1 + u_1 \\ v_2 + u_2 \\ \vdots \\ v_n + u_n \end{bmatrix} \right) = \begin{bmatrix} r(v_1 + u_1) \\ r(v_2 + u_2) \\ \vdots \\ r(v_n + u_n) \end{bmatrix}$$

$$= \begin{bmatrix} rv_1 + ru_1 \\ rv_2 + ru_2 \\ \vdots \\ rv_n + ru_n \end{bmatrix} = \begin{bmatrix} rv_1 \\ rv_2 \\ \vdots \\ rv_n \end{bmatrix} + \begin{bmatrix} ru_1 \\ ru_2 \\ \vdots \\ ru_n \end{bmatrix}.$$

Factoring an r out of both vectors gives us

$$\begin{bmatrix} rv_1 \\ rv_2 \\ \vdots \\ rv_n \end{bmatrix} + \begin{bmatrix} ru_1 \\ ru_2 \\ \vdots \\ ru_n \end{bmatrix} = r \begin{bmatrix} v_1 \\ v_2 \\ \vdots \\ v_n \end{bmatrix} + r \begin{bmatrix} u_1 \\ u_2 \\ \vdots \\ u_n \end{bmatrix} = r \cdot \vec{v} + r \cdot \vec{u}.$$

This means scalar multiplication distributes over vector addition.

To summarize these properties, we can say that vector addition is closed, commutative, associative, has an identity element and inverses, that scalar multiplication of vectors is closed, associative, and has an identity element, and that vector addition and scalar multiplication of vectors distribute nicely. These properties are expressed below as equations.

Theorem 1. For every $\vec{v}, \vec{w}, \vec{u}$ in $\mathbb{R}^n$ and any scalars r and s, we have

$\vec{v} + \vec{u}$ is in $\mathbb{R}^n$	$r \cdot \vec{v}$ is in $\mathbb{R}^n$
$\vec{v} + \vec{w} = \vec{w} + \vec{v}$	$r \cdot (s \cdot \vec{v}) = (rs) \cdot \vec{v}$
$(\vec{v} + \vec{u}) + \vec{w} = \vec{v} + (\vec{u} + \vec{w})$	$1 \cdot \vec{v} = \vec{v}$
$\vec{v} + \vec{0} = \vec{v}$	$(r + s) \cdot \vec{v} = r \cdot \vec{v} + s \cdot \vec{v}$
$-\vec{v} + \vec{v} = \vec{0}$ for some $-\vec{v}$ in $\mathbb{R}^n$	$r \cdot (\vec{v} + \vec{u}) = r \cdot \vec{v} + r \cdot \vec{u}$

In short, with these two operations, each $\mathbb{R}^n$ behaves a lot like $\mathbb{R}$.

Exercises 1.1.

1. For which n is the vector $\begin{bmatrix} -1 \\ 0 \\ 10 \\ -3 \end{bmatrix}$ in $\mathbb{R}^n$?

2. For which n is the vector $\begin{bmatrix} 7 \\ -8 \end{bmatrix}$ in $\mathbb{R}^n$?

3. Compute $-2 \begin{bmatrix} 1 \\ 2 \\ 3 \end{bmatrix}$.

4. Compute $\frac{1}{3} \begin{bmatrix} 12 \\ -9 \\ 1 \end{bmatrix}$

5. Compute $-4 \begin{bmatrix} 3 \\ 1 \end{bmatrix}$

6. Compute $-5 \begin{bmatrix} -1 \\ 2 \\ 1 \end{bmatrix}$

7. Compute $4 \begin{bmatrix} 3 \\ -2 \end{bmatrix}$.

8. Compute $-\frac{1}{2} \begin{bmatrix} -1 \\ 6 \\ 4 \end{bmatrix}$

9. Compute $3 \begin{bmatrix} -1 \\ 0 \\ 10 \\ -3 \end{bmatrix}$.

10. Is it possible to have a scalar r and a vector $\vec{v}$ so that $r\vec{v}$ doesn't make sense?

11. Compute each sum or explain why it is impossible.

(a) $\begin{bmatrix} 2 \\ -4 \end{bmatrix} + \begin{bmatrix} 1 \\ 2 \\ 3 \end{bmatrix}$

(b) $\begin{bmatrix} 2 \\ -4 \end{bmatrix} + \begin{bmatrix} 7 \\ 5 \end{bmatrix}$

(c) $\begin{bmatrix} 1 \\ 2 \\ 3 \end{bmatrix} + \begin{bmatrix} 7 \\ 5 \end{bmatrix}$.

12. Compute each sum or explain why it is impossible.

(a) $\begin{bmatrix} 1 \\ -3 \\ 0 \\ 4 \end{bmatrix} + \begin{bmatrix} 0 \\ 0 \end{bmatrix}$

(b) $\begin{bmatrix} -1 \\ 5 \end{bmatrix} + \begin{bmatrix} 2 \\ -3 \end{bmatrix}$

13. Compute each sum or explain why it is impossible.

(a) $\begin{bmatrix} 2 \\ -1 \\ 17 \end{bmatrix} + \begin{bmatrix} 1 \\ -3 \end{bmatrix}$

(b) $\begin{bmatrix} 0 \\ 5 \\ 10 \end{bmatrix} + \begin{bmatrix} -2 \\ 7 \\ -1 \end{bmatrix}$

14. Compute each sum or explain why it is impossible.

(a) $\begin{bmatrix} -1 \\ 2 \\ 1 \end{bmatrix} + \begin{bmatrix} 3 \\ -2 \end{bmatrix}$

(b) $\begin{bmatrix} -1 \\ 2 \\ 1 \end{bmatrix} + \begin{bmatrix} 11 \\ 5 \\ 0 \end{bmatrix}$

(c) $\begin{bmatrix} 3 \\ -2 \end{bmatrix} + \begin{bmatrix} 11 \\ 5 \\ 0 \end{bmatrix}$

15. Compute each sum or explain why it is impossible.

(a) $\begin{bmatrix} -1 \\ 2 \\ 1 \end{bmatrix} + \begin{bmatrix} 3 \\ -2 \end{bmatrix}$

(b) $\begin{bmatrix} -1 \\ 2 \\ 1 \end{bmatrix} + \begin{bmatrix} 11 \\ 5 \end{bmatrix}$

(c) $\begin{bmatrix} 3 \\ -2 \end{bmatrix} + \begin{bmatrix} 11 \\ 5 \end{bmatrix}$

16. Compute $\begin{bmatrix} 1 \\ 2 \\ 3 \end{bmatrix} + \begin{bmatrix} -1 \\ 6 \\ 4 \end{bmatrix} + \begin{bmatrix} 5 \\ -8 \\ -1 \end{bmatrix}$.

17. Let $\vec{v}_1 = \begin{bmatrix} -1 \\ 0 \\ 10 \\ -3 \end{bmatrix}$, $\vec{v}_2 = \begin{bmatrix} 2 \\ -3 \\ 1 \end{bmatrix}$, $\vec{v}_3 = \begin{bmatrix} 4 \\ 0 \\ -5 \end{bmatrix}$.

    (a) Which sums of these vectors make sense?

    (b) Compute those sums.

18. Use the picture of $\vec{v}$ and $\vec{w}$ below to draw a picture of $\vec{v} + \vec{w}$.

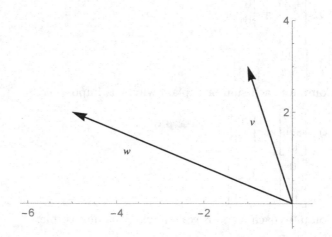

19. Use the picture of $\vec{v}$ below to draw a picture of $-2\vec{v}$.

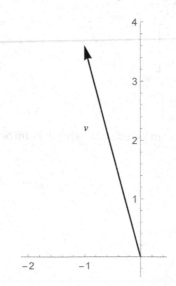

20. Use the picture of $\vec{v}$ and $\vec{w}$ below to draw a picture of $\vec{v} - \vec{w}$.

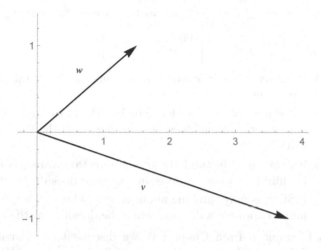

21. For each of the following scenarios, write a vector which records the given information. Be sure to describe what each entry of your vector represents.

    (a) The number of each type of atom in a molecule of glucose, $C_6H_{12}O_6$. As in Example 1 from Chapter 0, C stands for carbon, H for hydrogen, and O for oxygen.

    (b) The pirate treasure can be found two paces south and seven paces west of the only palm tree on the island.

    (c) A hospital patient is 68.5 inches tall, weighs 158 pounds, has a temperature of 98.7° F, and is 38 years old.

22. In the previous problem, imagine constructing the vector of a second molecule or treasure map or hospital patient. For each scenario, suppose you add together your two vectors. What is the practical meaning of your vector sum?

23. Suppose we work at the paint counter of a hardware store. A customer orders three gallons of lavender paint and half a gallon of sage green paint. To make a gallon of lavender paint we mix 1/4 gallon of red paint, 1/4 gallon of blue paint and 1/2 gallon of white paint. To make a gallon of sage green paint we mix 1/4 gallon of white paint, 1/4 gallon of yellow paint, and 1/2 gallon of blue paint. Use the method from Chapter 0's Example 3 to find a vector which gives the amounts of each color of paint needed to fill this order.

24. Adam and Steve are getting divorced, and they are trying to divide up their assets: a car, a boat, a house, and a ski cabin. Each of them has assigned each object a monetary and an emotional value. If the first value vector entry is an object's monetary value (in

thousands of dollars) and the second is its emotional value (on a 0 to 10 scale), Steve's value vectors are as follows: $\vec{v}_{car} = \begin{bmatrix} 12 \\ 3 \end{bmatrix}$, $\vec{v}_{boat} = \begin{bmatrix} 2 \\ 6 \end{bmatrix}$, $\vec{v}_{house} = \begin{bmatrix} 100 \\ 5 \end{bmatrix}$, $\vec{v}_{cabin} = \begin{bmatrix} 75 \\ 10 \end{bmatrix}$.

(a) What is the vector which represents Steve's total value of all four assets?

(b) What does Steve consider to be half the monetary value of these four assets? What does he consider to be half the emotional value?

(c) If Steve is only thinking about monetary value, is it possible for him to approve a division where he doesn't get the house?

(d) If Steve is only thinking about emotional value, is it possible for him to approve a division where he doesn't get the ski cabin?

25. In Example 5 from Chapter 0, we discussed the visual effect of multiplying both coordinates of a point in $\mathbb{R}^2$ by 2. What is the visual effect of multiplying all points in $\mathbb{R}^2$ by -2? (You may want to start by computing $-2\vec{v}$ for several specific vectors $\vec{v}$. Remember to try $\vec{v}$s from each quadrant of the plane.)

26. Can you think of a property of our familiar addition and multiplication in $\mathbb{R}$ which we didn't discuss in $\mathbb{R}^n$? Does $\mathbb{R}^n$ have that property?

27. Draw some examples of 3D vector addition and come up with your own description of how the sum relates to the shape formed by the two vectors. (You may want to use Mathematica.)

1.2 Span

So far, we've seen examples of problems where we want to add vectors or multiply a vector by a scalar. However, there are many situations where we'll want to see what vectors we can get by adding and scaling a whole set of vectors – perhaps to combine existing solutions to a problem into new solutions. For example, rather than simply adding together one molecule of carbon dioxide and one molecule of water, we may want to combine 15 molecules of carbon dioxide with 25 molecules of water. Since this idea comes up so often, we'll spend this section studying the set of new vectors we can get by adding and scaling a given set of vectors. Notice that as with the combination of carbon dioxide and water above, we will often want to use different scalar multiples of each vector. Let's start by considering a single combination of a set of vectors.

Definition. Let $\vec{v}_1, \ldots, \vec{v}_k$ be vectors in $\mathbb{R}^n$. A *linear combination* of these vectors is a vector of the form $a_1\vec{v}_1 + \cdots + a_k\vec{v}_k$ where $a_1, \ldots, a_k$ are scalars.

Note that if $\vec{v}_1, \ldots, \vec{v}_k$ are n-vectors, then any linear combination of them will also be an n-vector. Also notice that to create a linear combination of a given set of vectors we can choose whichever scalars we like, but must use the specified vectors.

Example 1. Create a linear combination of $\vec{v}_1 = \begin{bmatrix} -2 \\ 8 \\ 6 \end{bmatrix}$, $\vec{v}_2 = \begin{bmatrix} 1 \\ 0 \\ -5 \end{bmatrix}$, and $\vec{v}_3 = \begin{bmatrix} 3 \\ -1 \\ 2 \end{bmatrix}$.

To create a linear combination of these three vectors, we need to first select three scalars. I'll use $a_1 = \frac{1}{2}$, $a_2 = 2$, and $a_3 = -1$. Our linear combination is then

$$a_1\vec{v}_1 + a_2\vec{v}_2 + a_3\vec{v}_3 = \frac{1}{2}\begin{bmatrix} -2 \\ 8 \\ 6 \end{bmatrix} + 2\begin{bmatrix} 1 \\ 0 \\ -5 \end{bmatrix} + (-1)\begin{bmatrix} 3 \\ -1 \\ 2 \end{bmatrix}$$

$$= \begin{bmatrix} -1 \\ 4 \\ 3 \end{bmatrix} + \begin{bmatrix} 2 \\ 0 \\ -10 \end{bmatrix} + \begin{bmatrix} -3 \\ 1 \\ -2 \end{bmatrix} = \begin{bmatrix} -2 \\ 5 \\ -9 \end{bmatrix}.$$

Notice that since our original vectors were 3-vectors, our linear combination of them is also a 3-vector.

Geometrically, we can think of a linear combination as starting at the origin and taking a sequence of moves in $\mathbb{R}^n$ along each of the lines defined by $\vec{v}_i$. Through this point of view, our linear combination is the vector corresponding to the point where we end up at the end of our trip. Remember from Figure 1.2 in 1.1 that we can add vectors geometrically using the parallelogram pattern, and from Examples 9 and 10 of 1.1 that which direction we travel along $\vec{v}_i$'s line is determined by whether its scalar coefficient is positive or negative.

Example 2. Find the linear combination of $\vec{v}_1 = \begin{bmatrix} 1 \\ -2 \end{bmatrix}$ and $\vec{v}_2 = \begin{bmatrix} 4 \\ 4 \end{bmatrix}$ with $a_1 = 3$ and $a_2 = -\frac{1}{2}$.

Plugging these vectors and scalars into our definition gives us

$$a_1\vec{v}_1 + a_2\vec{v}_2 = 3\begin{bmatrix} 1 \\ -2 \end{bmatrix} - \frac{1}{2}\begin{bmatrix} 4 \\ 4 \end{bmatrix} = \begin{bmatrix} 3 \\ -6 \end{bmatrix} - \begin{bmatrix} 2 \\ 2 \end{bmatrix} = \begin{bmatrix} 1 \\ -8 \end{bmatrix}.$$

Alternately, we can construct $3\begin{bmatrix} 1 \\ -2 \end{bmatrix}$ and $-\frac{1}{2}\begin{bmatrix} 4 \\ 4 \end{bmatrix}$ geometrically.

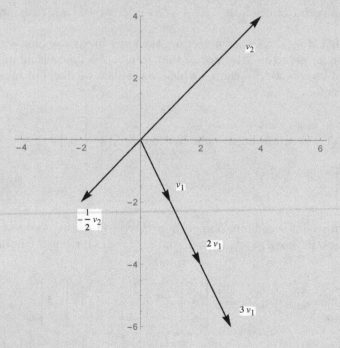

Now we can find the sum by drawing in the parallelogram formed by these vectors, which we can see exactly matches the result we computed algebraically above.

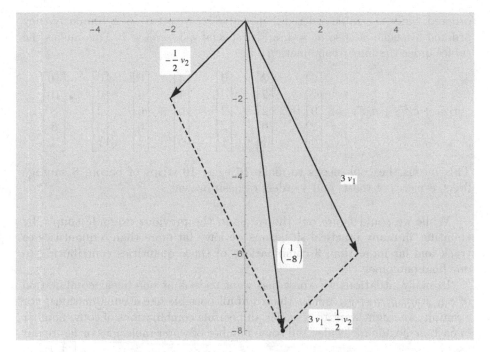

In a more practical sense, we can use linear combinations to create an ordered total amount of a variety of different quantities contributed by several vectors.

Example 3. Suppose there are three lunch specials at a restaurant. Special #1 is 2 eggs, 2 strips of bacon, and 2 pieces of toast, special #2 is 2 eggs, 2 strips of bacon, 2 sausage links, and an order of hash browns, and special #3 is 2 sausage links, an order of hash browns, and 2 pieces of toast. If a table of 6 people orders 2 special #1s, 3 special #2s, and 1 special #3, what should the cook make?

As with any practical example, we first have to decide how to assign vector entries to numeric pieces of information. There are 5 different food types in the three specials, so we can encode the information about each breakfast special's contents as a 5-vector whose entries are (in order): eggs, strips of bacon, sausage links, pieces of toast, and orders of hash browns. If we call special #1's vector $\vec{s}_1$, special #2's vector $\vec{s}_2$, and special #3's vector $\vec{s}_3$, we

$$\text{get } \vec{s}_1 = \begin{bmatrix} 2 \\ 2 \\ 0 \\ 2 \\ 0 \end{bmatrix}, \vec{s}_2 = \begin{bmatrix} 2 \\ 2 \\ 2 \\ 0 \\ 1 \end{bmatrix}, \text{ and } \vec{s}_3 = \begin{bmatrix} 0 \\ 0 \\ 2 \\ 2 \\ 1 \end{bmatrix}.$$

The table's order can now be thought of as a linear combination of these three vectors whose coefficients a_1, a_2, and a_3, are the numbers of each special ordered. Since the table ordered 2 special #1s, we set $a_1 = 2$. Similarly they ordered 3 special #2s so $a_2 = 3$, and 1 special #3 so $a_3 = 1$. This makes the tables order the linear combination

$$a_1 \vec{s}_1 + a_2 \vec{s}_2 + a_3 \vec{s}_3 = 2 \begin{bmatrix} 2 \\ 2 \\ 0 \\ 2 \\ 0 \end{bmatrix} + 3 \begin{bmatrix} 2 \\ 2 \\ 2 \\ 0 \\ 1 \end{bmatrix} + \begin{bmatrix} 0 \\ 0 \\ 2 \\ 2 \\ 1 \end{bmatrix} = \begin{bmatrix} 4 \\ 4 \\ 0 \\ 4 \\ 0 \end{bmatrix} + \begin{bmatrix} 6 \\ 6 \\ 6 \\ 0 \\ 3 \end{bmatrix} + \begin{bmatrix} 0 \\ 0 \\ 2 \\ 2 \\ 1 \end{bmatrix} = \begin{bmatrix} 10 \\ 10 \\ 8 \\ 6 \\ 4 \end{bmatrix}.$$

This means the cook needs to make 10 eggs, 10 strips of bacon, 8 sausage links, 6 pieces of toast, and 4 orders of hash browns.

While we could figure out the totals in the previous example simply by counting, in many practical situations we have far more than 5 quantities to track and far more than 3 combinations of those quantities contributing to our final outcome.

In many situations we won't just want to look at one linear combination of our starting vectors, but at the set of all possible linear combinations. For example, we might want to consider all possible combinations of some number of carbon dioxide molecules and some number of water molecules or figure out which points in the plane we can reach using linear combinations of $\begin{bmatrix} 1 \\ -2 \end{bmatrix}$ and $\begin{bmatrix} 4 \\ 4 \end{bmatrix}$. This motivates the following definition.

Definition. Let $\vec{v}_1, \ldots, \vec{v}_k$ be vectors in $\mathbb{R}^n$. The *span* of these vectors is the set of all possible linear combinations, which is written $\text{Span}\{\vec{v}_1, \ldots, \vec{v}_k\}$. The vectors $\vec{v}_1, \ldots, \vec{v}_k$ are called the *spanning set*.

In other words, the span is the set of vectors we get by making all possible choices of scalars in our linear combination of $\vec{v}_1, \ldots, \vec{v}_k$. Note that if $\vec{v}_1, \ldots, \vec{v}_k$ are n-vectors then all their linear combinations are also n-vectors, so their span is a subset of $\mathbb{R}^n$. Geometrically, we can visualize the span of a set of vectors by thinking of our spanning vectors $\vec{v}_1, \ldots, \vec{v}_k$ as giving possible directions of movement in $\mathbb{R}^n$. This makes the span all possible points in $\mathbb{R}^n$ that we can reach by traveling in some combination of those possible directions. Practically, you could imagine your spanning vectors as different types of coins and bills which makes the span the possible amounts of money you can give as change using those coins and bills. (Of course in this example the only coefficients that make sense are positive whole numbers, but we will learn to explore these situations in the most general case since that allows us to solve the greatest array of problems.)

Let's start by exploring the easiest possible case of a span: when we have only one vector in our spanning set.

Example 4. Find the span of $\vec{v} = \begin{bmatrix} 1 \\ -2 \end{bmatrix}$.

Since we only have one spanning vector, our span is just the set of all multiples of that vector. This means we have

$$\text{Span}\left\{\begin{bmatrix} 1 \\ -2 \end{bmatrix}\right\} = \left\{a\begin{bmatrix} 1 \\ -2 \end{bmatrix}\right\} = \left\{\begin{bmatrix} a \\ -2a \end{bmatrix}\right\}.$$

In other words, the span of $\begin{bmatrix} 1 \\ -2 \end{bmatrix}$ is the set of all vectors in $\mathbb{R}^2$ whose second entries are -2 times their first entries.

To visualize this span geometrically, recall from Examples 9 and 10 in 1.1 that positive multiples of a vector keep the direction of that vector but scale its length, while negative multiples of a vector scale its length and reverse its direction. This gives us the picture below.

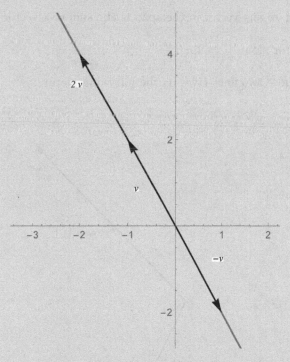

This shows that the span of this vector is the entire line through the origin which points in the same direction as $\begin{bmatrix} 1 \\ -2 \end{bmatrix}$.

Next, let's step up one level of complexity and add a second spanning vector.

Example 5. Find the span of $\begin{bmatrix} 1 \\ -2 \end{bmatrix}$ and $\begin{bmatrix} 4 \\ 4 \end{bmatrix}$.

By definition, our span is all vectors $\vec{b}$ which can be written as

$$\vec{b} = a_1 \begin{bmatrix} 1 \\ -2 \end{bmatrix} + a_2 \begin{bmatrix} 4 \\ 4 \end{bmatrix} = \begin{bmatrix} a_1 + 4a_2 \\ -2a_1 + 4a_2 \end{bmatrix}$$

for some a_1 and a_2.

This algebraic description of the span is harder to interpret than the previous example, so let's turn our attention to what the span looks like geometrically. From our discussion in Example 4, we know that vectors of the form $a_1 \begin{bmatrix} 1 \\ -2 \end{bmatrix}$ all lie along the line through the origin defined by $\begin{bmatrix} 1 \\ -2 \end{bmatrix}$ and that vectors of the form $a_2 \begin{bmatrix} 4 \\ 4 \end{bmatrix}$ all lie along the line through the origin defined by $\begin{bmatrix} 4 \\ 4 \end{bmatrix}$. Therefore any vector in the span is the sum of a vector along $\begin{bmatrix} 1 \\ -2 \end{bmatrix}$'s line and a vector along $\begin{bmatrix} 4 \\ 4 \end{bmatrix}$'s line.

Drawing those two lines gives us the following picture.

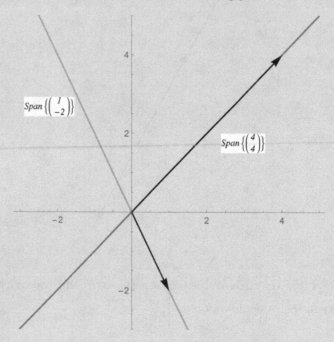

We saw in 1.1 that we can visualize the sum of two vectors as the diagonal of the parallelogram formed by the two vectors we're adding. Thus our span consists of all points in the plane which can be thought of as the fourth corner of a parallelogram with one corner at the origin and two sides along the two lines defined by our two spanning vectors. This is shown for a generic point in the plane in the picture below.

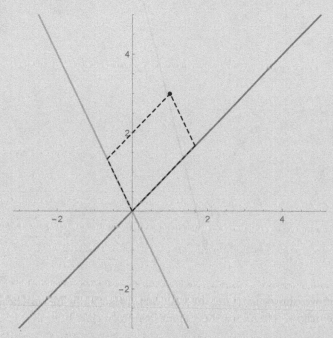

Hopefully you can convince yourself that no matter which point in the plane we pick, we can always draw a parallelogram to show it is an element of the span. This means $\text{Span}\left\{\begin{bmatrix} 1 \\ -2 \end{bmatrix}, \begin{bmatrix} 4 \\ 4 \end{bmatrix}\right\} = \mathbb{R}^2$.

Example 6. Find the span of $\begin{bmatrix} 2 \\ 8 \end{bmatrix}$ and $\begin{bmatrix} -1 \\ -4 \end{bmatrix}$.

We know

$$\text{Span}\left\{\begin{bmatrix} 2 \\ 8 \end{bmatrix}, \begin{bmatrix} -1 \\ -4 \end{bmatrix}\right\} = \left\{a_1 \begin{bmatrix} 2 \\ 8 \end{bmatrix} + a_2 \begin{bmatrix} -1 \\ -4 \end{bmatrix}\right\} = \left\{\begin{bmatrix} 2a_1 - a_2 \\ 8a_1 - 4a_2 \end{bmatrix}\right\}.$$

As in the previous example, this algebraic description is not very illuminating, so we turn to a geometric interpretation. We can visualize all vectors of the form $a_1 \begin{bmatrix} 2 \\ 8 \end{bmatrix}$ as the line through the origin along $\begin{bmatrix} 2 \\ 8 \end{bmatrix}$ and all vectors of the form $a_2 \begin{bmatrix} -1 \\ -4 \end{bmatrix}$ as the line through the origin along $\begin{bmatrix} -1 \\ -4 \end{bmatrix}$.

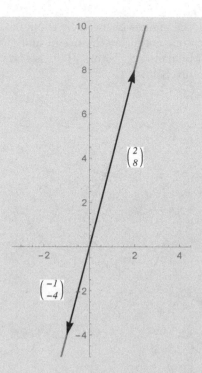

This picture shows us that our two vectors actually lie on the same line through the origin! Therefore their span is simply that line, i.e.,

$$\text{Span}\left\{\begin{bmatrix}2\\8\end{bmatrix}, \begin{bmatrix}-1\\-4\end{bmatrix}\right\} = \text{Span}\left\{\begin{bmatrix}2\\8\end{bmatrix}\right\} = \text{Span}\left\{\begin{bmatrix}-1\\-4\end{bmatrix}\right\}.$$

Notice that the previous example shows that it is possible for multiple different spanning sets to have the same span.

While geometric descriptions are fairly straightforward in $\mathbb{R}^2$, as we work with larger vectors we may start to rely more on the algebraic description from the definition. (This is certainly true for $\mathbb{R}^n$ with $n > 3$!)

Example 7. Describe the span of $\vec{v}_1 = \begin{bmatrix}1\\1\\1\end{bmatrix}$ and $\vec{v}_2 = \begin{bmatrix}-1\\0\\-1\end{bmatrix}$.

By definition, $\text{Span}\{\vec{v}_1, \vec{v}_2\} = \{a_1\vec{v}_1 + a_2\vec{v}_2\}$ for all scalars a_1 and a_2. Plugging in our two vectors gives us

$$\text{Span}\{\vec{v}_1, \vec{v}_2\} = \left\{a_1\begin{bmatrix}1\\1\\1\end{bmatrix} + a_2\begin{bmatrix}-1\\0\\-1\end{bmatrix}\right\} = \left\{\begin{bmatrix}a_1\\a_1\\a_1\end{bmatrix} + \begin{bmatrix}-a_2\\0\\-a_2\end{bmatrix}\right\} = \left\{\begin{bmatrix}a_1 - a_2\\a_1\\a_1 - a_2\end{bmatrix}\right\}.$$

Since we can let a_1 be whatever number we like, it is clear that our middle entry can be anything. We also get a free choice of a_2, so we can make $a_1 - a_2$ be anything we want as well. This means our span is any 3-vector whose first and last entries are the same, i.e.,

$$\text{Span}\left\{\begin{bmatrix}1\\1\\1\end{bmatrix}, \begin{bmatrix}-1\\0\\-1\end{bmatrix}\right\} = \left\{\begin{bmatrix}x\\y\\x\end{bmatrix}\right\},$$

which can be visualized geometrically as the plane $z - x$. Notice that since our spanning vectors were from $\mathbb{R}^3$, their span is a subset of $\mathbb{R}^3$.

We can also use spans to model practical problems, even those which may not initially look like they have anything to do with linear combinations.

Example 8. Suppose we want to swap produce at a farmer's market. There are four goods available: bacon, beans, cornmeal, and turnips. We can exchange 4 lbs of turnips for 1 lb of bacon, 4 lbs of beans for 3 lbs of cornmeal, and 1 lb of turnips for 1 lb of beans. What possible trades can we make?

Before we can even start thinking about how spans might be relevant here, we need to reinterpret this situation using vectors. We have four different goods, so we will use 4-vectors whose entries give the quantity, in pounds, of the various goods. I'll use the first entry for bacon, the second for beans, the third for cornmeal, and the fourth for turnips. Normally it would be very strange to have a negative quantity of something like bacon, but here we will view negative quantities as how much of a good we are giving to someone else and positive quantities as how much we are getting.

Using this interpretation of positive vector entries as getting and negative entries as giving, we can model an accepted exchange as a 4-vector. We know that we can give someone 4 lbs of turnips in exchange for getting 1 lb of bacon, which we can now view as the vector $\begin{bmatrix}1\\0\\0\\-4\end{bmatrix}$. (If you're worried that I chose to view exchanging turnips and bacon as giving turnips and getting bacon rather than the other way around, we'll get to that in the next paragraph where we discuss the scalars in our linear combinations.) Similarly, giving someone 4 lbs of beans and getting 3 lbs of cornmeal has the vector $\begin{bmatrix}0\\-4\\3\\0\end{bmatrix}$, and giving someone 1 lb of turnips and getting 1 lb of beans has the vector $\begin{bmatrix}0\\1\\0\\-1\end{bmatrix}$.

Multiplying one of these exchange vectors by -1 simply changes the signs of all the entries, i.e., which goods are being given or gotten. In other words, multiplying by -1 changes the direction of the exchange. For example, the vector $-1 \begin{bmatrix} 1 \\ 0 \\ 0 \\ -4 \end{bmatrix} = \begin{bmatrix} -1 \\ 0 \\ 0 \\ 4 \end{bmatrix}$ means you're giving someone 1 lb of bacon and getting 4 lbs of turnips, the opposite direction of our initial model of exchanging turnips for bacon. Multiplying an exchange vector by a positive scalar changes the amounts of the goods involved, but keeps the ratio of their quantities, their "exchange rate", the same. For example, $3 \begin{bmatrix} 1 \\ 0 \\ 0 \\ -4 \end{bmatrix} = \begin{bmatrix} 3 \\ 0 \\ 0 \\ -12 \end{bmatrix}$ means we're giving 12 lbs of turnips and getting 3 lbs of bacon, but we're still giving 4 times as much turnips as we get bacon. Putting these two ideas together means that our scalar multiples allow us to trade different amounts of goods in both directions, but keeps the relative values of the goods the same.

Using this model, we can view each linear combination of our exchange vectors as an overall trade at the market which is made up of some combination of our three given exchange ratios. For example, suppose we arrive at the market with 8 lbs of beans and swap 4 lbs of beans for 3 lbs of cornmeal, swap another 4 lbs of beans for 4 lbs of turnips, and then swap 2 lbs of our turnips for 1/2 lb of bacon. Our first exchange is $\begin{bmatrix} 0 \\ -4 \\ 3 \\ 0 \end{bmatrix} = 1 \begin{bmatrix} 0 \\ -4 \\ 3 \\ 0 \end{bmatrix}$. Our second exchange is $\begin{bmatrix} 0 \\ -4 \\ 0 \\ 4 \end{bmatrix} = -4 \begin{bmatrix} 0 \\ 1 \\ 0 \\ -1 \end{bmatrix}$. Our final exchange is $\begin{bmatrix} 1/2 \\ 0 \\ 0 \\ -2 \end{bmatrix} = (1/2) \begin{bmatrix} 1 \\ 0 \\ 0 \\ -4 \end{bmatrix}$ This means our overall exchange vector is the linear combination

$$(1/2) \begin{bmatrix} 1 \\ 0 \\ 0 \\ -4 \end{bmatrix} + 1 \begin{bmatrix} 0 \\ -4 \\ 3 \\ 0 \end{bmatrix} - 4 \begin{bmatrix} 0 \\ 1 \\ 0 \\ -1 \end{bmatrix} = \begin{bmatrix} 1/2 \\ -8 \\ 3 \\ 2 \end{bmatrix}$$

which correctly tells us overall we exchanged our 8 lbs of beans for 1/2 lb of bacon, 3 lbs of cornmeal, and 2 lbs of turnips.

If we view our possible overall trades as exchange vectors, the set of all possible trades is therefore the span of $\begin{bmatrix} 1 \\ 0 \\ 0 \\ -4 \end{bmatrix}$, $\begin{bmatrix} 0 \\ -4 \\ 3 \\ 0 \end{bmatrix}$, and $\begin{bmatrix} 0 \\ 1 \\ 0 \\ -1 \end{bmatrix}$.

One basic question we can ask about spans is whether or not a given vector $\vec{b}$ is in the span of a set of vectors $\vec{v}_1, \ldots, \vec{v}_k$. Geometrically, this can be reinterpreted as asking whether we can get to the point in $\mathbb{R}^n$ corresponding to $\vec{b}$ via a combination of movements along the vectors $\vec{v}_1, \ldots, \vec{v}_k$. Practically, this can be reinterpreted as asking whether we can create the output represented by $\vec{b}$ using only the inputs represented by the $\vec{v}_i$'s. If $\vec{b}$ is not the same size as the $\vec{v}$'s then the answer is clearly no, since a linear combination of vectors has the same size as the vectors in the spanning set. If $\vec{b}$ is the same size as the $\vec{v}$'s, then we can answer this question by trying to find scalars $a_1, \ldots, a_k$ so that $a_1 \vec{v}_1 + \cdots + a_k \vec{v}_k = \vec{b}$. Since we're used to solving for x, this type of equation is often written $x_1 \vec{v}_1 + \cdots + x_k \vec{v}_k = \vec{b}$ to help remind us which variables we're solving for. These equations come up a lot, so they have their own name.

Definition. A *vector equation* is an equation of the form $x_1 \vec{v}_1 + \cdots + x_k \vec{v}_k = \vec{b}$ where $\vec{v}_1, \ldots, \vec{v}_k, \vec{b}$ are n-vectors and we are solving for scalars $x_1, \ldots, x_k$.

Example 9. Solve the vector equation $x_1 \begin{bmatrix} 2 \\ -1 \\ 3 \end{bmatrix} + x_2 \begin{bmatrix} 1 \\ -1 \\ 0 \end{bmatrix} + x_3 \begin{bmatrix} 0 \\ 2 \\ 1 \end{bmatrix} = \begin{bmatrix} 6 \\ -9 \\ 1 \end{bmatrix}$.

We can start by combining the left-hand side of our equation into a single vector to give us

$$\begin{bmatrix} 2x_1 + x_2 \\ -x_1 - x_2 + 2x_3 \\ 3x_1 + x_3 \end{bmatrix} = \begin{bmatrix} 6 \\ -9 \\ 1 \end{bmatrix}.$$

Setting corresponding entries equal gives us three equations: $2x_1 + x_2 = 6$, $-x_1 - x_2 + 2x_3 = -9$, and $3x_1 + x_3 = 1$. We can use the first and third equations to solve for x_2 and x_3 in terms of x_1, which gives us $x_2 = 6 - 2x_1$ and $x_3 = 1 - 3x_1$. Plugging these back into our middle equation above gives $-x_1 - (6 - 2x_1) + 2(1 - 3x_1) = -9$, which simplifies to $-5x_1 - 4 = -9$ or $x_1 = 1$. This means $x_2 = 6 - 2 = 4$ and $x_3 = 1 - 3 = -2$. Therefore the solution to our vector equation is $x_1 = 1$, $x_2 = 4$, and $x_3 = -2$.

Since $\vec{b}$ is in $\text{Span}\{\vec{v}_1, \ldots, \vec{v}_k\}$ exactly when we can find scalars $a_1, \ldots, a_k$ so that $a_1 \vec{v}_1 + \cdots + a_k \vec{v}_k = \vec{b}$, figuring out whether or not $\vec{b}$ is in the span is equivalent to deciding whether or not the vector equation $x_1 \vec{v}_1 + \cdots + x_k \vec{v}_k = \vec{b}$ has a solution.

Example 10. Is $\vec{b} = \begin{bmatrix} 1 \\ 2 \\ 3 \end{bmatrix}$ in the span of $\vec{v}_1 = \begin{bmatrix} 3 \\ 0 \\ 1 \end{bmatrix}$ and $\vec{v}_2 = \begin{bmatrix} 0 \\ -1 \\ 0 \end{bmatrix}$?

We can answer this question by trying to solve the vector equation

$$x_1 \begin{bmatrix} 3 \\ 0 \\ 1 \end{bmatrix} + x_2 \begin{bmatrix} 0 \\ -1 \\ 0 \end{bmatrix} = \begin{bmatrix} 1 \\ 2 \\ 3 \end{bmatrix}.$$

As in the previous example, we can simplify the left-hand side of this equation to get

$$\begin{bmatrix} 3x_1 \\ -x_2 \\ x_1 \end{bmatrix} = \begin{bmatrix} 1 \\ 2 \\ 3 \end{bmatrix}.$$

Setting corresponding entries equal gives us the three equations $3x_1 = 1$, $-x_2 = 2$, and $x_1 = 3$. Unfortunately, the first equation says $x_1 = \frac{1}{3}$, while the third equation says $x_1 = 3$. Since we can't satisfy both of these conditions at once, there is no solution to our vector equation. Therefore $\vec{b}$ is not in the span of $\vec{v}_1$ and $\vec{v}_2$.

Notice that this vector equation version of checking whether or not a vector is in a span doesn't require any geometric understanding of what that span looks like, so it is particularly useful in $\mathbb{R}^n$ for $n > 3$.

Asking whether or not a vector is in a span also comes up in practical situations, as in the example below.

Example 11. With the same basic setup of bartering goods from Example 8, can we trade 3 lbs of cornmeal and 4 lbs of turnips for 1/2 lb of bacon and 6 lbs of beans?

Since we saw in Example 8 that the set of all possible exchanges is the span of $\begin{bmatrix} 1 \\ 0 \\ 0 \\ -4 \end{bmatrix}$, $\begin{bmatrix} 0 \\ -4 \\ 3 \\ 0 \end{bmatrix}$, and $\begin{bmatrix} 0 \\ 1 \\ 0 \\ -1 \end{bmatrix}$, this is really just asking if our trade is in this span. Giving 3 lbs of cornmeal and 4 lbs of turnips and getting 1/2 lb of bacon and 6 lbs of beans corresponds to the vector $\begin{bmatrix} 1/2 \\ 6 \\ -3 \\ -4 \end{bmatrix}$. Therefore this question can be rephrased as asking if $\begin{bmatrix} 1/2 \\ 6 \\ -3 \\ -4 \end{bmatrix}$ is in the span of $\begin{bmatrix} 1 \\ 0 \\ 0 \\ -4 \end{bmatrix}$, $\begin{bmatrix} 0 \\ -4 \\ 3 \\ 0 \end{bmatrix}$, and $\begin{bmatrix} 0 \\ 1 \\ 0 \\ -1 \end{bmatrix}$.

We can figure this out using the vector equation

$$x_1 \begin{bmatrix} 1 \\ 0 \\ 0 \\ -4 \end{bmatrix} + x_2 \begin{bmatrix} 0 \\ -4 \\ 3 \\ 0 \end{bmatrix} + x_3 \begin{bmatrix} 0 \\ 1 \\ 0 \\ -1 \end{bmatrix} = \begin{bmatrix} 1/2 \\ 6 \\ -3 \\ -4 \end{bmatrix}.$$

Simplifying the left-hand side gives us

$$\begin{bmatrix} x_1 \\ -4x_2 + x_3 \\ 3x_2 \\ -4x_1 - x_3 \end{bmatrix} = \begin{bmatrix} 1/2 \\ 6 \\ -3 \\ -4 \end{bmatrix}.$$

Setting corresponding entries equal gives us the four equations $x_1 = 1/2$, $-4x_2 + x_3 = 6$, $3x_2 = -3$, and $-4x_1 - x_3 = -4$. The first equation gives us the value of x_1, and we can use the third equation to see that $x_2 = -1$. Plugging $x_2 = -1$ into the second equation and solving for x_3 gives us $4 + x_3 = 6$, so $x_3 = 2$. Finally, we can check that these values satisfy the fourth equation.

Since we have a solution to our vector equation, our vector is in the span of possible exchanges. This means that it is possible to trade 3 lbs of cornmeal and 4 lbs of turnips for 1/2 lb bacon and 6 lbs of beans.

Now that we have an initial understanding of spans, let's look at a few of their nice properties. These will often be quite helpful, so we'll happily generalize them in later chapters. The first of these properties is that the zero vector is always in any span. To see this, just choose zero as the coefficient on every spanning vector. The second property is that spans are closed under addition. This means that if we take any two vectors from a span and add them together, that sum will also be in our span. To see this, suppose our spanning vectors are $\vec{v}_1, \ldots, \vec{v}_k$. Our two vectors in the span of the $\vec{v}$s can therefore be written $\vec{w} = a_1\vec{v}_1 + \cdots + a_k\vec{v}_k$ and $\vec{u} = b_1\vec{v}_1 + \cdots + b_k\vec{v}_k$ for some scalars $a_1, \ldots, a_k$ and $b_1, \ldots, b_k$. Their sum is

$$\begin{aligned} \vec{w} + \vec{u} &= a_1\vec{v}_1 + \cdots + a_k\vec{v}_k + b_1\vec{v}_1 + \cdots + b_k\vec{v}_k \\ &= (a_1\vec{v}_1 + b_1\vec{v}_1) + \cdots + (a_k\vec{v}_k + b_k\vec{v}_k) \\ &= (a_1 + b_1)\vec{v}_1 + \cdots + (a_k + b_k)\vec{v}_k \end{aligned}$$

which is clearly in the span of $\vec{v}_1, \ldots, \vec{v}_k$. Finally, the third property is that spans are closed under scalar multiplication. This means that if we take any vector from our span and multiply it by any scalar we'll get a vector which is also in our span. To see this, again suppose our spanning vectors are $\vec{v}_1, \ldots, \vec{v}_k$. Our vector in the span must have the form $\vec{w} = a_1\vec{v}_1 + \cdots + a_k\vec{v}_k$ for some scalars $a_1, \ldots, a_k$. If we multiply this vector by a scalar r, we get

$$r\vec{w} = r(a_1\vec{v}_1 + \cdots + a_k\vec{v}_k) = ra_1\vec{v}_1 + \cdots + ra_k\vec{v}_k$$

which is clearly still in the span of $\vec{v}_1, \ldots, \vec{v}_k$.

These three properties are important characteristics of $\mathbb{R}^n$, i.e., n-dimensional space, so we'll use them to make the following definition.

Definition. A subset of $\mathbb{R}^n$ is a *subspace* if it contains $\vec{0}$, is closed under addition, and is closed under scalar multiplication.

Example 12. The span of any set of vectors in $\mathbb{R}^n$ is a subspace of $\mathbb{R}^n$.

We showed in the previous paragraph that any span in $\mathbb{R}^n$ satisfies the three properties to be a subspace.

Exercises 1.2.

1. Compute the linear combination $2\begin{bmatrix} 1 \\ -3 \end{bmatrix} + 7\begin{bmatrix} -1 \\ 0 \end{bmatrix} - \begin{bmatrix} 2 \\ 5 \end{bmatrix}$.

2. Compute the linear combination $4\begin{bmatrix} 1 \\ 2 \\ 3 \end{bmatrix} - 2\begin{bmatrix} 4 \\ -1 \\ 6 \end{bmatrix} + 5\begin{bmatrix} 1 \\ 0 \\ -1 \end{bmatrix}$.

3. Compute the linear combination

$$x_1\begin{bmatrix} 2 \\ 1 \end{bmatrix} + x_2\begin{bmatrix} 0 \\ -4 \end{bmatrix} + x_3\begin{bmatrix} 3 \\ -1 \end{bmatrix} + x_4\begin{bmatrix} 5 \\ 0 \end{bmatrix}.$$

4. Compute the linear combination $x_1\begin{bmatrix} -2 \\ 4 \\ 1 \end{bmatrix} + x_2\begin{bmatrix} 3 \\ -6 \\ 0 \end{bmatrix} + x_3\begin{bmatrix} 1 \\ -1 \\ -2 \end{bmatrix}$.

5. Find a value of h so that $x_1\begin{bmatrix} -2 \\ h \end{bmatrix} + x_2\begin{bmatrix} 4 \\ -3 \end{bmatrix} = \begin{bmatrix} 0 \\ 1 \end{bmatrix}$.

6. Find a value of k so that $x_1\begin{bmatrix} 1 \\ -1 \\ 2 \end{bmatrix} + x_2\begin{bmatrix} 0 \\ k \\ 5 \end{bmatrix} = \begin{bmatrix} 4 \\ -5 \\ 3 \end{bmatrix}$.

7. Rust is formed by an initial reaction of iron, Fe, oxygen gas, O_2, and water, H_2O to form iron hydroxide, $Fe(OH)_3$. As in Chapter 0's Example 1, we can write each molecule in this reaction as a vector which counts the number of each type of atom in that molecule. Let's agree that the first entry in our vectors will count iron atoms (Fe), the second will count oxygen (O), and the third will count hydrogen (H). If our reaction combines x_1 molecules of iron, x_2 molecules of oxygen, and x_3 molecules of water to create x_4 molecules of rust, give the vector equation of this chemical reaction.

8. Sodium chlorate, $NaClO_3$, is produced by electrolysis of sodium chloride, $NaCl$, and water, H_2O. The reaction also produces hydrogen gas, H_2. As in Chapter 0's Example 1, we can write each molecule in this reaction as a vector which counts the number

of each type of atom in that molecule. Let's agree that the first entry in our vectors will count sodium atoms (Na), the second will count chlorine (Cl), the third will count oxygen (O), and the fourth will count hydrogen (H). If our reaction combines x_1 molecules of sodium chloride with x_2 molecules of water to produce x_3 molecules of sodium chlorate and x_4 molecules of hydrogen gas, give the vector equation of this chemical reaction.

9. Does the span of $\begin{bmatrix} 1 \\ -1 \\ 2 \end{bmatrix}$ and $\begin{bmatrix} 3 \\ 1 \\ 4 \end{bmatrix}$ include the vector $\begin{bmatrix} 1 \\ 3 \\ 5 \end{bmatrix}$?

10. Does the span of $\begin{bmatrix} -2 \\ 0 \\ 5 \end{bmatrix}$ and $\begin{bmatrix} 6 \\ 3 \\ 9 \end{bmatrix}$ include the vector $\begin{bmatrix} 0 \\ 1 \\ 8 \end{bmatrix}$?

11. Is $\vec{b} = \begin{bmatrix} 1 \\ 1 \\ 1 \\ 1 \end{bmatrix}$ in Span $\left\{ \begin{bmatrix} 1 \\ 0 \\ -1 \\ 2 \end{bmatrix}, \begin{bmatrix} 2 \\ -3 \\ 0 \\ 1 \end{bmatrix}, \begin{bmatrix} -3 \\ 0 \\ 6 \\ -2 \end{bmatrix} \right\}$?

12. Is $\vec{b} = \begin{bmatrix} 3 \\ 0 \\ 3 \end{bmatrix}$ in Span $\left\{ \begin{bmatrix} 2 \\ -2 \\ 0 \end{bmatrix}, \begin{bmatrix} 1 \\ 1 \\ 1 \end{bmatrix}, \begin{bmatrix} -1 \\ 0 \\ 1 \end{bmatrix} \right\}$?

13. Sketch a picture of the span of $\begin{bmatrix} 4 \\ -2 \end{bmatrix}$.

14. Sketch a picture of the span of $\begin{bmatrix} 2 \\ 0 \end{bmatrix}$ and $\begin{bmatrix} 1 \\ 1 \end{bmatrix}$.

15. Sketch a picture of the span of $\begin{bmatrix} 2 \\ 0 \\ 0 \end{bmatrix}$ and $\begin{bmatrix} 1 \\ 1 \\ 0 \end{bmatrix}$.

16. Do $\begin{bmatrix} 1 \\ 0 \\ -1 \end{bmatrix}, \begin{bmatrix} 0 \\ 1 \\ 2 \end{bmatrix}$, and $\begin{bmatrix} -4 \\ 0 \\ 6 \end{bmatrix}$ span all of $\mathbb{R}^3$? Briefly say why or why not.

17. Do $\begin{bmatrix} 1 \\ 0 \\ 1 \end{bmatrix}, \begin{bmatrix} 2 \\ 0 \\ 2 \end{bmatrix}$ and $\begin{bmatrix} 0 \\ 1 \\ 1 \end{bmatrix}$ span all of $\mathbb{R}^3$? Briefly say why or why not.

18. Develop a strategy to figure out whether or not a set of 4-vectors span all of $\mathbb{R}^4$.

19. Using the setup of Example 3, is it possible to order a combination of specials so that the cook needs to make 12 eggs, 10 strips of bacon, 6 sausage links, 10 pieces of toast, and 5 orders of hash browns? If it is possible, explain how. (Remember that the coefficients in your solution must be positive whole numbers!)

20. Using the setup of Example 8, can you trade 3 lbs of bacon and 8 lbs of beans for 3 lbs of cornmeal and 16 lbs of turnips?

21. Is the graph of the line $y = x + 1$ a subspace of $\mathbb{R}^2$?

22. Is $W = \left\{ \begin{bmatrix} x \\ 0 \\ z \end{bmatrix} \right\}$ a subspace of $\mathbb{R}^3$?

23. What is the smallest subspace of $\mathbb{R}^n$? What is the largest subspace of $\mathbb{R}^n$?

24. Let S be the subset of $\mathbb{R}^3$ consisting of the x-axis, y-axis and z-axis. Show that S is not a subspace of $\mathbb{R}^3$.

1.3 Linear Independence

In this section we'll develop a way to figure out how big the span of a given set of vectors is. Since we'll answer that question by giving the dimension of the span, we first need to discuss the idea of dimension. Most of us have a geometric idea of dimension as long as we stay in $\mathbb{R}^2$ (the plane) or $\mathbb{R}^3$ (3D space). This tells us that a line has dimension 1, a plane has dimension 2, and a three-dimensional object like a solid ball has dimension 3.

One way of articulating this geometric notion of dimension is to say that a k-dimensional object is one that contains k possible independent directions of travel from any given point where by "independent" we mean the different directions are all at right angles to each other. (Think of up/down and right/left in a plane or up/down, right/left, forward/back in 3-space.) We'll develop a more linear algebra flavored definition of dimension which can be applied much more broadly (including in dimensions greater than 3) in Chapter 3.

If we are working in $\mathbb{R}^2$ or $\mathbb{R}^3$, this visual idea of dimension allows us to figure out the dimension of a span by computing the span and drawing a picture. However, in $\mathbb{R}^n$ for $n > 3$ this approach won't work, so we'd like to develop a way to figure out the dimension of a span without using pictures. This alternate strategy will also help with problems in $\mathbb{R}^2$ or $\mathbb{R}^3$ where drawing a picture becomes overly complicated.

We saw in Examples 4 and 5 of 1.2 that the dimension of a span sometimes equals the number of vectors in our spanning set. However, Example 6 shows us that this isn't always true. It shouldn't be too hard to convince yourself that the dimension of a span can't be larger than the number of spanning vectors, i.e., the dimension of $\text{Span}\{\vec{v}_1, \ldots, \vec{v}_k\}$ is at most k. If we think of our spanning vectors as giving possible directions of travel in $\mathbb{R}^n$, we can see that there are at most k possible independent directions included in the span. Thus $\dim(\text{Span}\{\vec{v}_1, \ldots, \vec{v}_k\}) \leq k$.

This reduces our problem of finding the dimension of a span to figuring out how to tell when the dimension is less than the number of spanning vectors. Since we saw that in Example 6 of 1.2, let's think about what happened there to see if we can identify and generalize what happened in that situation. In that example, our two vectors lay along the same line through the origin. In other words, the second vector lay along the line spanned by the first vector. This kept their span as a one-dimensional line instead of expanding it to a two-dimensional plane. We explore a similar situation with more vectors in the next example.

Example 1. Give a geometric description of the span of $\begin{bmatrix} 1 \\ 0 \\ 0 \end{bmatrix}$, $\begin{bmatrix} 1 \\ 1 \\ 0 \end{bmatrix}$, and $\begin{bmatrix} 2 \\ 3 \\ 0 \end{bmatrix}$.

From the definition of the span we get

$$\text{Span}\left\{\begin{bmatrix}1\\0\\0\end{bmatrix},\begin{bmatrix}1\\1\\0\end{bmatrix},\begin{bmatrix}2\\3\\0\end{bmatrix}\right\}=\left\{x_1\begin{bmatrix}1\\0\\0\end{bmatrix}+x_2\begin{bmatrix}1\\1\\0\end{bmatrix}+x_3\begin{bmatrix}2\\3\\0\end{bmatrix}\right\}$$

which simplifies to

$$\left\{\begin{bmatrix}x_1+x_2+2x_3\\x_2+3x_3\\0\end{bmatrix}\right\}.$$

We can choose any values we like for x_1, x_2, and x_3, so we can make the first two entries of our vector equal whatever we want. To see this, suppose we want to get the vector $\begin{bmatrix}a\\b\\0\end{bmatrix}$. We can do this by choosing $x_1=a-b$, $x_2=b$, and $x_3=0$ to get

$$\begin{bmatrix}x_1+x_2+2x_3\\x_2+3x_3\\0\end{bmatrix}=\begin{bmatrix}(a-b)+b+2(0)\\b+3(0)\\0\end{bmatrix}=\begin{bmatrix}a\\b\\0\end{bmatrix}$$

as desired. However, no matter what scalars we choose, we'll always get 0 as our third entry. Therefore geometrically, this span is the plane $\left\{\begin{bmatrix}a\\b\\0\end{bmatrix}\right\}$ inside $\mathbb{R}^3$. Notice that even though we have three spanning vectors, our span is two-dimensional not three-dimensional.

In the previous example, none of the vectors lay along the lines spanned by the other vectors. You can check this geometrically using a 3D picture or computationally by noticing that none of them is a multiple of one of the other vectors. However, the third vector is in the plane spanned by the first two vectors which we can check computationally by noticing that

$$-1\begin{bmatrix}1\\0\\0\end{bmatrix}+3\begin{bmatrix}1\\1\\0\end{bmatrix}=\begin{bmatrix}2\\3\\0\end{bmatrix}.$$

This made the span two-dimensional instead of three-dimensional. We can generalize this idea by saying that the span of $\vec{v}_1,\ldots,\vec{v}_k$ has dimension less than k only if some of the spanning vectors don't add anything to the overall span because they are already in the span of the other vectors. This motivates the following definition.

Definition. The vectors $\vec{v}_1, \ldots, \vec{v}_k$ in $\mathbb{R}^n$ are *linearly dependent* if one of the vectors is in the span of the others. Otherwise $\vec{v}_1, \ldots, \vec{v}_k$ are *linearly independent*.

Example 2. Show that $\vec{v}_1 = \begin{bmatrix} 2 \\ -1 \\ 3 \end{bmatrix}$, $\vec{v}_2 = \begin{bmatrix} 1 \\ 1 \\ 1 \end{bmatrix}$, and $\vec{v}_3 = \begin{bmatrix} 3 \\ 0 \\ 4 \end{bmatrix}$ are linearly dependent.

Here we can observe that $\vec{v}_1 + \vec{v}_2 = \vec{v}_3$. This means $\vec{v}_3$ is an element of Span $\{\vec{v}_1, \vec{v}_2\}$. Since one vector is in the span of the other two, these three vectors are linearly dependent.

This definition is great for showing a set of vectors is linearly dependent, but can be harder to use if we want to show they are linearly independent. To give ourselves a more computational option to check linear independence or dependence, we use the following theorem.

Theorem 1. The vectors $\vec{v}_1, \ldots, \vec{v}_k$ in $\mathbb{R}^n$ are *linearly dependent* if the vector equation $x_1\vec{v}_1 + \cdots + x_k\vec{v}_k = \vec{0}$ has a solution where at least one of the x_is is nonzero. Otherwise $\vec{v}_1, \ldots, \vec{v}_k$ are *linearly independent*.

To see that this is equivalent to our definition, suppose we have $v_1, \ldots, \vec{v}_k$ in $\mathbb{R}^n$ and that one of these vectors is in the span of the others. Let's renumber our vectors if necessary so that $\vec{v}_1$ is in Span$\{\vec{v}_2, \ldots, \vec{v}_k\}$. This means that $\vec{v}_1$ is a linear combination of $\vec{v}_2, \ldots, \vec{v}_k$, so we can find scalars $a_2, \ldots, a_k$ for which

$$\vec{v}_1 = a_2\vec{v}_2 + \cdots + a_k\vec{v}_k.$$

Subtracting $\vec{v}_1$ from both sides of this vector equation gives us

$$\vec{0} = -\vec{v}_1 + a_2\vec{v}_2 + \cdots + a_k\vec{v}_k.$$

The coefficient on $\vec{v}_1$ is -1, so we have a solution to the vector equation $x_1\vec{v}_1 + \cdots + x_k\vec{v}_k = \vec{0}$ where at least one coefficient is nonzero. Therefore if $\vec{v}_1, \ldots, \vec{v}_k$ are linearly dependent according to our first definition, they are linearly dependent according to our second definition.

Now suppose that we know the vector equation $x_1\vec{v}_1 + \cdots + x_k\vec{v}_k = \vec{0}$ has a solution with at least one nonzero coefficient. This means we have scalars $a_1, \ldots, a_k$ with

$$a_1\vec{v}_1 + \cdots + a_k\vec{v}_k = \vec{0}$$

and at least one a_i nonzero. Again, let's renumber our vectors if necessary so that $a_1 \neq 0$. Subtracting $a_1\vec{v}_1$ from both sides of our equation gives us

$$a_2\vec{v}_2 + \cdots + a_k\vec{v}_k = -a_1\vec{v}_1.$$

Multiplying both sides of this equation by $-\frac{1}{a_1}$ gives us

$$-\frac{a_2}{a_1}\vec{v}_2 - \cdots - \frac{a_k}{a_1}\vec{v}_k = \vec{v}_1.$$

This shows us that $\vec{v}_1$ is in the span of the other vectors. Therefore if $\vec{v}_1, \ldots, \vec{v}_k$ are linearly dependent according to our second definition, they are linearly dependent according to our first definition.

We've now shown that our two definitions of linear dependence are the same. Since linear independence was defined to be "not linearly dependent", this means that our two definitions of linear independence are also the same.

Example 3. Show that $\begin{bmatrix} 1 \\ 0 \\ 1 \end{bmatrix}$, $\begin{bmatrix} 2 \\ 3 \\ 0 \end{bmatrix}$, and $\begin{bmatrix} 1 \\ 1 \\ -1 \end{bmatrix}$ are linearly independent.

Let's show this using our second definition, which says our vectors are linearly independent if the vector equation

$$x_1 \begin{bmatrix} 1 \\ 0 \\ 1 \end{bmatrix} + x_2 \begin{bmatrix} 2 \\ 3 \\ 0 \end{bmatrix} + x_3 \begin{bmatrix} 1 \\ 1 \\ -1 \end{bmatrix} = \begin{bmatrix} 0 \\ 0 \\ 0 \end{bmatrix}$$

can only be solved by letting $x_1 = x_2 = x_3 = 0$.

We can simplify the left-hand side of this equation to get

$$\begin{bmatrix} x_1 + 2x_2 + x_3 \\ 3x_2 + x_3 \\ x_1 - x_3 \end{bmatrix} = \begin{bmatrix} 0 \\ 0 \\ 0 \end{bmatrix}.$$

Setting corresponding entries equal to each other gives us the three equations $x_1 + 2x_2 + x_3 = 0$, $3x_2 + x_3 = 0$, and $x_1 - x_3 = 0$. The third equation says $x_1 = x_3$. Solving the second equation for x_2 gives us $x_2 = -\frac{1}{3}x_3$. Plugging both of those into the first equation gives us $x_3 + 2(-\frac{1}{3}x_3) + x_3 = 0$ which simplifies to $\frac{4}{3}x_3 = 0$. This means $x_3 = 0$, so $x_1 = x_3 = 0$ and $x_2 = -\frac{1}{3}x_3 = 0$. Since our only solution is to have all coefficients equal to 0, our three vectors are linearly independent.

Example 4. Are $\begin{bmatrix} 2 \\ -1 \\ 3 \end{bmatrix}$, $\begin{bmatrix} 1 \\ 1 \\ 1 \end{bmatrix}$, and $\begin{bmatrix} 5 \\ -4 \\ 8 \end{bmatrix}$ linearly independent or linearly dependent?

If you have a particularly clever moment, you might observe that

$$3 \begin{bmatrix} 2 \\ -1 \\ 3 \end{bmatrix} + (-1) \begin{bmatrix} 1 \\ 1 \\ 1 \end{bmatrix} = \begin{bmatrix} 5 \\ -4 \\ 8 \end{bmatrix}.$$

This means our third vector is in the span of the other two, so our three vectors are linearly dependent.

However, I am not confident that I'd notice that relationship between these vectors immediately. In that case, we can always solve the vector equation

$$x_1 \begin{bmatrix} 2 \\ -1 \\ 3 \end{bmatrix} + x_2 \begin{bmatrix} 1 \\ 1 \\ 1 \end{bmatrix} + x_3 \begin{bmatrix} 5 \\ -4 \\ 8 \end{bmatrix} = \begin{bmatrix} 0 \\ 0 \\ 0 \end{bmatrix}$$

to see if there is a solution with at least one nonzero coefficient.

This equation can be simplified to

$$\begin{bmatrix} 2x_1 + x_2 + 5x_3 \\ -x_1 + x_2 - 4x_3 \\ 3x_1 + x_2 + 8x_3 \end{bmatrix} = \begin{bmatrix} 0 \\ 0 \\ 0 \end{bmatrix}$$

which gives us three equations: $2x_1 + x_2 + 5x_3 = 0$, $-x_1 + x_2 - 4x_3 = 0$, and $3x_1 + x_2 + 8x_3 = 0$. Subtracting our first equation from our third equation gives us $x_1 + 3x_3 = 0$, so $x_1 = -3x_3$. Plugging that into our second equation gives us $3x_3 + x_2 - 4x_3 = 0$, which simplifies to $x_2 - x_3 = 0$, so $x_2 = x_3$. However, no matter which of our three original equations we plug $x_1 = -3x_3$ and $x_2 = x_3$ into, the terms all cancel out. This means any value of x_3 will work. (In particular we can have $x_3 = -1$ so $x_2 = -1$ and $x_1 = 3$ to get the linear combination at the start of this example.) Since it is possible to have $x_3 \neq 0$, these three vectors are linearly dependent.

This is a good example of why having two different but equivalent ways to check something can be very helpful!

Going back to our original motivation for defining linear independence and linear dependence, we can now state the following.

Theorem 2. Let $\vec{v}_1, \ldots, \vec{v}_k$ be in $\mathbb{R}^n$. Span$\{\vec{v}_1, \ldots, \vec{v}_k\}$ is k-dimensional if and only if $\vec{v}_1, \ldots, \vec{v}_k$ are linearly independent.

If our spanning set is linearly independent, this allows us to find the dimension of the span by simply counting the number of spanning vectors. This doesn't require pictures, so it can be used in $\mathbb{R}^n$ for any n. This provides an alternate way to compute the dimension of the span from Example 5 in 1.2.

Example 5. Find the dimension of the span of $\begin{bmatrix} 1 \\ -2 \end{bmatrix}$ and $\begin{bmatrix} 4 \\ 4 \end{bmatrix}$.

From the theorem above, we know that if our two spanning vectors are linearly independent, the span has dimension 2. If they are linearly dependent, the span has dimension less than 2.

Checking linear independence or dependence means solving the vector equation

$$x_1 \begin{bmatrix} 1 \\ -2 \end{bmatrix} + x_2 \begin{bmatrix} 4 \\ 4 \end{bmatrix} = \begin{bmatrix} 0 \\ 0 \end{bmatrix}$$

which can be simplified to

$$\begin{bmatrix} x_1 + 4x_2 \\ -2x_1 + 4x_2 \end{bmatrix} = \begin{bmatrix} 0 \\ 0 \end{bmatrix}.$$

This gives us the equations $x_1 + 4x_2 = 0$ and $-2x_1 + 4x_2 = 0$. The first equation tells us $x_1 = -4x_2$. Plugging this into the second equation, we get $8x_2 + 4x_2 = 0$ or $12x_2 = 0$. This means $x_2 = 0$, so $x_1 = 0$ as well.

Since our only solution was both coefficients equal to zero, our two spanning vectors are linearly independent. This means their span has dimension 2, which matches up with the dimension of the picture we drew in 1.2.

Looking at Example 6 from 1.2 where our two spanning vectors are linearly dependent shows us Theorem 1's other conclusion.

Example 6. Find the dimension of the span of $\begin{bmatrix} 2 \\ 8 \end{bmatrix}$ and $\begin{bmatrix} -1 \\ -4 \end{bmatrix}$.

From the theorem above, we know that if our two spanning vectors are linearly independent, the span has dimension 2. If they are linearly dependent, the span has dimension less than 2.

Checking linear independence or dependence means solving the vector equation

$$x_1 \begin{bmatrix} 2 \\ 8 \end{bmatrix} + x_2 \begin{bmatrix} -1 \\ -4 \end{bmatrix} = \begin{bmatrix} 0 \\ 0 \end{bmatrix}$$

which can be simplified to

$$\begin{bmatrix} 2x_1 - x_2 \\ 8x_1 - 4x_2 \end{bmatrix} = \begin{bmatrix} 0 \\ 0 \end{bmatrix}.$$

This gives us the equations $2x_1 - x_2 = 0$ and $8x_1 - 4x_2 = 0$. The first equation tells us $x_2 = 2x_1$. Plugging this into the second equation, we get $8x_1 - 8x_1 = 0$ or $0 = 0$. This means we can have any value of x_1 as long as we set $x_2 = 2x_1$.

Since we have solutions where the coefficients don't equal zero, our two spanning vectors are linearly dependent. This means their span has dimension less than 2, which again matches up with the dimension of the picture we drew in 1.2.

If our set of vectors is linearly dependent, we can still use Theorem 1 to find the dimension of their span. During the check that our vectors are linearly dependent, we will have identified at least one which is in the span of the others (or has a nonzero coefficient in our other definition). That vector doesn't contribute anything new to the overall span, so we can remove it from our spanning set without changing either the span or its dimension. If our new smaller spanning set is now linearly independent, we can find the dimension by counting our new set of spanning vectors. If not, we repeat our identification and removal of a redundant spanning vector. Since we started with finitely many vectors, this process will eventually stop, and the dimension of the original span will be the number of remaining spanning vectors.

Example 7. Find the dimension of the span of $\begin{bmatrix} 1 \\ 0 \\ 1 \\ 0 \end{bmatrix}$, $\begin{bmatrix} 1 \\ 0 \\ 1 \\ 1 \end{bmatrix}$, and $\begin{bmatrix} 2 \\ 0 \\ 0 \\ 2 \end{bmatrix}$.

We have three spanning vectors, so the dimension of their span is at most 3. If our vectors are linearly independent, their span is three-dimensional. If they are linearly dependent, their span has dimension less than 3, and we can use the algorithm outlined above to find the dimension. Let's start by checking whether our vectors are linearly independent or linearly dependent.

Since our spanning vectors are in $\mathbb{R}^4$, we can't tackle this problem geometrically, so let's look at the vector equation

$$x_1 \begin{bmatrix} 1 \\ 0 \\ 1 \\ 0 \end{bmatrix} + x_2 \begin{bmatrix} 1 \\ 0 \\ 1 \\ 1 \end{bmatrix} + x_3 \begin{bmatrix} 2 \\ 0 \\ 0 \\ 2 \end{bmatrix} = \begin{bmatrix} 0 \\ 0 \\ 0 \\ 0 \end{bmatrix}.$$

This simplifies to

$$\begin{bmatrix} x_1 + x_2 + 2x_3 \\ 0 \\ x_1 + x_2 \\ x_2 + 2x_3 \end{bmatrix} = \begin{bmatrix} 0 \\ 0 \\ 0 \\ 0 \end{bmatrix}$$

which gives us $x_1 + x_2 + 2x_3 = 0$, $0 = 0$, $x_1 + x_2 = 0$, and $x_2 + 2x_3 = 0$. The third equation tells us $x_1 = -x_2$, and the fourth equation tells us $x_3 = -\frac{1}{2}x_2$. Plugging these into the first equation gives us $-x_2 + x_2 - x_2 = 0$ so $x_2 = 0$. This also means $x_1 = 0$ and $x_3 = 0$.

Since all three coefficients in our vector equation must be 0, our vectors are linearly independent which means their span has dimension 3.

Example 8. Find the dimension of the span of $\begin{bmatrix} 4 \\ 2 \\ 0 \\ -1 \end{bmatrix}$, $\begin{bmatrix} 6 \\ 6 \\ 0 \\ 1 \end{bmatrix}$, and $\begin{bmatrix} -3 \\ 0 \\ 0 \\ 2 \end{bmatrix}$.

As in the previous example we have three spanning vectors, so the maximum dimension of this span is 3. Again, we'll start by determining if these vectors are linearly independent or linearly dependent using the vector equation

$$x_1 \begin{bmatrix} 4 \\ 2 \\ 0 \\ -1 \end{bmatrix} + x_2 \begin{bmatrix} 6 \\ 6 \\ 0 \\ 1 \end{bmatrix} + x_3 \begin{bmatrix} -3 \\ 0 \\ 0 \\ 2 \end{bmatrix} = \begin{bmatrix} 0 \\ 0 \\ 0 \\ 0 \end{bmatrix}.$$

This simplifies to

$$\begin{bmatrix} 4x_1 + 6x_2 - 3x_3 \\ 2x_1 + 6x_2 \\ 0 \\ -x_1 + x_2 + 2x_3 \end{bmatrix} = \begin{bmatrix} 0 \\ 0 \\ 0 \\ 0 \end{bmatrix}$$

which gives us the equations $4x_1 + 6x_2 - 3x_3 = 0$, $2x_1 + 6x_2 = 0$, $0 = 0$, and $-x_1 + x_2 + 2x_3 = 0$. The second equation gives us $x_1 = -3x_2$. Plugging this into the fourth equation gives us $3x_2 + x_2 + 2x_3 = 0$ so $4x_2 + 2x_3 = 0$ or $x_3 = -2x_2$. Plugging $x_1 = -3x_2$ and $x_3 = -2x_2$ into any of our original equations simplifies down to $0 = 0$, so x_2 can be whatever we choose. In particular, we can choose to have $x_2 \neq 0$, so our three vectors are linearly dependent. This means their span has dimension less than 3.

Following our algorithm, we now need to remove one of our spanning vectors which is in the span of the other two. The clue here is that this vector will have a nonzero coefficient in our vector equation. Since we said we could choose any value for x_2, let's remove our second vector. (Actually if $x_2 \neq 0$ we also have x_1 and x_3 nonzero, so here we could also have chosen to remove either of the other two vectors as well.)

Since $\begin{bmatrix} 6 \\ 6 \\ 0 \\ 1 \end{bmatrix}$ is in the span of $\begin{bmatrix} 4 \\ 2 \\ 0 \\ -1 \end{bmatrix}$ and $\begin{bmatrix} -3 \\ 0 \\ 0 \\ 2 \end{bmatrix}$, we know

$$\text{Span} \left\{ \begin{bmatrix} 4 \\ 2 \\ 0 \\ -1 \end{bmatrix}, \begin{bmatrix} 6 \\ 6 \\ 0 \\ 1 \end{bmatrix}, \begin{bmatrix} -3 \\ 0 \\ 0 \\ 2 \end{bmatrix} \right\} = \text{Span} \left\{ \begin{bmatrix} 4 \\ 2 \\ 0 \\ -1 \end{bmatrix}, \begin{bmatrix} -3 \\ 0 \\ 0 \\ 2 \end{bmatrix} \right\}.$$

In particular, we know these two spans have the same dimension. To find this dimension, let's check whether our remaining two spanning vectors are linearly independent or linearly dependent by solving the vector equation

$$x_1 \begin{bmatrix} 4 \\ 2 \\ 0 \\ -1 \end{bmatrix} + x_2 \begin{bmatrix} -3 \\ 0 \\ 0 \\ 2 \end{bmatrix} = \begin{bmatrix} 0 \\ 0 \\ 0 \\ 0 \end{bmatrix}.$$

This simplifies to

$$\begin{bmatrix} 4x_1 - 3x_2 \\ 2x_1 \\ 0 \\ -x_1 + 2x_2 \end{bmatrix} = \begin{bmatrix} 0 \\ 0 \\ 0 \\ 0 \end{bmatrix}$$

which gives us the equations $4x_1 - 3x_2 = 0$, $2x_1 = 0$, $0 = 0$, and $-x_1 + 2x_2 = 0$. The second equation tells us that $x_1 = 0$, and plugging that back into the fourth equation tells us $x_2 = 0$. (This also satisfies the first equation.) Since our only solution is to have both coefficients equal to 0, these two vectors are linearly independent and that their span has dimension 2.

Thus the span of $\begin{bmatrix} 4 \\ 2 \\ 0 \\ -1 \end{bmatrix}$, $\begin{bmatrix} 6 \\ 6 \\ 0 \\ 1 \end{bmatrix}$, and $\begin{bmatrix} -3 \\ 0 \\ 0 \\ 2 \end{bmatrix}$ is two-dimensional. (Geometri-

cally, this means the span is a plane inside $\mathbb{R}^4$, which is fun to think about even if we can't draw a good picture of it.)

Exercises 1.3.

1. Are $\begin{bmatrix} 1 \\ -1 \\ 1 \end{bmatrix}$, $\begin{bmatrix} 3 \\ 0 \\ 6 \end{bmatrix}$, and $\begin{bmatrix} 1 \\ 2 \\ 4 \end{bmatrix}$ linearly independent or linearly dependent?

2. Are $\begin{bmatrix} 2 \\ 2 \\ 2 \end{bmatrix}$, $\begin{bmatrix} 1 \\ 0 \\ -1 \end{bmatrix}$, and $\begin{bmatrix} 3 \\ 2 \\ 0 \end{bmatrix}$ linearly independent or linearly dependent?

3. Are $\begin{bmatrix} 2 \\ 1 \\ 0 \\ 1 \end{bmatrix}$, $\begin{bmatrix} 3 \\ 0 \\ 2 \\ 1 \end{bmatrix}$, and $\begin{bmatrix} 1 \\ 0 \\ 2 \\ 0 \end{bmatrix}$ linearly independent or linearly dependent?

4. Are $\begin{bmatrix} -1 \\ 1 \\ 0 \\ 0 \end{bmatrix}$, $\begin{bmatrix} 3 \\ 0 \\ 1 \\ 3 \end{bmatrix}$, and $\begin{bmatrix} 2 \\ 4 \\ 2 \\ 6 \end{bmatrix}$ linearly independent or linearly dependent?

5. Are the two vectors pictured below linearly independent or linearly dependent?

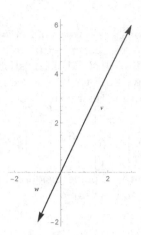

6. Are the two vectors pictured below linearly independent or linearly dependent?

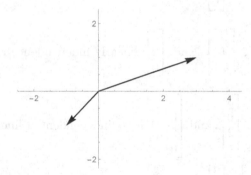

7. (a) Sketch two vectors in $\mathbb{R}^2$ which are linearly independent.
   (b) Sketch two vectors in $\mathbb{R}^2$ which are linearly dependent.

8. (a) Sketch three vectors in $\mathbb{R}^3$ which are linearly independent.
   (b) Sketch three vectors in $\mathbb{R}^3$ which are linearly dependent.

9. Let $\vec{v}_1 = \begin{bmatrix} 3 \\ 1 \\ 0 \\ -1 \end{bmatrix}$, $\vec{v}_2 = \begin{bmatrix} 5 \\ -2 \\ 1 \\ 2 \end{bmatrix}$, and $\vec{v}_3 = \begin{bmatrix} 4 \\ 5 \\ -1 \\ 2 \end{bmatrix}$.

   (a) Are $\vec{v}_1, \vec{v}_2, \vec{v}_3$ linearly dependent or linearly independent? Show work to support your answer.
   (b) What does your answer to (a) tell you about the dimension of Span$\{\vec{v}_1, \vec{v}_2, \vec{v}_3\}$?

10. Let $\vec{v}_1 = \begin{bmatrix} 1 \\ -1 \\ 0 \\ 0 \end{bmatrix}$, $\vec{v}_2 = \begin{bmatrix} 2 \\ -1 \\ 1 \\ 0 \end{bmatrix}$, and $\vec{v}_3 = \begin{bmatrix} 0 \\ -1 \\ 1 \\ 0 \end{bmatrix}$.

(a) Are $\vec{v}_1$, $\vec{v}_2$, $\vec{v}_3$ linearly dependent or linearly independent? Show work to support your answer.

(b) What does your answer to (a) tell you about the dimension of Span$\{\vec{v}_1, \vec{v}_2, \vec{v}_3\}$?

11. Find the dimension of Span $\left\{ \begin{bmatrix} -3 \\ 12 \\ 9 \end{bmatrix}, \begin{bmatrix} 1 \\ -4 \\ -3 \end{bmatrix}, \begin{bmatrix} -2 \\ 8 \\ 6 \end{bmatrix} \right\}$.

12. Find the dimension of Span $\left\{ \begin{bmatrix} -1 \\ 1 \\ 0 \\ 0 \end{bmatrix}, \begin{bmatrix} 3 \\ 0 \\ 1 \\ 3 \end{bmatrix}, \begin{bmatrix} 2 \\ 4 \\ 2 \\ 6 \end{bmatrix} \right\}$.

13. Briefly explain why two 3-vectors cannot span all of $\mathbb{R}^3$.

14. Is it possible to span all of $\mathbb{R}^6$ with five 6-vectors, i.e., five vectors from $\mathbb{R}^6$? Briefly say why or why not.

15. Use linear independence to decide whether or not $\begin{bmatrix} 1 \\ 0 \\ -1 \end{bmatrix}$, $\begin{bmatrix} 0 \\ 1 \\ 2 \end{bmatrix}$, and $\begin{bmatrix} -4 \\ 0 \\ 6 \end{bmatrix}$ span all of $\mathbb{R}^3$.

16. Use linear independence to decide whether or not $\begin{bmatrix} 1 \\ 0 \\ 1 \end{bmatrix}$, $\begin{bmatrix} 2 \\ 0 \\ 2 \end{bmatrix}$ and $\begin{bmatrix} 0 \\ 1 \\ 1 \end{bmatrix}$ span all of $\mathbb{R}^3$.

17. Briefly explain how you would use the idea of linear independence to figure out whether or not a set of 4-vectors span all of $\mathbb{R}^4$.

18. Using the setup of 1.2's Example 8, are there trades that are impossible? Explain why or why not using the idea of linear independence. Why does your answer make sense from a practical perspective?

2

Functions of Vectors

2.1 Linear Functions

Now that we've explored vectors in $\mathbb{R}^n$ in Chapter 1, let's explore functions of vectors. Think of this as a parallel to a first calculus course's study of functions from $\mathbb{R}$ to $\mathbb{R}$ after earlier mathematical explorations of the real numbers. By functions of vectors, we simply mean functions whose inputs and outputs are vectors instead of real numbers. Since both their inputs and outputs are vectors, these are often called vector-valued functions. For example, we might consider the following map from $\mathbb{R}^2$ to itself.

Example 1. $f\left(\begin{bmatrix} x_1 \\ x_2 \end{bmatrix}\right) = \begin{bmatrix} \frac{1}{2}x_1 \\ 2x_2 \end{bmatrix}$.

This function acts on $\mathbb{R}^2$ by multiplying the first component of each vector by $1/2$ and the second component by 2.

Unlike a first calculus course, we don't have to use the same kind of input and output.

Example 2. $f\left(\begin{bmatrix} x_1 \\ x_2 \end{bmatrix}\right) = \begin{bmatrix} x_1 \\ x_2 \\ 3 \end{bmatrix}$.

In this example our inputs are 2-vectors and our outputs are 3-vectors. We can imagine a practical interpretation of this function as taking a position vector on a 2D table top and changing it into a position vector in a 3D dining room using the fact that the table top has a height of 3 feet.

In the previous example, as in many applications, we needed to use input and output vectors of different sizes. This doesn't really make our lives too much more complicated, but it does mean we'll need to be careful to think about whether a vector makes sense as an input or an output (or neither) of a given function f. To help keep this straight, we have special names for these two spaces of vectors.

Definition. Suppose $f : \mathbb{R}^n \to \mathbb{R}^m$. The *domain* of f is $\mathbb{R}^n$ where n is the size of the input vectors, and the *codomain* of f is $\mathbb{R}^m$ where m is the size of the output vectors.

In other words, the set from which the function maps is its domain, and the set it maps to is its codomain, as shown in Figure 2.1.

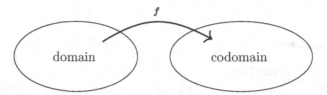

Figure 2.1: Visualizing a function

Example 3. Find the domain and codomain of $f\left(\begin{bmatrix} x_1 \\ x_2 \\ x_3 \\ x_4 \end{bmatrix} \right) = \begin{bmatrix} x_1 + x_4 \\ x_2 - x_3 \end{bmatrix}$.

This function's inputs are 4-vectors and its outputs are 2-vectors, so we could write $f : \mathbb{R}^4 \to \mathbb{R}^2$. Therefore f's domain is $\mathbb{R}^4$ and its codomain is $\mathbb{R}^2$.

Since these vector-valued functions are still functions, we can ask all the same sorts of questions about them that we asked about the functions in calculus. These include things like computing $f(\vec{x})$ for a given input vector, solving $f(\vec{x}) = \vec{b}$ for $\vec{x}$, and doing basic function operations like adding two functions, multiplying a function by a scalar, doing function composition, and finding inverse functions. We will eventually tackle all of these questions, both for general mathematical interest and because we will need their answers to solve practical problems involving vectors. However, we will not discuss all vector-valued functions, but instead restrict our attention to a special class of functions.

Recall that calculus quickly narrows its focus to study only continuous functions. This is because calculus relies so heavily on limits of real numbers that it wants functions to respect and play well with limits. In our case, we don't care about limits, but do care about addition and scalar multiplication of vectors. Therefore, we will restrict our attention to functions that respect and play well with vector addition and scalar multiplication. Putting this more precisely, we define a linear function as follows.

Definition. A function $f : \mathbb{R}^n \to \mathbb{R}^m$ is *linear* if $f(\vec{v} + \vec{u}) = f(\vec{v}) + f(\vec{u})$ and $f(r \cdot \vec{v}) = r \cdot f(\vec{v})$ for all vectors $\vec{v}, \vec{u}$ in $\mathbb{R}^n$ and all scalars r.

Intuitively, this is saying that our function f respects addition and scalar multiplication because we'll get the same answer whether we add or scale our vectors before or after applying the function f. This often gives us two choices on how to tackle a computation, which can be very helpful if one is easier than the other.

Additionally, a linear function preserves lines in the sense that if f is linear then the image of a line in $\mathbb{R}^n$ is a line in $\mathbb{R}^m$. To see this, recall that one way to write the equation of a line in $\mathbb{R}^n$ is $t\vec{v} + \vec{b}$ where $\vec{b}$ is any vector on that line, $\vec{v}$ is any vector parallel to that line, and t is any scalar. If we apply a linear function f, then the image is the set of all points $f(t\vec{v} + \vec{b})$, which by linearity can be broken up to give $f(t\vec{v} + \vec{b}) = tf(\vec{v}) + f(\vec{b})$. This is also a line, in particular the line in $\mathbb{R}^m$ containing the vector $f(\vec{b})$ and parallel to $f(\vec{v})$.

Let's move from the realm of theory to the realm of computations and look at a few examples.

Example 4. Show $f : \mathbb{R}^2 \to \mathbb{R}^3$ by $f\left(\begin{bmatrix} x \\ y \end{bmatrix}\right) = \begin{bmatrix} x \\ y \\ 0 \end{bmatrix}$ is linear.

This map is one of the three standard ways to map $\mathbb{R}^2$ into $\mathbb{R}^3$ as what is often called the xy-plane. To check that it is a linear map, we need to check the two conditions from the definition. To do that, we'll need to write down two generic vectors $\vec{v}$ and $\vec{u}$ in $\mathbb{R}^2$ and a generic scalar r in $\mathbb{R}$. I'll use $\vec{v} = \begin{bmatrix} x \\ y \end{bmatrix}$, $\vec{u} = \begin{bmatrix} z \\ w \end{bmatrix}$.

First we need to check that $f(\vec{v} + \vec{u}) = f(\vec{v}) + f(\vec{u})$.
Computing the left-hand side gives us

$$f(\vec{v} + \vec{u}) = f\left(\begin{bmatrix} x \\ y \end{bmatrix} + \begin{bmatrix} z \\ w \end{bmatrix}\right) = f\left(\begin{bmatrix} x + z \\ y + w \end{bmatrix}\right) = \begin{bmatrix} x + z \\ y + w \\ 0 \end{bmatrix}.$$

Computing the right-hand side gives us

$$f(\vec{v}) + f(\vec{u}) = f\left(\begin{bmatrix} x \\ y \end{bmatrix}\right) + f\left(\begin{bmatrix} z \\ w \end{bmatrix}\right) = \begin{bmatrix} x \\ y \\ 0 \end{bmatrix} + \begin{bmatrix} z \\ w \\ 0 \end{bmatrix} = \begin{bmatrix} x + z \\ y + w \\ 0 \end{bmatrix}.$$

Since our two answers are equal, it's clear that f respects addition.
Next we need to check that $f(r \cdot \vec{v}) = r \cdot f(\vec{v})$.
Computing the left-hand side gives us

$$f(r \cdot \vec{v}) = f\left(r \cdot \begin{bmatrix} x \\ y \end{bmatrix}\right) = f\left(\begin{bmatrix} rx \\ ry \end{bmatrix}\right) = \begin{bmatrix} rx \\ ry \\ 0 \end{bmatrix}.$$

Computing the right-hand side gives us

$$r \cdot f(\vec{v}) = r \cdot f\left(\begin{bmatrix} x \\ y \end{bmatrix}\right) = r \cdot \begin{bmatrix} x \\ y \\ 0 \end{bmatrix} = \begin{bmatrix} rx \\ ry \\ 0 \end{bmatrix}.$$

Again, since our answers are equal, it's clear that f respects scalar multiplication.

Since both conditions hold, f is a linear function.

Example 5. Show $g : \mathbb{R}^2 \to \mathbb{R}^2$ by $g\left(\begin{bmatrix} x \\ y \end{bmatrix}\right) = \begin{bmatrix} 2x - y \\ x + 3y \end{bmatrix}$ is linear.

As in Example 4, I'll use the generic vectors $\vec{v} = \begin{bmatrix} x \\ y \end{bmatrix}$ and $\vec{u} = \begin{bmatrix} z \\ w \end{bmatrix}$ and the scalar r to show that g respects addition and scalar multiplication.

Plugging $\vec{v}$ and $\vec{u}$ into the left-hand side of $g(\vec{v} + \vec{u}) = g(\vec{v}) + g(\vec{u})$ gives us

$$g(\vec{v} + \vec{u}) = g\left(\begin{bmatrix} x \\ y \end{bmatrix} + \begin{bmatrix} z \\ w \end{bmatrix}\right) = g\left(\begin{bmatrix} x + z \\ y + w \end{bmatrix}\right)$$
$$= \begin{bmatrix} 2(x + z) - (y + w) \\ (x + z) + 3(y + w) \end{bmatrix} = \begin{bmatrix} 2x + 2z - y - w \\ x + z + 3y + 3w \end{bmatrix}.$$

Computing the right-hand side gives us

$$g(\vec{v}) + g(\vec{u}) = g\left(\begin{bmatrix} x \\ y \end{bmatrix}\right) + f\left(\begin{bmatrix} z \\ w \end{bmatrix}\right) = \begin{bmatrix} 2x - y \\ x + 3y \end{bmatrix} + \begin{bmatrix} 2z - w \\ z + 3w \end{bmatrix}$$
$$= \begin{bmatrix} 2x - y + 2z - w \\ x + 3y + z + 3w \end{bmatrix} = \begin{bmatrix} 2x + 2z - y - w \\ x + z + 3y + 3w \end{bmatrix}.$$

Since our two answers are equal, it's clear that g respects addition.

Computing the left-hand side of $g(r \cdot \vec{v}) = r \cdot g(\vec{v})$ gives us

$$g(r \cdot \vec{v}) = g\left(r \cdot \begin{bmatrix} x \\ y \end{bmatrix}\right) = g\left(\begin{bmatrix} rx \\ ry \end{bmatrix}\right) = \begin{bmatrix} 2(rx) - (ry) \\ (rx) + 3(ry) \end{bmatrix} = \begin{bmatrix} 2rx - ry \\ rx + 3ry \end{bmatrix}.$$

Computing the right-hand side gives us

$$r \cdot g(\vec{v}) = r \cdot g\left(\begin{bmatrix} x \\ y \end{bmatrix}\right) = r \cdot \begin{bmatrix} 2x - y \\ x + 3y \end{bmatrix} = \begin{bmatrix} r(2x - y) \\ r(x + 3y) \end{bmatrix} = \begin{bmatrix} 2rx - ry \\ rx + 3ry \end{bmatrix}.$$

Again, since our answers are equal, g respects scalar multiplication.

Both our conditions hold, so g is a linear function.

Example 6. Show $h : \mathbb{R}^2 \to \mathbb{R}^2$ by $h\left(\begin{bmatrix} x \\ y \end{bmatrix}\right) = \begin{bmatrix} x^2 \\ y - 1 \end{bmatrix}$ is not linear.

As in the previous two examples, we start by checking whether or not $h(\vec{v} + \vec{u}) = h(\vec{v}) + h(\vec{u})$ using $\vec{v} = \begin{bmatrix} x \\ y \end{bmatrix}$ and $\vec{u} = \begin{bmatrix} z \\ w \end{bmatrix}$.

$$h(\vec{v} + \vec{u}) = h\left(\begin{bmatrix} x \\ y \end{bmatrix} + \begin{bmatrix} z \\ w \end{bmatrix}\right) = h\left(\begin{bmatrix} x + z \\ y + w \end{bmatrix}\right)$$
$$= \begin{bmatrix} (x + z)^2 \\ (y + w) - 1 \end{bmatrix} = \begin{bmatrix} x^2 + 2xz + z^2 \\ y + w - 1 \end{bmatrix}.$$

However,

$$h(\vec{v}) + h(\vec{u}) = h\left(\begin{bmatrix} x \\ y \end{bmatrix}\right) + h\left(\begin{bmatrix} z \\ w \end{bmatrix}\right) = \begin{bmatrix} x^2 \\ y - 1 \end{bmatrix} + \begin{bmatrix} z^2 \\ w - 1 \end{bmatrix}$$
$$= \begin{bmatrix} x^2 + z^2 \\ y - 1 + w - 1 \end{bmatrix} = \begin{bmatrix} x^2 + z^2 \\ y + w - 2 \end{bmatrix}.$$

These are clearly not equal (in both components!), so this function is not linear.

If I were just concerned with checking whether or not h was linear, I'd stop here. However, in the interest of practice, let's check the other condition as well using the same $\vec{v}$ and the scalar r.

$$h(r \cdot \vec{v}) = h\left(r \cdot \begin{bmatrix} x \\ y \end{bmatrix}\right) = h\left(\begin{bmatrix} rx \\ ry \end{bmatrix}\right) = \begin{bmatrix} (rx)^2 \\ (ry) - 1 \end{bmatrix} = \begin{bmatrix} r^2 x^2 \\ ry - 1 \end{bmatrix}.$$

Computing the right-hand side gives us

$$r \cdot h(\vec{v}) = r \cdot h\left(\begin{bmatrix} x \\ y \end{bmatrix}\right) = r \cdot \begin{bmatrix} x^2 \\ y - 1 \end{bmatrix} = \begin{bmatrix} rx^2 \\ r(y - 1) \end{bmatrix} = \begin{bmatrix} rx^2 \\ ry - r \end{bmatrix}.$$

Again, these are not equal (in both components), so h doesn't respect scalar multiplication. This is also enough, even without the check on addition, to show h isn't linear.

Have you noticed a pattern in the functions above? One possible pattern is that for both of our linear functions, each component of the output vector was a linear combination of the variables from the input vector. Our function that wasn't linear had both an exponent and a constant term in its output vector's components. This is another, probably easier way to determine if a function from $\mathbb{R}^n$ to $\mathbb{R}^m$ is linear. Why didn't we adopt that as our definition of a linear map? If we were only going to talk about $\mathbb{R}^n$, we could have. However, in 2.4 and Chapter 3, we'll want to expand our focus to other types of spaces and our formal definition will be easier to generalize to those situations than this second idea. In many areas of math and science you'll see this happening; collecting many ways of describing an idea and sorting through them later to figure out which one ends up working best. That decision will depend heavily on what you want to do later, so you may find yourself changing your mind. One nice thing about studying an older mathematical subject like basic linear algebra is that it has had time to settle down to a set of definitions that work best for what we want to do.

Exercises 2.1.

1. Compute $f\left(\begin{bmatrix} -3 \\ 5 \end{bmatrix}\right)$ where $f\left(\begin{bmatrix} x_1 \\ x_2 \end{bmatrix}\right) = \begin{bmatrix} x_2 \\ x_1 + x_2 \end{bmatrix}$.

2. Compute $f\left(\begin{bmatrix} 1 \\ 2 \\ 3 \end{bmatrix}\right)$ where $f\left(\begin{bmatrix} x_1 \\ x_2 \\ x_3 \end{bmatrix}\right) = \begin{bmatrix} -x_3 \\ x_1 - x_2 + x_3 \end{bmatrix}$.

3. Compute $f\left(\begin{bmatrix} 8 \\ -2 \end{bmatrix}\right)$ where $f\left(\begin{bmatrix} x_1 \\ x_2 \end{bmatrix}\right) = \begin{bmatrix} 0 \\ 2x_1 + 4x_2 \\ \frac{1}{4}x_1 \\ x_1 + x_2 \end{bmatrix}$.

4. Compute $f\left(\begin{bmatrix} 6 \\ 2 \\ 0 \\ -1 \end{bmatrix}\right)$ where $f\left(\begin{bmatrix} x_1 \\ x_2 \\ x_3 \\ x_4 \end{bmatrix}\right) = \begin{bmatrix} x_1 + x_3 + x_4 \\ -5x_2 \\ 4x_3 - x_4 \end{bmatrix}$.

5. Give the domain and codomain of $f\left(\begin{bmatrix} x_1 \\ x_2 \end{bmatrix}\right) = \begin{bmatrix} 2x_1 - x_2 \\ 0 \\ -x_1 + 4x_2 \end{bmatrix}$.

6. Give the domain and codomain of $f\left(\begin{bmatrix} x_1 \\ x_2 \\ x_3 \end{bmatrix}\right) = \begin{bmatrix} x_1 + x_2 + x_3 \\ x_1 - x_3 \end{bmatrix}$.

7. Give the domain and codomain of $f\left(\begin{bmatrix} x_1 \\ x_2 \\ x_3 \\ x_4 \\ x_5 \end{bmatrix}\right) = \begin{bmatrix} x_2 - x_4 \\ 5x_1 + x_5 \\ -x_3 \\ x_1 + x_3 \end{bmatrix}$.

8. Give the domain and codomain of $f\left(\begin{bmatrix} x_1 \\ x_2 \\ x_3 \\ x_4 \end{bmatrix}\right) = \begin{bmatrix} 3x_2 + x_4 \\ -x_1 \end{bmatrix}$.

9. Find the formula of the function $f : \mathbb{R}^2 \to \mathbb{R}^2$ which switches the order of a 2-vector's entries.

10. Find the formula of the function $f : \mathbb{R}^2 \to \mathbb{R}^3$ which multiplies the

first entry by -1, keeps the second entry the same, and has a zero in the third entry.

11. Show that $f\left(\begin{bmatrix} x_1 \\ x_2 \end{bmatrix}\right) = \begin{bmatrix} 2x_1 - x_2 \\ x_1 + 2 \end{bmatrix}$ is not a linear map.

12. Show that $f\left(\begin{bmatrix} x \\ y \end{bmatrix}\right) = \begin{bmatrix} 2y \\ x \\ 1 \end{bmatrix}$ is not a linear map.

13. Show that $f\left(\begin{bmatrix} x_1 \\ x_2 \end{bmatrix}\right) = \begin{bmatrix} x_2 \\ x_1 + x_2 \end{bmatrix}$ is a linear map.

14. Show that $f\left(\begin{bmatrix} x_1 \\ x_2 \end{bmatrix}\right) = \begin{bmatrix} x_2 \\ x_1 - x_2 \\ -x_1 \end{bmatrix}$ is a linear map.

15. If $f : \mathbb{R}^n \to \mathbb{R}^m$ is linear, explain why $f(\vec{0}_n) = \vec{0}_m$ where $\vec{0}_n$ is the zero vector in $\mathbb{R}^n$ and $\vec{0}_m$ is the zero vector in $\mathbb{R}^m$. (This can be restated as saying that a linear function always maps the zero vector of the domain to the zero vector of the codomain.)

16. If $f : \mathbb{R}^n \to \mathbb{R}^m$ is linear, explain why $f(-\vec{v}) = -f(v)$ for every $\vec{v}$ in $\mathbb{R}^n$. (This can be restated as saying that a linear function always maps the additive inverse of $\vec{v}$ to the additive inverse of $f(\vec{v})$.)

17. Use $r = 2$ and $\vec{v} = \begin{bmatrix} 0 \\ 1/2 \end{bmatrix}$ to show that the function $f : \mathbb{R}^2 \to \mathbb{R}^2$ which reflects the plane across the line $y = x + 1$ is not a linear function.

2.2   Matrices

We saw in the last section that a linear map f from $\mathbb{R}^n$ to $\mathbb{R}^m$ had a certain pattern to its vector of outputs: each entry was a linear combination of the entries of the input vector. This means if we want to describe a particular linear map to someone, we really only need to tell them three pieces of information: the size of the input vectors, the size of the output vectors, and the coefficients on each input entry that appear in each of our output entries. Instead of writing down the whole function with all the variables, we can keep track of of all three pieces of information by writing down our coefficients in a grid of numbers called a matrix.

Definition.   A *matrix* is an ordered grid of real numbers written $\begin{bmatrix} a_{11} & a_{12} & \cdots & a_{1n} \\ a_{21} & a_{22} & \cdots & a_{2n} \\ \vdots & & \ddots & \\ a_{m1} & a_{m2} & \cdots & a_{mn} \end{bmatrix}$. A matrix with m rows and n columns is called $m \times n$.

Note that when talking about matrices, our notation always puts row information before column information. Thus an $m \times n$ matrix has m rows and n columns, and a_{ij} is the entry in the ith row and jth column. As with vectors, two matrices are equal exactly when they are the same size and all corresponding pairs of entries are equal.

Example 1. Find the size of A and give a_{32} where $A = \begin{bmatrix} -3 & 0 & 1 & 8 \\ 2 & -4 & 3 & 1 \\ 5 & -1 & 0 & 7 \end{bmatrix}$.

This matrix has three rows and four columns, so it is 3×4. The entry a_{32} is in the third row and second column, so $a_{32} = -1$.

Now that we have some understanding of what a matrix is and how to write it down, let's explore the idea of using matrices to record the coefficients from our linear maps. In the first part of this section we'll use algebraic techniques and in the second part we'll use geometry.

Example 2. What matrix should we use to encode the sizes of input and output vectors and the coefficients used to build the function

$$f\left(\begin{bmatrix} x_1 \\ x_2 \end{bmatrix}\right) = \begin{bmatrix} 6x_1 + 4x_2 \\ 5x_1 + 2x_2 \\ -3x_1 + 7x_2 \end{bmatrix}?$$

Let's start by focusing on the coefficients on the variables in each entry of our output vector. Those numbers already look like they form a grid if we ignore the variables and plus signs. The rows of this grid pattern are the entries of our output vector, and the columns are for our input variables x_1 and x_2 which are the entries of our input vector. If we write down just these coefficients in this grid pattern, we get the matrix $\begin{bmatrix} 6 & 4 \\ 5 & 2 \\ -3 & 7 \end{bmatrix}$.

This matrix definitely tells us everything we need to know about the coefficients (as long as we remember that the first column contains coefficients of x_1 and the second column contains coefficients of x_2). As a bonus, it also tells us the size of the input and output vectors. Each column came from one of our variables, and those variables were the entries of our input vector. Since our matrix has two columns, this means our inputs must be 2-vectors. Similarly each row of our matrix came from the coefficients in one entry of the output vector, so we can tell our output vectors are 3-vectors because our matrix has three rows.

If we look at the pattern revealed in this last example, we get the following.

Definition. The matrix A corresponding to a linear map $f : \mathbb{R}^n \to \mathbb{R}^m$ is the $m \times n$ matrix where a_{ij} is the coefficient of x_j in the ith entry of $f(\vec{x})$.

Notice that the dimensions of our matrix are the sizes of the vectors from the domain and codomain of f, but in reverse order. As we saw in the previous example, this is because each column of f's matrix contains the coefficients on one variable from $\vec{x}$, so the number of columns is the same as the size of a vector from f's domain. Conversely, each row of f's matrix contains the coefficients from one entry of $f(\vec{x})$, so the number of rows is the same as the size of a vector from f's codomain. In fact, this is a very important idea to keep in mind: rows of f's matrix correspond to output entries while columns correspond to input entries.

Finding the matrix of a linear function may seem complicated at first, but after a few repetitions it will quickly become routine.

Example 3. Find the matrix of the function $f\left(\begin{bmatrix} x_1 \\ x_2 \\ x_3 \end{bmatrix}\right) = \begin{bmatrix} -x_1 + x_3 \\ 4x_1 - x_2 + 3x_3 \end{bmatrix}$.

The input vector has three entries, so our matrix A must have three columns. The output vector has two entries, so our matrix A must have two rows. Another way of thinking about this is that f's domain is $\mathbb{R}^3$ and its codomain is $\mathbb{R}^2$. Therefore $f : \mathbb{R}^3 \to \mathbb{R}^2$, so A is 2×3.

The first column of A contains the coefficients of x_1. In the first component of $f(\vec{x})$ we have $-x_1$, so the first entry in this column is -1. In the second output component we have $4x_1$, so the second entry in A's first column is 4. Similarly, the second column of A contains the coefficients on x_2. These are 0 in the first component of $f(\vec{x})$ and -1 in the second component. Finally, A's third column contains the coefficients on x_3. The first component of $f(\vec{x})$ has coefficient 1 and the second has coefficient 3.

Putting this all together we get that A is the 2×3 matrix $\begin{bmatrix} -1 & 0 & 1 \\ 4 & -1 & 3 \end{bmatrix}$.

Example 4. Find the matrix of the identity function $f : \mathbb{R}^n \to \mathbb{R}^n$ with $f(\vec{x}) = \vec{x}$.

Since the domain and codomain of f are both $\mathbb{R}^n$, this matrix will be $n \times n$. Each variable x_i appears only in the ith entry of $f(\vec{x})$ with coefficient 1, so the jth column of f's matrix has all entries equal to zero except for a 1 in the jth spot. Thus f has matrix

$$I_n = \begin{bmatrix} 1 & 0 & 0 & \cdots & 0 \\ 0 & 1 & 0 & \cdots & 0 \\ \vdots & \vdots & \ddots & & \vdots \\ 0 & 0 & & 1 & 0 \\ 0 & 0 & \cdots & 0 & 1 \end{bmatrix}.$$

Since f is the identity map, its matrix I_n is called the identity matrix.

In biology, linear functions are often used to model population dynamics. The matrices of these functions are called demographic matrices.

Example 5. In Example 6 from Chapter 0, we discussed setting up a vector to model the five stages (seedling, small plant, medium plant, large plant, and flowering) of the life cycle of the smooth coneflower. A recent study measured how plants moved between these stages after a year of growth. It found that seedlings have a 0.35 probability of becoming a small plant. Small plants have a 0.78 probability of staying small plants, a 0.18 probability of becoming medium plants, a 0.03 probability of becoming large plants, and a 0.01 probability of flowering. Medium plants have a 0.24 probability of becoming small plants, a 0.49 probability of staying medium plants, a 0.17 probability of becoming large plants, and a 0.10 probability of flowering. Large plants have a 0.07 probability of becoming small plants, a 0.21 probability of becoming medium plants, a 0.38 probability of staying large plants, and a 0.33 probability of flowering. Flowering plants have a 0.43 probability of becoming small plants, a 0.28 probability of becoming medium plants, a 0.18 probability of becoming large plants, and a 0.11 probability of flowering. Flowering plants also have a 17.76 probability of producing seedlings. (Normally this wouldn't be an acceptable number for a probability because it is greater than 1, but one flowering plant is capable of producing multiple seedlings.) Find the matrix of the function that takes a given year's population vector and gives the next year's population vector.

Let's start by thinking about the size of this matrix. Our input and output vectors both have 5 entries, one for each stage in the coneflower's life cycle. This means we have $f : \mathbb{R}^5 \to \mathbb{R}^5$ and our matrix is 5×5.

As in Chapter 0's Example 6, let's order the entries in our population vectors $\begin{bmatrix} x_1 \\ x_2 \\ x_3 \\ x_4 \\ x_5 \end{bmatrix}$ so that x_1 counts seedlings, x_2 counts small plants, x_3 counts medium plants, x_4 counts large plants, and x_5 counts flowering plants.

The first row of our matrix is the coefficients on these x_is that make up the first entry of $f(\vec{x})$, in other words, the coefficients on the current year's life cycle counts which tell us how many seedlings there will be next year. The only life cycle stage that produces seedlings is flowering plants, and each flowering plant produces 17.76 seedlings. This means that the first entry of $f(\vec{x})$ is $17.76x_5$.

The second row of our matrix is the coefficients that tell us how to compute the number of small plants there will be next year. Each seedling has a 0.35 chance to become a small plant, each small plant has a 0.78 chance to stay a small plant, each medium plant has a 0.24 chance to become a small plant, each large plant has a 0.07 chance to become a small plant, and each flowering plant has a 0.43 chance to become a small plant. This means the second entry of $f(\vec{x})$ is $0.35x_1 + 0.78x_2 + 0.24x_3 + 0.07x_4 + 0.43x_5$.

The third row of our matrix is the coefficients that tell us how to compute the number of medium plants there will be next year. No seedlings become medium plants, each small plant has a 0.18 chance to become a medium plant, each medium plant has a 0.49 chance to stay a medium plant, each large plant has a 0.21 chance to become a medium plant, and each flowering plant has a 0.28 chance to become a medium plant. This means the third entry of $f(\vec{x})$ is $0.18x_2 + 0.49x_3 + 0.21x_4 + 0.28x_5$.

The fourth row of our matrix is the coefficients that tell us how to compute the number of large plants there will be next year. No seedlings become large plants, each small plant has a 0.03 chance to become a large plant, each medium plant has a 0.17 chance to become a large plant, each large plant has a 0.38 chance to stay a large plant, and each flowering plant has a 0.18 chance to become a large plant. This means the fourth entry of $f(\vec{x})$ is $0.03x_2 + 0.17x_3 + 0.38x_4 + 0.18x_5$.

The fifth row of our matrix is the coefficients that tell us how to compute the number of flowering plants there will be next year. No seedlings become flowering plants, each small plant has a 0.01 chance to become a flowering

64 Functions of Vectors

plant, each medium plant has a 0.10 chance to become a flowering plant, each large plant has a 0.33 chance to become a flowering plant, and each flowering plant has a 0.11 chance to stay a flowering plant. This means the fifth entry of $f(\vec{x})$ is $0.01x_2 + 0.10x_3 + 0.33x_4 + 0.11x_5$.

Putting the coefficients from each entry of $f(\vec{x})$ into the corresponding rows of f's matrix gives us

$$\begin{bmatrix} 0 & 0 & 0 & 0 & 17.76 \\ 0.35 & 0.78 & 0.24 & 0.07 & 0.43 \\ 0 & 0.18 & 0.49 & 0.21 & 0.28 \\ 0 & 0.03 & 0.17 & 0.38 & 0.18 \\ 0 & 0.01 & 0.10 & 0.33 & 0.11 \end{bmatrix}.$$

The idea of building a matrix whose entries are probabilities of going from one stage in a process to another is part of a type of modeling called a discrete Markov chain. These are used in many other applications where we're transitioning from one state to another. For example, in game theory the states may be squares on a game board and the probabilities in the associated matrix will encode how likely a player is to move between a given pair of squares.

Now that we understand how to find the matrix of a given function, we can turn this process around to find the function of a given matrix.

Example 6. Find the equation of the function corresponding to the matrix
$A = \begin{bmatrix} 4 & 0 \\ 1 & -3 \\ 2 & 7 \end{bmatrix}.$

Our matrix A has three rows and two columns, so is 3×2. This means it corresponds to a function $f : \mathbb{R}^2 \to \mathbb{R}^3$. This means we're trying to fill in the gaps in the equation

$$f\left(\begin{bmatrix} x_1 \\ x_2 \end{bmatrix}\right) = \begin{bmatrix} - \\ - \\ - \end{bmatrix}.$$

Each row of our matrix contains the coefficients (on x_1 in the first column and x_2 in the second column) of an entry of $f(\vec{x})$. The first row gives us $4x_1 + 0x_2$, the second row gives us $x_1 - 3x_2$, and the third row gives us $2x_1 + 7x_2$. Plugging this into our formula above means

$$f\left(\begin{bmatrix} x_1 \\ x_2 \end{bmatrix}\right) = \begin{bmatrix} 4x_1 \\ x_1 - 3x_2 \\ x_1 + 7x_2 \end{bmatrix}.$$

Suppose we are given the matrix A of a linear function $f : \mathbb{R}^n \to \mathbb{R}^m$, and we want to find $f(\vec{x})$ for some vector $\vec{x}$ from $\mathbb{R}^n$. We could find the function's equation as in Example 6 and then plug in $\vec{x}$, but that seems a bit complicated to do every time – especially if we want to compute $f(\vec{x})$ for multiple vectors! Instead, let's figure out a way to go directly from A and $\vec{x}$ to $f(\vec{x})$. We'll write this as the product of a matrix and a vector, so $A\vec{x} = f(\vec{x})$.

Before we figure out how to compute the entries of the vector $A\vec{x}$, we need to talk about how to make sure the sizes of A and $\vec{x}$ are compatible. For $f(\vec{x})$ to make sense, we need $\vec{x}$ to be in the domain of f. We can connect this to A by remembering that the size of f's input vectors is the same as the number of columns of A. Therefore $A\vec{x}$ only makes sense if A has n columns and $\vec{x}$ is an n-vector. The product $A\vec{x}$ is $f(\vec{x})$, which means it is in the codomain of f. Since the size of f's codomain is the number of rows of A, this means $A\vec{x}$ is a vector with as many entries as A has rows.

Personally, I've always remembered this by restating it as "the adjacent numbers match", i.e., $(m \times n$ matrix)$(n$-vector), "and cancelling the matching numbers leaves the size of the product", i.e., removing the n's leaves m.

Example 7. What are the sizes of $\vec{x}$ and $A\vec{x}$ if A is a 5×8 matrix?

Because A is 5×8, we know $f : \mathbb{R}^8 \to \mathbb{R}^5$. Since $\vec{x}$ is an input of f, this means it must be an 8-vector. Similarly, since $A\vec{x}$ is an output of f, it must be a 5-vector.

Alternately we need $(5 \times 8$ matrix)$(8$-vector) to get the adjacent numbers to match. The leftover number 5 is the size of the vector $A\vec{x}$.

Example 8. Explain why we can't multiply the matrix $A = \begin{bmatrix} -1 & 0 & 1 \\ 2 & 4 & -1 \end{bmatrix}$ by the vector $\vec{x} = \begin{bmatrix} 0 \\ 1 \end{bmatrix}$.

Here our matrix A is 2×3 but $\vec{x}$ is a 2-vector. Since the size of $\vec{x}$ doesn't match the number of columns of A, we know $\vec{x}$ isn't in the domain of A's function. Therefore this product isn't defined.

Now that we understand how to ensure $A\vec{x}$ is defined and what size vector it is, let's figure out how to compute its entries from those of A and $\vec{x}$.

Example 9. Let $A = \begin{bmatrix} 2 & 3 \\ 4 & -2 \end{bmatrix}$ and $\vec{x} = \begin{bmatrix} 5 \\ -8 \end{bmatrix}$. Use the fact that $A\vec{x} = f(\vec{x})$ to compute $A\vec{x}$.

Using the same process as Example 6, we find that A is the matrix of the function $f\left(\begin{bmatrix} x_1 \\ x_2 \end{bmatrix}\right) = \begin{bmatrix} 2x_1 + 3x_2 \\ 4x_1 - 2x_2 \end{bmatrix}$. Plugging in $\vec{x} = \begin{bmatrix} 5 \\ -8 \end{bmatrix}$, we get

$$f\left(\begin{bmatrix} 5 \\ -8 \end{bmatrix}\right) = \begin{bmatrix} 2(5) + 3(-8) \\ 4(5) - 2(-8) \end{bmatrix}$$

so since $A\vec{x} = f(\vec{x})$ means we want

$$\begin{bmatrix} 2 & 3 \\ 4 & -2 \end{bmatrix} \begin{bmatrix} 5 \\ -8 \end{bmatrix} = \begin{bmatrix} 2(5) + 3(-8) \\ 4(5) - 2(-8) \end{bmatrix}.$$

Looking at this last equation, we can start to see a pattern emerge. To get the top entry of $A\vec{x}$, we added together the first entry in the top row of A times the first entry of $\vec{x}$ and the second entry in the top row of A times the second entry of $\vec{x}$. To get the bottom entry of $A\vec{x}$, we added together the first entry in the bottom row of A times the first entry of $\vec{x}$ and the second entry in the bottom row of A times the second entry of $\vec{x}$. This should seem reasonable, since it means we're using the entries of A as coefficients on the entries of $\vec{x}$.

We can generalize the pattern we observed in the previous example as follows: to find the ith entry of $A\vec{x}$ we combine the entries from the ith row of the matrix A with the entries of $\vec{x}$. Each matrix row is combined with $\vec{x}$ by adding together the pairwise products of their corresponding entries. Pairing up the entries along a row with the entries of $\vec{x}$ in this way should make sense, because the jth entry of each row of A is a coefficient on the jth entry, x_j, of $\vec{x}$.

This motivates the following definition.

Definition. The $m \times n$ matrix $A = \begin{bmatrix} a_{11} & a_{12} & \cdots & a_{1n} \\ a_{21} & a_{22} & \cdots & a_{2n} \\ \vdots & & \ddots & \\ a_{m1} & a_{m2} & \cdots & a_{mn} \end{bmatrix}$ and n-vector

$\vec{x} = \begin{bmatrix} x_1 \\ x_2 \\ \vdots \\ x_n \end{bmatrix}$ have m-vector product $A\vec{x} = \begin{bmatrix} a_{11}x_1 + a_{12}x_2 + \cdots + a_{1n}x_n \\ a_{21}x_1 + a_{22}x_2 + \cdots + a_{2n}x_n \\ \vdots \\ a_{m1}x_1 + a_{m2}x_2 + \cdots + a_{mn}x_n \end{bmatrix}.$

Notice that since a_{ij} is the coefficient on x_j in the ith component of $A\vec{x}$, if A is the matrix of a linear map f we have $A\vec{x} = f(\vec{x})$ as planned.

This definition's notation is quite complicated, but once you have the overall pattern down, it becomes more routine. Let's walk through a few examples.

Example 10. Compute $A\vec{x}$ where $A = \begin{bmatrix} -1 & 0 & 1 \\ 2 & 4 & -1 \end{bmatrix}$ and $\vec{x} = \begin{bmatrix} 2 \\ -1 \\ 3 \end{bmatrix}$.

First of all, notice that A is 2×3 and $\vec{x}$ is a 3-vector, so this product makes sense. It also tells us $A\vec{x}$ is a 2-vector.

To find the first entry of $A\vec{x}$, we add up the pairwise products across the first row of A and down $\vec{x}$.

$$\begin{bmatrix} \boxed{-1 \quad 0 \quad 1} \\ 2 \quad 4 \quad -1 \end{bmatrix} \begin{bmatrix} 2 \\ -1 \\ 3 \end{bmatrix}.$$

This gives us $(-1)(2) + (0)(-1) + (1)(3) = 1$. The second entry of $A\vec{x}$ is the same format, but uses the second row of A.

$$\begin{bmatrix} -1 \quad 0 \quad 1 \\ \boxed{2 \quad 4 \quad -1} \end{bmatrix} \begin{bmatrix} 2 \\ -1 \\ 3 \end{bmatrix}.$$

This gives us $(2)(2) + (4)(-1) + (-1)(3) = -3$.

Therefore $A\vec{x} = \begin{bmatrix} 1 \\ -3 \end{bmatrix}$.

Now that we understand how to multiply a vector by a matrix, we can use our new skill to compute $f(\vec{x})$.

Example 11. Compute $f\left(\begin{bmatrix} 2 \\ 1 \\ 1 \end{bmatrix}\right)$ where f has matrix $A = \begin{bmatrix} -1 & 0 & 1 \\ 4 & -5 & 3 \end{bmatrix}$.

By our construction of matrix-vector multiplication, we know

$$f\left(\begin{bmatrix} 2 \\ 1 \\ 1 \end{bmatrix}\right) = \begin{bmatrix} -1 & 0 & 1 \\ 4 & -5 & 3 \end{bmatrix} \begin{bmatrix} 2 \\ 1 \\ 1 \end{bmatrix}.$$

Adding up the pairwise products along the rows of our matrix and down our vector gives us

$$\begin{bmatrix} -1 & 0 & 1 \\ 4 & -5 & 3 \end{bmatrix} \begin{bmatrix} 2 \\ 1 \\ 1 \end{bmatrix} = \begin{bmatrix} -1(2) + 0(1) + 1(1) \\ 4(2) + (-5)(1) + 3(1) \end{bmatrix} = \begin{bmatrix} -1 \\ 6 \end{bmatrix}.$$

Thus $f\left(\begin{bmatrix} 2 \\ 1 \\ 1 \end{bmatrix}\right) = \begin{bmatrix} -1 \\ 6 \end{bmatrix}.$

The fact that multiplying $\vec{x}$ by a matrix A is the same as plugging $\vec{x}$ into the corresponding linear function f means that this type of multiplication is linear in the sense that $A(\vec{v} + \vec{w}) = A\vec{v} + A\vec{w}$ and $A(r \cdot \vec{v}) = r \cdot A\vec{v}$ since

$$A(\vec{v} + \vec{w}) = f(\vec{v} + \vec{w}) = f(\vec{v}) + f(\vec{w}) = A\vec{v} + A\vec{w}$$

and
$$A(r \cdot \vec{v}) = f(r \cdot \vec{v}) = r \cdot f(\vec{v}) = r \cdot A\vec{v}.$$

We can relate this new multiplication to the linear combinations of vectors discussed in 1.2. There we were discussing the span of a set of vectors $\vec{v}_1, \vec{v}_2, \ldots, \vec{v}_n$, which had the form $x_1\vec{v}_1 + x_2\vec{v}_2 + \cdots + x_n\vec{v}_n$. If we have an $m \times n$ matrix A and we think of its n columns as the m-vectors $\vec{a}_1, \vec{a}_2, \ldots, \vec{a}_n$, we can rewrite $A\vec{x}$ as

$$A\vec{x} = \begin{bmatrix} a_{11}x_1 + a_{12}x_2 + \cdots + a_{1n}x_n \\ a_{21}x_1 + a_{22}x_2 + \cdots + a_{2n}x_n \\ \vdots \\ a_{m1}x_1 + a_{m2}x_2 + \cdots + a_{mn}x_n \end{bmatrix}$$

$$= \begin{bmatrix} a_{11}x_1 \\ a_{21}x_1 \\ \vdots \\ a_{m1}x_1 \end{bmatrix} + \begin{bmatrix} a_{12}x_2 \\ a_{22}x_2 \\ \vdots \\ a_{m2}x_2 \end{bmatrix} + \cdots + \begin{bmatrix} a_{1n}x_n \\ a_{2n}x_n \\ \vdots \\ a_{mn}x_n \end{bmatrix}$$

$$= x_1 \begin{bmatrix} a_{11} \\ a_{21} \\ \vdots \\ a_{m1} \end{bmatrix} + x_2 \begin{bmatrix} a_{12} \\ a_{22} \\ \vdots \\ a_{m2} \end{bmatrix} + \cdots + x_n \begin{bmatrix} a_{1n} \\ a_{2n} \\ \vdots \\ a_{mn} \end{bmatrix}$$

$$= x_1\vec{a}_1 + x_2\vec{a}_2 + \cdots + x_n\vec{a}_n.$$

This means the vector equation $x_1\vec{v}_1 + x_2\vec{v}_2 + \cdots + x_n\vec{v}_n = \vec{b}$ can also be thought of as the matrix equation $A\vec{x} = \vec{b}$ or the equation $f(\vec{x}) = \vec{b}$ where f is the linear function whose matrix is A. This connection shouldn't be totally surprising, since we constructed A so that the entries in its jth column were the coefficients of x_j in $f(\vec{x})$.

Example 12. Rewrite the matrix-vector product $A\vec{x}$ as a linear combination of vectors where $A = \begin{bmatrix} -1 & 0 & 1 \\ 2 & 4 & -1 \end{bmatrix}$.

To do this, we need to split A up into column vectors and multiply each column by the appropriate variable. Since A has three columns, $\vec{x}$ is a 3-vector so our variables are x_1, x_2, and x_3. Thus

$$A\vec{x} = \begin{bmatrix} -1 & 0 & 1 \\ 2 & 4 & -1 \end{bmatrix} \begin{bmatrix} x_1 \\ x_2 \\ x_3 \end{bmatrix} = x_1 \begin{bmatrix} -1 \\ 2 \end{bmatrix} + x_2 \begin{bmatrix} 0 \\ 4 \end{bmatrix} + x_3 \begin{bmatrix} 1 \\ -1 \end{bmatrix}.$$

Example 13. Write $x_1 \begin{bmatrix} 3 \\ -2 \\ 1 \\ 0 \end{bmatrix} + x_2 \begin{bmatrix} -1 \\ 4 \\ 2 \\ 6 \end{bmatrix} = \begin{bmatrix} 1 \\ 5 \\ 0 \\ -3 \end{bmatrix}$ as a matrix equation.

To write this as a matrix equation, we simply need to rewrite the left-hand side as $A\vec{x}$ for some matrix A. This is the opposite of what we did in the previous example, so A is the matrix whose columns are the vectors on the left of our equation. This means we can rewrite our vector equation as

$$\begin{bmatrix} 3 & -1 \\ -2 & 4 \\ 1 & 2 \\ 0 & 6 \end{bmatrix} \vec{x} = \begin{bmatrix} 1 \\ 5 \\ 0 \\ -3 \end{bmatrix}.$$

Example 14. Let $f : \mathbb{R}^3 \to \mathbb{R}^3$ by $f\left(\begin{bmatrix} x_1 \\ x_2 \\ x_3 \end{bmatrix} \right) = \begin{bmatrix} 7x_1 - x_2 + 6x_3 \\ 2x_1 + 2x_2 - 4x_3 \\ 5x_1 - 3x_2 + 8x_3 \end{bmatrix}$. Find $\vec{x}$ so that $f(\vec{x}) = \vec{b}$ where $\vec{b} = \begin{bmatrix} 5 \\ -2 \\ 5 \end{bmatrix}$.

We now have three options on how to tackle this problem, because we have three ways of writing this equation: as $f(\vec{x}) = \vec{b}$, as $A\vec{x} = \vec{b}$ where A is f's matrix, or as the vector equation $x_1\vec{a}_1 + x_2\vec{a}_2 + x_3\vec{a}_3 = \vec{b}$ where the vectors $\vec{a}_1$, $\vec{a}_2$, and $\vec{a}_3$ are the columns of A. All of these methods will give us the same answer for $\vec{x}$, so we can choose whichever one seems easiest to us.

Solving for $\vec{x}$ directly from $f(\vec{x}) = \vec{b}$ doesn't seem easy, so let's explore our two other options by rewriting this equation as both a matrix and vector equation. In both cases, we need to find the matrix A of our function f. Since $f : \mathbb{R}^3 \to \mathbb{R}^3$, we know A is a 3×3 matrix. Picking off the coefficients from f, gives us

$$A = \begin{bmatrix} 7 & -1 & 6 \\ 2 & 2 & -4 \\ 5 & -3 & 8 \end{bmatrix}.$$

This means our matrix equation $A\vec{x} = \vec{b}$ is

$$\begin{bmatrix} 7 & -1 & 6 \\ 2 & 2 & -4 \\ 5 & -3 & 8 \end{bmatrix} \begin{bmatrix} x_1 \\ x_2 \\ x_3 \end{bmatrix} = \begin{bmatrix} 5 \\ -2 \\ 5 \end{bmatrix}.$$

Unfortunately, this doesn't seem easier, so let's try the vector equation version. The three columns of A are $\vec{a}_1 = \begin{bmatrix} 7 \\ 2 \\ 5 \end{bmatrix}$, $\vec{a}_2 = \begin{bmatrix} -1 \\ 2 \\ -3 \end{bmatrix}$, and $\vec{a}_3 = \begin{bmatrix} 6 \\ -4 \\ 8 \end{bmatrix}$. This means our vector equation $x_1\vec{a}_1 + x_2\vec{a}_2 + x_3\vec{a}_3 = \vec{b}$ is

$$x_1\begin{bmatrix}7\\2\\5\end{bmatrix} + x_2\begin{bmatrix}-1\\2\\-3\end{bmatrix} + x_3\begin{bmatrix}6\\-4\\8\end{bmatrix} = \begin{bmatrix}5\\-2\\5\end{bmatrix}.$$

In this format it is easier to realize that $\vec{b} = \vec{a}_2 + \vec{a}_3$. This means our solution

is $x_1 = 0$, $x_2 = 1$, and $x_3 = 1$ or $\vec{x} = \begin{bmatrix}0\\1\\1\end{bmatrix}$. (You can go back and check that

this vector satisfies the other two formats as well.)

If you're worried you wouldn't have noticed this relationship on your own, don't worry. We'll be discussing how to solve $f(\vec{x}) = \vec{b}$ in more detail soon.

Let's switch gears for a moment and explore how to use the geometric action of a function on $\mathbb{R}^n$ to find its matrix. Since we don't yet know how to do that, we'll start in the opposite direction by exploring the geometric effects of a matrix in hopes that once we understand this process we can reverse it.

Example 15. Suppose f has matrix $A = \begin{bmatrix}0 & 1\\-1 & 0\end{bmatrix}$. What does f do geometrically?

The first thing to notice here is that since A is 2×2 we know $f : \mathbb{R}^2 \to \mathbb{R}^2$. This means we're looking at the geometric effect f has on the plane.

One way to explore the impact of f geometrically is to start with a picture of a shape in the plane and compare that starting shape with its image after we apply our map f. Let's start with the unit square pictured below.

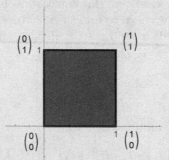

If we plug each of the corners of our unit square into f, we get

$$f\left(\begin{bmatrix}0\\0\end{bmatrix}\right) = \begin{bmatrix}0 & 1\\-1 & 0\end{bmatrix}\begin{bmatrix}0\\0\end{bmatrix} = \begin{bmatrix}0\\0\end{bmatrix},$$

$$f\left(\begin{bmatrix}1\\0\end{bmatrix}\right) = \begin{bmatrix}0 & 1\\-1 & 0\end{bmatrix}\begin{bmatrix}1\\0\end{bmatrix} = \begin{bmatrix}0\\-1\end{bmatrix},$$

$$f\left(\begin{bmatrix} 1 \\ 1 \end{bmatrix}\right) = \begin{bmatrix} 0 & 1 \\ -1 & 0 \end{bmatrix} \begin{bmatrix} 1 \\ 1 \end{bmatrix} = \begin{bmatrix} 1 \\ -1 \end{bmatrix},$$

and

$$f\left(\begin{bmatrix} 0 \\ 1 \end{bmatrix}\right) = \begin{bmatrix} 0 & 1 \\ -1 & 0 \end{bmatrix} \begin{bmatrix} 0 \\ 1 \end{bmatrix} = \begin{bmatrix} 1 \\ 0 \end{bmatrix}.$$

These image points outline a new square as shown below. (The dotted outline is the original unit square.)

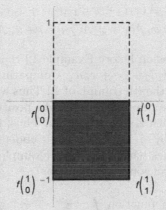

Looking at this image of the unit square, we can guess that f is the function which rotates the plane clockwise by 90°. However, we also saw an interesting thing: $f\left(\begin{bmatrix} 1 \\ 0 \end{bmatrix}\right)$ was the first column of f's matrix and $f\left(\begin{bmatrix} 0 \\ 1 \end{bmatrix}\right)$ was the second column of f's matrix.

Why should the images of the $\begin{bmatrix} 1 \\ 0 \end{bmatrix}$ and $\begin{bmatrix} 0 \\ 1 \end{bmatrix}$ have given us the columns of our function's matrix in the example above? If we think geometrically about a vector $\vec{x}$ in $\mathbb{R}^n$, the entries of $\vec{x}$ tell us that vector's position along the n axes of $\mathbb{R}^n$. We can think of each axis as being the span of a vector of length 1 which lies along that axis in the positive direction. These are the n special n-vectors

$$\vec{e}_1 = \begin{bmatrix} 1 \\ 0 \\ 0 \\ \vdots \\ 0 \end{bmatrix}, \vec{e}_2 = \begin{bmatrix} 0 \\ 1 \\ 0 \\ \vdots \\ 0 \end{bmatrix}, \ldots, \vec{e}_n = \begin{bmatrix} 0 \\ 0 \\ \vdots \\ 0 \\ 1 \end{bmatrix},$$

which are sometimes called the standard unit vectors.

This makes

$$\vec{x} = \begin{bmatrix} x_1 \\ x_2 \\ \vdots \\ x_n \end{bmatrix} = x_1 \vec{e}_1 + x_2 \vec{e}_2 + \cdots + x_n \vec{e}_n.$$

Multiplying a vector by a matrix splits up over linear combinations, so

$$\begin{aligned} A\vec{x} &= A\left(x_1 \vec{e}_1 + x_2 \vec{e}_2 + \cdots + x_n \vec{e}_n\right) \\ &= x_1 A\vec{e}_1 + x_2 A\vec{e}_2 + \cdots + x_n A\vec{e}_n. \end{aligned}$$

Recall from the discussion before Example 12 that if $\vec{a}_i$ is the ith column of A, then $A\vec{x} = x_1\vec{a}_1 + x_2\vec{a}_2 + \cdots + x_n\vec{a}_n$. Comparing this with the equation above tells us that $A\vec{e}_i$ is the ith column of A! Thus we can interpret a matrix geometrically by using the rule that A maps the positive unit vector along the ith axis to the vector which is its ith column. Thus, to understand the effect on $\mathbb{R}^n$ of multiplication by A, it is enough to understand the effect of A on $\vec{e}_1, \vec{e}_2, \ldots, \vec{e}_n$. This matches what we saw in Example 15. Next, let's use this idea to find the matrices of some 2D functions.

Example 16. Find the function $f : \mathbb{R}^2 \to \mathbb{R}^2$ which rotates the plane counterclockwise by $45°$.

We can do this by figuring out where f sends each of the positive unit vectors along the axes and using them to create f's matrix. Put more concretely, we need to use some geometry to find $f\left(\begin{bmatrix} 1 \\ 0 \end{bmatrix}\right)$ and $f\left(\begin{bmatrix} 0 \\ 1 \end{bmatrix}\right)$. Let's start by visualizing f's effect on the plane.

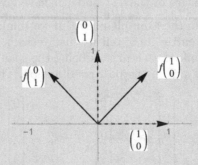

The image of $\begin{bmatrix} 1 \\ 0 \end{bmatrix}$ is the vector of length 1 along the line $y = x$. We can use the Pythagorean Theorem to figure out its coordinates.

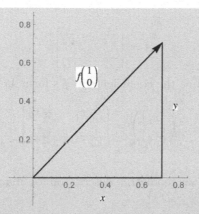

Rotating $\begin{bmatrix} 1 \\ 0 \end{bmatrix}$ doesn't change its length, so from the picture above, we can see that $x^2 + y^2 = 1$ and $x = y$. This means $x = y = \frac{1}{\sqrt{2}}$, so

$$f\left(\begin{bmatrix} 1 \\ 0 \end{bmatrix}\right) = \begin{bmatrix} \frac{1}{\sqrt{2}} \\ \frac{1}{\sqrt{2}} \end{bmatrix}.$$

Similarly, we can use the image of $\begin{bmatrix} 0 \\ 1 \end{bmatrix}$ in the picture below to figure out its exact coordinates.

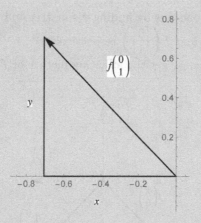

Here $x = -\frac{1}{\sqrt{2}}$ and $y = \frac{1}{\sqrt{2}}$, so

$$f\left(\begin{bmatrix} 0 \\ 1 \end{bmatrix}\right) = \begin{bmatrix} -\frac{1}{\sqrt{2}} \\ \frac{1}{\sqrt{2}} \end{bmatrix}.$$

Using these two images (in order) as the columns of f's matrix, we see that $f(\vec{x}) = A\vec{x}$ where

$$A = \begin{bmatrix} \frac{1}{\sqrt{2}} & -\frac{1}{\sqrt{2}} \\ \frac{1}{\sqrt{2}} & \frac{1}{\sqrt{2}} \end{bmatrix}.$$

Using the entries of A as the coefficients in f's formula gives

$$f\left(\begin{bmatrix} x_1 \\ x_2 \end{bmatrix}\right) = \begin{bmatrix} \frac{1}{\sqrt{2}}x_1 - \frac{1}{\sqrt{2}}x_2 \\ \frac{1}{\sqrt{2}}x_1 + \frac{1}{\sqrt{2}}x_2 \end{bmatrix}.$$

Example 17. Give a geometric description of map $f : \mathbb{R}^2 \to \mathbb{R}^2$ by
$$f\left(\begin{bmatrix} x_1 \\ x_2 \end{bmatrix}\right) = \begin{bmatrix} -\frac{1}{\sqrt{2}}x_1 + \frac{1}{\sqrt{2}}x_2 \\ -\frac{1}{\sqrt{2}}x_1 - \frac{1}{\sqrt{2}}x_2 \end{bmatrix}.$$

To understand what f does to $\mathbb{R}^2$ geometrically, let's start by computing the images of the two positive unit vectors along the axes. This gives us

$$f\left(\begin{bmatrix} 1 \\ 0 \end{bmatrix}\right) = \begin{bmatrix} -\frac{1}{\sqrt{2}} \\ -\frac{1}{\sqrt{2}} \end{bmatrix}$$

and

$$f\left(\begin{bmatrix} 0 \\ 1 \end{bmatrix}\right) = \begin{bmatrix} \frac{1}{\sqrt{2}} \\ -\frac{1}{\sqrt{2}} \end{bmatrix}.$$

(We could also have done this by finding f's matrix and remembering that its columns are $f\left(\begin{bmatrix} 1 \\ 0 \end{bmatrix}\right)$ and $f\left(\begin{bmatrix} 0 \\ 1 \end{bmatrix}\right)$ respectively.)

Visually, this means we get the following picture of f's effect on the plane.

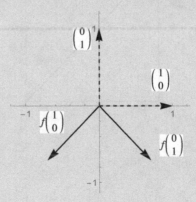

Looking at the picture, we can see that f rotates $\mathbb{R}^2$ clockwise $135°$.

Our geometric and equation oriented approaches to finding the matrix of a linear function give us the option to find a function's matrix whichever way seems easier for our particular function. The option to view a problem as $f(\vec{x}) = \vec{b}$, $A\vec{x} = \vec{b}$, or $x_1\vec{a}_1 + \cdots + x_n\vec{a}_n = \vec{b}$ gives us several different ways to solve for $\vec{x}$.

Exercises 2.2.

1. Consider $A\vec{x} = \vec{b}$, where A is a 4×6 matrix.

   (a) What size vector is $\vec{x}$?

   (b) What size vector is $\vec{b}$?

2. Consider $A\vec{x} = \vec{b}$, where A is a 3×8 matrix.

   (a) What size vector is $\vec{x}$?

   (b) What size vector is $\vec{b}$?

3. Consider $A\vec{x} = \vec{b}$, where A is a 5×4 matrix.

   (a) What size vector is $\vec{x}$?

   (b) What size vector is b?

4. Consider $A\vec{x} = \vec{b}$, where A is a 7×2 matrix.

   (a) What size vector is $\vec{x}$?

   (b) What size vector is $\vec{b}$?

5. Let $f(\vec{x}) = A\vec{x}$ where $A = \begin{bmatrix} 1 & 0 & 0 & -2 \\ 0 & 0 & 1 & 3 \end{bmatrix}$.

   (a) What is the domain of f?

   (b) What is the codomain of f?

6. Let $f(\vec{x}) = A\vec{x}$ where $A = \begin{bmatrix} 1 & -3 & 0 \\ 0 & 0 & 1 \\ 0 & 0 & 0 \\ 0 & 0 & 0 \end{bmatrix}$.

   (a) What is the domain of f?

   (b) What is the codomain of f?

7. Let f be the linear map whose matrix is $\begin{bmatrix} 1 & -1 \\ 4 & -3 \\ 0 & 3 \end{bmatrix}$.

   (a) What is the domain of f?

   (b) What is the codomain of f?

8. Let f be the linear map whose matrix is $A = \begin{bmatrix} 1 & 1 & 0 \\ 0 & 0 & -7 \\ 2 & 5 & 3 \end{bmatrix}$.

(a) What is the domain of f?

(b) What is the codomain of f?

9. Let $f\left(\begin{bmatrix} x_1 \\ x_2 \\ x_3 \end{bmatrix}\right) = \begin{bmatrix} 3x_1 - x_3 \\ -x_1 + x_2 + 2x_3 \end{bmatrix}$

    (a) What is the domain of this map?

    (b) What is the codomain of this map?

    (c) What is the size of f's matrix?

10. Let $f\left(\begin{bmatrix} x_1 \\ x_2 \end{bmatrix}\right) = \begin{bmatrix} -x_1 + 4x_2 \\ x_2 \\ -2x_1 + 3x_2 \end{bmatrix}$

    (a) What is the domain of this map?

    (b) What is the codomain of this map?

    (c) What is the size of f's matrix?

11. Let $f\left(\begin{bmatrix} x_1 \\ x_2 \\ x_3 \\ x_4 \end{bmatrix}\right) = \begin{bmatrix} -x_1 + 4x_2 - 3x_4 \\ x_2 + x_3 \\ -2x_1 + 3x_2 \end{bmatrix}$

    (a) What is the domain of this map?

    (b) What is the codomain of this map?

    (c) What is the size of f's matrix?

12. Let $f\left(\begin{bmatrix} x_1 \\ x_2 \end{bmatrix}\right) = \begin{bmatrix} x_1 + 2x_2 \\ 2x_1 + 5x_2 \end{bmatrix}$

    (a) What is the domain of this map?

    (b) What is the codomain of this map?

    (c) What is the size of f's matrix?

13. Let $f : \mathbb{R}^3 \to \mathbb{R}^2$ by $f\left(\begin{bmatrix} x \\ y \\ z \end{bmatrix}\right) = \begin{bmatrix} 2x - y + z \\ -x + 10z \end{bmatrix}$. Find the matrix A so that $f(\vec{x}) = A\vec{x}$.

14. Let $f : \mathbb{R}^3 \to \mathbb{R}^3$ by $f\left(\begin{bmatrix} x \\ y \\ z \end{bmatrix}\right) = \begin{bmatrix} 2x - y \\ 3z \\ y + 8z - x \end{bmatrix}$. Find the matrix A so that $f(\vec{x}) = A\vec{x}$.

15. Find the matrix of the linear map $f\left(\begin{bmatrix} x \\ y \\ z \end{bmatrix}\right) = \begin{bmatrix} x + y \\ 2x + y \\ 2y - z \end{bmatrix}$.

16. Find the matrix of the linear map $f\left(\begin{bmatrix} x_1 \\ x_2 \\ x_3 \end{bmatrix}\right) = \begin{bmatrix} -x_2 + x_3 \\ 2x_1 + 4x_2 + 2x_3 \\ x_1 - 3x_2 + x_3 \\ x_2 + x_3 \end{bmatrix}$.

17. Find the matrix of the linear map $f\left(\begin{bmatrix} x_1 \\ x_2 \\ x_3 \\ x_4 \end{bmatrix}\right) = \begin{bmatrix} 3x_1 - 7x_3 + x_4 \\ x_1 + x_3 + x_4 \\ -x_1 + 8x_3 \end{bmatrix}$.

18. Find the matrix of the linear map $f\left(\begin{bmatrix} x_1 \\ x_2 \\ x_3 \end{bmatrix}\right) = \begin{bmatrix} 2x_1 - 3x_2 + x_3 \\ x_1 - x_3 \end{bmatrix}$.

19. Find the matrix of the linear map $f : \mathbb{R}^2 \to \mathbb{R}^2$ which rotates the plane 270° counterclockwise.

20. Find the matrix of the linear map $f : \mathbb{R}^2 \to \mathbb{R}^2$ which reflects the plane about the line $y = -x$.

21. The picture below shows the effect of the map f on $\begin{bmatrix} 1 \\ 0 \end{bmatrix}$ and $\begin{bmatrix} 0 \\ 1 \end{bmatrix}$. Use this to find f's matrix.

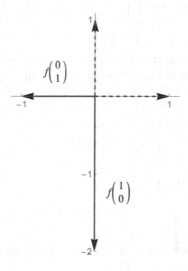

22. The picture below shows the effect of the map f on $\begin{bmatrix} 1 \\ 0 \end{bmatrix}$ and $\begin{bmatrix} 0 \\ 1 \end{bmatrix}$. Use this to find f's matrix.

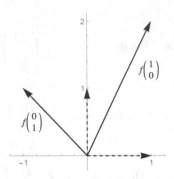

23. Compute $\begin{bmatrix} 2 & 1 & 4 \\ 0 & 4 & -1 \end{bmatrix} \begin{bmatrix} 1 \\ 2 \\ 3 \end{bmatrix}$.

24. Compute $\begin{bmatrix} 1 & -1 \\ 4 & -3 \\ 0 & 3 \end{bmatrix} \begin{bmatrix} 2 \\ 5 \end{bmatrix}$.

25. Compute each of the following or explain why it isn't possible.

    (a) $\begin{bmatrix} 1 & 2 & 5 \\ 3 & -1 & 1 \end{bmatrix} \cdot \begin{bmatrix} -2 \\ 4 \\ 1 \end{bmatrix}$

    (b) $\begin{bmatrix} 1 & 2 & 5 \\ 3 & -1 & 1 \end{bmatrix} \cdot \begin{bmatrix} -1 \\ 2 \end{bmatrix}$

26. Compute each of the following or explain why it isn't possible.

    (a) $\begin{bmatrix} 1 & 2 \\ -1 & 4 \\ 0 & -5 \end{bmatrix} \cdot \begin{bmatrix} -1 \\ 2 \end{bmatrix}$

    (b) $\begin{bmatrix} 1 & 2 \\ -1 & 4 \\ 0 & -5 \end{bmatrix} \cdot \begin{bmatrix} -5 \\ 7 \\ 9 \end{bmatrix}$

27. Let $\vec{v}_1 = \begin{bmatrix} -1 \\ 2 \\ 1 \end{bmatrix}$, $\vec{v}_2 = \begin{bmatrix} 3 \\ -2 \end{bmatrix}$, $\vec{v}_3 = \begin{bmatrix} 11 \\ 5 \end{bmatrix}$, and $A = \begin{bmatrix} -1 & 3 & 4 \\ 0 & -2 & 9 \end{bmatrix}$.
    Compute whichever of $A\vec{v}_1$, $A\vec{v}_2$ and $A\vec{v}_3$ are possible.

28. Let $\vec{v}_1 = \begin{bmatrix} 2 \\ -4 \end{bmatrix}$, $\vec{v}_2 = \begin{bmatrix} 1 \\ 2 \\ 3 \end{bmatrix}$, $\vec{v}_3 = \begin{bmatrix} 7 \\ 5 \end{bmatrix}$, and $A = \begin{bmatrix} 1 & -2 \\ -4 & 2 \end{bmatrix}$. Compute
    whichever of $A\vec{v}_1$, $A\vec{v}_2$ and $A\vec{v}_3$ are possible.

29. Let $\vec{v}_1 = \begin{bmatrix} -1 \\ 0 \\ 10 \\ -3 \end{bmatrix}$, $\vec{v}_2 = \begin{bmatrix} 2 \\ -3 \\ 1 \end{bmatrix}$, $\vec{v}_3 = \begin{bmatrix} 4 \\ 0 \\ -5 \end{bmatrix}$ and $A = \begin{bmatrix} 1 & 4 & 0 & 2 \\ 0 & 1 & 2 & 0 \end{bmatrix}$.
    Compute whichever of $A\vec{v}_1$, $A\vec{v}_2$ and $A\vec{v}_3$ are possible.

30. Let $\vec{v}_1 = \begin{bmatrix} 4 \\ -2 \\ 8 \end{bmatrix}$, $\vec{v}_2 = \begin{bmatrix} 6 \\ -5 \end{bmatrix}$, $\vec{v}_3 = \begin{bmatrix} -1 \\ 9 \end{bmatrix}$ and $A = \begin{bmatrix} -1 & 2 \\ 4 & -3 \\ 7 & 1 \end{bmatrix}$.

Compute whichever of $A\vec{v}_1$, $A\vec{v}_2$ and $A\vec{v}_3$ are possible.

31. Write $x_1 \begin{bmatrix} 2 \\ 1 \end{bmatrix} + x_2 \begin{bmatrix} 4 \\ 2 \end{bmatrix} + x_3 \begin{bmatrix} -2 \\ -1 \end{bmatrix} + x_4 \begin{bmatrix} 4 \\ 0 \end{bmatrix} = \begin{bmatrix} -1 \\ 9 \end{bmatrix}$ as a matrix equation.

32. Write $\begin{bmatrix} 1 \\ -1 \end{bmatrix} x_1 + \begin{bmatrix} -2 \\ 3 \end{bmatrix} x_2 + \begin{bmatrix} 3 \\ 1 \end{bmatrix} x_3 = \begin{bmatrix} -2 \\ 4 \end{bmatrix}$ as a matrix equation.

33. Write $\begin{bmatrix} -10 & 3 \\ 1 & 0 \\ 7 & -2 \end{bmatrix} \vec{x} = \begin{bmatrix} 8 \\ -2 \\ 3 \end{bmatrix}$ as a vector equation.

34. Write $\begin{bmatrix} 2 & 1/2 & 4 \\ -1 & 0 & 5 \\ 7 & 8 & -4 \end{bmatrix} \vec{x} = \begin{bmatrix} 10 \\ 0 \\ -9 \end{bmatrix}$ as a vector equation.

35. (a) Show that $f_\theta : \mathbb{R}^2 \to \mathbb{R}^2$ with matrix $R_\theta = \begin{bmatrix} \cos(\theta) & -\sin(\theta) \\ \sin(\theta) & \cos(\theta) \end{bmatrix}$ rotates the plane counterclockwise by the angle θ by computing $f_\theta \left(\begin{bmatrix} 1 \\ 0 \end{bmatrix} \right)$ and $f_\theta \left(\begin{bmatrix} 0 \\ 1 \end{bmatrix} \right)$ and arguing that these are the images of $\begin{bmatrix} 1 \\ 0 \end{bmatrix}$ and $\begin{bmatrix} 0 \\ 1 \end{bmatrix}$ after a counterclockwise rotation by θ.

   (b) We can compute the coordinates of k holes evenly spaced around a circle of radius r centered at the origin by starting with a single point on the perimeter of the circle and applying f_θ repeatedly, where θ is $\frac{1}{k}$th of a complete circle. Use this method to find the coordinates of 12 evenly spaced holes on the perimeter of a circle of radius 10.

36. The general formula for the matrix of the map $f : \mathbb{R}^2 \to \mathbb{R}^2$ which reflects the plane across the line $y = mx$ is $R_m = \begin{bmatrix} \frac{1-m^2}{1+m^2} & \frac{2m}{1+m^2} \\ \frac{2m}{1+m^2} & \frac{m^2-1}{1+m^2} \end{bmatrix}$.

   (a) Check your answer from Exercise 20 by plugging in $m = -1$.

   (b) Use this formula to find the matrix of the map which reflects $\mathbb{R}^2$ across the line $y = x$. Check your answer by using geometry to find the images of $\begin{bmatrix} 1 \\ 0 \end{bmatrix}$ and $\begin{bmatrix} 0 \\ 1 \end{bmatrix}$.

   (c) Find the reflection of $\begin{bmatrix} 2 \\ 5 \end{bmatrix}$ across the line $y = 3x$.

2.3   Matrix Operations

In the last section, we learned how to rewrite a linear function from $\mathbb{R}^n$ to $\mathbb{R}^m$ as an $m \times n$ matrix. From our studies of functions in other math courses, we have a number of function operations. Since matrices are just another way of viewing linear functions, we should be able to use our function operations to define matrix operations. We'll do this for three basic function operations: addition, multiplication by a scalar, and composition. Let's start with function addition.

If we have two functions f and g, then their sum should be another function $f + g$ which satisfies $(f + g)(\vec{x}) = f(\vec{x}) + g(\vec{x})$. This immediately makes it clear that our functions f and g must have the same domain and codomain. If f and g don't have the same domain, then we won't be able to plug the same vector $\vec{x}$ into both functions which means one of $f(\vec{x})$ and $g(\vec{x})$ won't make sense. If f and g don't have the same codomain then their output vectors won't have the same size, so $f(\vec{x}) + g(\vec{x})$ will be undefined. Thus $f + g$ only makes sense if we have $f : \mathbb{R}^n \to \mathbb{R}^m$ and $g : \mathbb{R}^n \to \mathbb{R}^m$ for the same n and m.

Before we try to translate this function behavior into matrix behavior, we need to be sure that the function $f + g$ will have a matrix. Since only linear functions have matrices, this means we need the following theorem.

Theorem 1. If f and g are linear functions for which $f + g$ is defined, then $f + g$ is a linear function.

To check that $f + g$ is a linear function, we need to show that it obeys both the additive and scalar multiplication conditions from 2.1's definition of a linear function. Let $\vec{v}$ and $\vec{u}$ be vectors in $f + g$'s domain and r be any scalar.

Function addition dictates that

$$(f + g)(\vec{v} + \vec{u}) = f(\vec{v} + \vec{u}) + g(\vec{v} + \vec{u})$$

and since both f and g are linear functions we know

$$f(\vec{v} + \vec{u}) = f(\vec{v}) + f(\vec{u}) \text{ and } g(\vec{v} + \vec{u}) = g(\vec{v}) + g(\vec{u}).$$

Putting these three equations together, we get

$$(f + g)(\vec{v} + \vec{u}) = f(\vec{v}) + f(\vec{u}) + g(\vec{v}) + g(\vec{u}).$$

Rearranging the terms in the right-hand side of this equation gives us

$$(f + g)(\vec{v} + \vec{u}) = f(\vec{v}) + g(\vec{v}) + f(\vec{u}) + g(\vec{u})$$

which can be rewritten as

$$(f + g)(\vec{v} + \vec{u}) = (f + g)(\vec{v}) + (f + g)(\vec{u}).$$

This shows $f + g$ satisfies the additive condition of a linear function.

Moving on to scalar multiplication, we have

$$(f + g)(r \cdot \vec{v}) = f(r \cdot \vec{v}) + g(r \cdot \vec{v}).$$

Again, since f and g are linear we have

$$f(r \cdot \vec{v}) = r \cdot f(\vec{v}) \text{ and } g(r \cdot \vec{v}) = r \cdot g(\vec{v}).$$

Plugging this back into our initial equation gives us

$$(f + g)(r \cdot \vec{v}) = r \cdot f(\vec{v}) + r \cdot g(\vec{v}).$$

This shows that $f + g$ satisfies the scalar multiplication property of a linear function, and therefore $f + g$ is a linear function whenever f and g are linear.

Thinking about function addition in the context of matrices, if f has matrix A and g has matrix B we want to create another matrix which we'll write $A + B$ so that $A + B$ is the matrix of the linear map $f + g$. We can only add f and g when they have the same domain and codomain, i.e., $f, g : \mathbb{R}^n \rightarrow \mathbb{R}^m$. In matrix terms, this means we can only add f and g when they both have $m \times n$ matrices. Therefore matrix addition only makes sense for matrices of the same size, which mirrors what we saw for vector addition. Additionally, since $(f + g)(\vec{x}) = f(\vec{x}) + g(\vec{x})$, we need our new matrix $A + B$ to satisfy $(A + B)\vec{x} = A\vec{x} + B\vec{x}$.

Example 1. Find $A + B$ where $A = \begin{bmatrix} 9 & -1 \\ 3 & 7 \end{bmatrix}$ and $B = \begin{bmatrix} 4 & 2 \\ 6 & 5 \end{bmatrix}$.

Since we don't know yet how to combine these matrices directly, let's instead find the formulas for their maps, add those maps, and then find the matrix of that new map. I'll call A's map f and B's map g, so $A + B$ is the matrix of $(f + g)(\vec{x}) = f(\vec{x}) + g(\vec{x})$.

Using the techniques we practiced in 2.2, we know

$$f\left(\begin{bmatrix} x_1 \\ x_2 \end{bmatrix}\right) = \begin{bmatrix} 9x_1 - x_2 \\ 3x_1 + 7x_2 \end{bmatrix} \text{ and } g\left(\begin{bmatrix} x_1 \\ x_2 \end{bmatrix}\right) = \begin{bmatrix} 4x_1 + 2x_2 \\ 6x_1 + 5x_2 \end{bmatrix}$$

so

$$f(\vec{x}) + g(\vec{x}) = \begin{bmatrix} 9x_1 - x_2 \\ 3x_1 + 7x_2 \end{bmatrix} + \begin{bmatrix} 4x_1 + 2x_2 \\ 6x_1 + 5x_2 \end{bmatrix} = \begin{bmatrix} (9x_1 - x_2) + (4x_1 + 2x_2) \\ (3x_1 + 7x_2) + (6x_1 + 5x_2) \end{bmatrix}$$

$$= \begin{bmatrix} (9 + 4)x_1 + (-1 + 2)x_2 \\ (3 + 6)x_1 + (7 + 5)x_2 \end{bmatrix}.$$

Since $A + B$ is the matrix of $f + g$, we must have

$$A + B = \begin{bmatrix} 9 + 4 & -1 + 2 \\ 3 + 6 & 7 + 5 \end{bmatrix}.$$

Comparing this to $A = \begin{bmatrix} 9 & -1 \\ 3 & 7 \end{bmatrix}$ and $B = \begin{bmatrix} 4 & 2 \\ 6 & 5 \end{bmatrix}$, we can see that each entry of $A + B$ is the sum of the corresponding entries of A and B.

The pattern from our previous example motivates the following definition.

Definition. The $m \times n$ matrices $A = \begin{bmatrix} a_{11} & a_{12} & \cdots & a_{1n} \\ a_{21} & a_{22} & \cdots & a_{2n} \\ \vdots & & \ddots & \\ a_{m1} & a_{m2} & \cdots & a_{mn} \end{bmatrix}$ and

$B = \begin{bmatrix} b_{11} & b_{12} & \cdots & b_{1n} \\ b_{21} & b_{22} & \cdots & b_{2n} \\ \vdots & & \ddots & \\ b_{m1} & b_{m2} & \cdots & b_{mn} \end{bmatrix}$ have sum

$A + B = \begin{bmatrix} a_{11} + b_{11} & a_{12} + b_{12} & \cdots & a_{1n} + b_{1n} \\ a_{21} + b_{21} & a_{22} + b_{22} & \cdots & a_{2n} + b_{2n} \\ \vdots & & \ddots & \\ a_{m1} + b_{m1} & a_{m2} + b_{m2} & \cdots & a_{mn} + b_{mn} \end{bmatrix}$.

Since the ijth entry of a matrix is the coefficient of x_j in its function's output vector's ith entry, we get $(A + B)\vec{x} = A\vec{x} + B\vec{x}$ as desired. To see that the matrix $A + B$ corresponds to the function $f + g$, we can compute

$$(A + B)\vec{x} = A\vec{x} + B\vec{x} = f(\vec{x}) + g(\vec{x}) = (f + g)(\vec{x})$$

Therefore our newly defined addition for matrices corresponds to our older notion of addition of functions.

Example 2. Compute $\begin{bmatrix} -1 & 0 & 4 \\ 10 & -5 & 2 \end{bmatrix} + \begin{bmatrix} 2 & 6 & 1 \\ -8 & 1 & 2 \end{bmatrix}$.

Since these two matrices are the same size (both are 2×3), we can add them together by adding corresponding pairs of entries. This gives us

$$\begin{bmatrix} -1 & 0 & 4 \\ 10 & -5 & 2 \end{bmatrix} + \begin{bmatrix} 2 & 6 & 1 \\ -8 & 1 & 2 \end{bmatrix} = \begin{bmatrix} -1+2 & 0+6 & 4+1 \\ 10+(-8) & -5+1 & 2+2 \end{bmatrix}$$

$$= \begin{bmatrix} 1 & 6 & 5 \\ 2 & 4 & 4 \end{bmatrix}.$$

Example 3. Why can't we compute $\begin{bmatrix} 3 & -1 & 0 & 4 \\ 7 & 10 & -5 & 2 \end{bmatrix} + \begin{bmatrix} 6 & 1 & -2 \\ 9 & 0 & -3 \end{bmatrix}$?

Our first matrix is 2×4 while the second is 2×3. Since they don't have the same number of columns, they aren't the same size. This means that if we tried to add their corresponding entries, we'd have entries of the first matrix without partners in the second matrix. Thus it is impossible to add these two matrices.

Next let's create a matrix operation to mirror multiplying a function by a constant. Here there are no restrictions on the size of the domain and codomain of our function, so we won't have any restrictions on the number of rows or columns of our matrix.

Example 4. Find $4A$ where $A = \begin{bmatrix} 9 & -1 \\ 3 & 7 \end{bmatrix}$.

Since we don't know yet how to do this directly, let's instead find the formula for A's map, multiply that map by 4, and then find the matrix of that new function. I'll call A's map f, so $4A$ is the matrix of $4f(\vec{x})$.

From Example 1, we know $f\left(\begin{bmatrix} x_1 \\ x_2 \end{bmatrix}\right) = \begin{bmatrix} 9x_1 - x_2 \\ 3x_1 + 7x_2 \end{bmatrix}$, so

$$4f(\vec{x}) = 4\begin{bmatrix} 9x_1 - x_2 \\ 3x_1 + 7x_2 \end{bmatrix} = \begin{bmatrix} 4(9x_1 - x_2) \\ 4(3x_1 + 7x_2) \end{bmatrix} = \begin{bmatrix} 4(9)x_1 + 4(-1)x_2 \\ 4(3)x_1 + 4(7)x_2 \end{bmatrix}.$$

Since $4A$ is the matrix of $4f$, we must have

$$4A = \begin{bmatrix} 4(9) & 4(-1) \\ 4(3) & 4(7) \end{bmatrix}.$$

Clearly this is just our matrix A with every entry multiplied by 4.

The pattern we saw in the previous example holds for any matrix and any scalar, because when we multiply a function f by a scalar r, we're really creating a new function $r \cdot f$ where $(r \cdot f)(\vec{x}) = r \cdot f(\vec{x})$. Thus to multiply a function by a scalar we're really just multiplying our function's output vector by that scalar. Since multiplying a vector by a scalar means multiplying each entry of the vector by the scalar, we can find $r \cdot f$ by multiplying each entry of $f(\vec{x})$ by r. Distributing this multiplication by r through each entry of $f(\vec{x})$ means multiplying the coefficient of each variable in each entry of $f(\vec{x})$ by r, so the coefficients for $r \cdot f$ are just r times the coefficients of f. We want to define multiplication of a matrix by a scalar so that if A is the matrix of f then $r \cdot A$ is the matrix of $r \cdot f$. This prompts the following definition.

Definition. Let $A = \begin{bmatrix} a_{11} & a_{12} & \cdots & a_{1n} \\ a_{21} & a_{22} & \cdots & a_{2n} \\ \vdots & & \ddots & \\ a_{m1} & a_{m2} & \cdots & a_{mn} \end{bmatrix}$ be an $m \times n$ matrix and r be any scalar. Then $r \cdot A = \begin{bmatrix} ra_{11} & ra_{12} & \cdots & ra_{1n} \\ ra_{21} & ra_{22} & \cdots & ra_{2n} \\ \vdots & & \ddots & \\ ra_{m1} & ra_{m2} & \cdots & ra_{mn} \end{bmatrix}$.

Our discussion above explains why the coefficient on x_j in the ith entry of $r \cdot f$ is r times the corresponding coefficient from f. Since those coefficients are the ijth entries of the matrices corresponding to f and $r \cdot f$, we see that $r \cdot A$ is the matrix of $r \cdot f$.

Example 5. Compute $-2 \cdot \begin{bmatrix} 5 & 3 \\ 0 & -1 \\ -4 & 2 \end{bmatrix}$.

To multiply this matrix by -2, we simply multiply each of its entries by -2. This gives us

$$-2 \cdot \begin{bmatrix} 5 & 3 \\ 0 & -1 \\ -4 & 2 \end{bmatrix} = \begin{bmatrix} -2(5) & -2(3) \\ -2(0) & -2(-1) \\ -2(-4) & -2(2) \end{bmatrix} = \begin{bmatrix} -10 & -6 \\ 0 & 2 \\ 8 & -4 \end{bmatrix}.$$

Finally, let's tackle figuring out how to model function composition. If we have two functions f and g, then $f \circ g$ is the function with $(f \circ g)(\vec{x}) = f(g(\vec{x}))$. (Notice that even though we read from left to right most of the time, functions in a composition act right to left since the rightmost function is the one applied to $\vec{x}$ first.)

As with addition, before we start to translate function composition into matrices, we need to establish that $f \circ g$ has a matrix.

Theorem 2. If f and g are linear functions and $f \circ g$ is defined, then $f \circ g$ is a linear function.

As with Theorem 1, we need to show that $f \circ g$ satisfies both the additive and scalar multiplication property of a linear function. Let $\vec{v}$ and $\vec{u}$ be vectors in $f \circ g$'s domain and r be any scalar.

Since we know g is linear, we have $g(\vec{v} + \vec{u}) = g(\vec{v}) + g(\vec{u})$ which means

$$(f \circ g)(\vec{v} + \vec{u}) = f(g(\vec{v} + \vec{u})) = f(g(\vec{v}) + g(\vec{u})).$$

Since f is also linear, the right-hand side can be split up to give us

$$(f \circ g)(\vec{v} + \vec{u}) = f(g(\vec{v})) + f(g(\vec{u}))$$

which can be rewritten as

$$(f \circ g)(\vec{v} + \vec{u}) = (f \circ g)(\vec{v}) + (f \circ g)(\vec{u}).$$

Thus $f \circ g$ splits up over vector addition as required.

Similarly, because g is linear we have $g(r \cdot \vec{v}) = r \cdot g(\vec{v})$ so

$$(f \circ g)(r \cdot \vec{v}) = f(g(r \cdot \vec{v})) = f(r \cdot g(\vec{v})).$$

Since f is also linear, we can pull the scalar r out of the right-hand side to get

$$(f \circ g)(r \cdot \vec{v}) = r \cdot f(g(\vec{v}))$$

which can be rewritten as

$$(f \circ g)(r \cdot \vec{v}) = r \cdot (f \circ g)(\vec{v}).$$

This means $f \circ g$ also splits up over scalar multiplication and therefore is a linear function.

If f has matrix A and g has matrix B, then $(f \circ g)(\vec{x}) = f(g(\vec{x}))$ can be rewritten as

$$(f \circ g)(\vec{x}) = f(g(\vec{x})) = A(B\vec{x}) = (AB)\vec{x}.$$

This means we need to define matrix multiplication so that AB is the matrix of the linear map $f \circ g$.

For $f \circ g$ to make sense, we must be able to plug $g(\vec{x})$ into f. This means the codomain of g must be the same as the domain of f. In other words, we need $f : \mathbb{R}^k \to \mathbb{R}^m$ and $g : \mathbb{R}^n \to \mathbb{R}^k$. This gives us $f \circ g : \mathbb{R}^n \to \mathbb{R}^m$. Translating this into conditions on our matrices A and B, we get that A is $m \times k$ and B is $k \times n$. In other words, the product AB only makes sense if the number of columns of A equals the number of rows of B. Since AB is the matrix of $f \circ g$, it is an $m \times n$ matrix.

Personally, I remember $(m \times k)(k \times n)$ as "the touching dimensions match" and "to get the dimensions of the product, cancel out the matching dimensions" i.e., removing the k's leaves $m \times n$. (This is similar to my method for remembering sizes when multiplying a matrix and a vector.)

Example 6. Is it possible to compute AB if A is 4×6 and B is 6×9? If it is possible, give the dimensions of the product matrix AB.

Remember that for AB to make sense we need the number of columns in A to equal the number of rows in B. Here A has 6 columns and B has 6 rows, so it is possible to compute their product AB. The product AB will have as many rows as A and as many columns as B. Since A has 4 rows and B has 9 columns, our product AB is a 4×9 matrix.

If you prefer to think of this in terms of functions, this means A's function has domain $\mathbb{R}^6$ while B's function has codomain $\mathbb{R}^6$. This means the output of B's map is the correct size to plug into A's map, so the composition is possible. The domain of the composition is the domain of B's map, i.e., $\mathbb{R}^9$, while the codomain of the composition is the codomain of A's map, i.e., $\mathbb{R}^4$. Therefore the matrix of the composition, AB, is 4×9.

Alternately, we can plug the dimensions of A and B into the product to get $(4 \times 6)(6 \times 9)$. Since the middle numbers match, the product is possible. If we remove our matching 6s the remaining numbers are the dimensions of the product, so 4×9.

Example 7. Is it possible to compute AB if A is 5×3 and B is 5×2? If it is possible, give the dimensions of the product matrix AB.

Here A has 3 columns and B has 5 rows, so this product is impossible.

In terms of maps this means B's map has outputs which are 5-vectors, while A's map has inputs which are 3-vectors. Therefore the composition of these two maps is impossible.

Alternately, when we plug in the dimensions of A and B we get $(5 \times 3)(5 \times 2)$ and the middle numbers don't match up, so this can't be done.

To make AB the matrix of $f \circ g$, we need the ijth entry of AB to be the coefficient on x_j in the ith entry of $f(g(\vec{x}))$. Before we dive into the general case, let's look at a small concrete example.

Example 8. Find AB where $A = \begin{bmatrix} 9 & -1 \\ 3 & 7 \end{bmatrix}$ and $B = \begin{bmatrix} 4 & 2 \\ 6 & 5 \end{bmatrix}$.

Since we don't know yet how to combine these matrices directly, let's instead find the formulas for their maps, compose those maps, and then find the matrix of that composition. I'll call A's map f and B's map g, so AB is the matrix of $(f \circ g)(\vec{x}) = f(g(\vec{x}))$.

Using the techniques we practiced in 2.2, we see $f\left(\begin{bmatrix} x_1 \\ x_2 \end{bmatrix} \right) = \begin{bmatrix} 9x_1 - x_2 \\ 3x_1 + 7x_2 \end{bmatrix}$ and $g\left(\begin{bmatrix} x_1 \\ x_2 \end{bmatrix} \right) = \begin{bmatrix} 4x_1 + 2x_2 \\ 6x_1 + 5x_2 \end{bmatrix}$. This means

$$f(g(\vec{x})) = f\left(\begin{bmatrix} 4x_1 + 2x_2 \\ 6x_1 + 5x_2 \end{bmatrix} \right) = \begin{bmatrix} 9(4x_1 + 2x_2) - (6x_1 + 5x_2) \\ 3(4x_1 + 2x_2) + 7(6x_1 + 5x_2) \end{bmatrix}.$$

Simplifying (although not all the way because we're looking for the pattern) gives us

$$f(g(\vec{x})) = \begin{bmatrix} 9(4)x_1 + 9(2)x_2 + (-1)(6)x_1 + (-1)(5)x_2 \\ 3(4)x_1 + 3(2)x_2 + 7(6)x_1 + (7)5x_2 \end{bmatrix}$$

$$= \begin{bmatrix} (9(4) + (-1)(6))x_1 + (9(2) + (-1)(5))x_2 \\ (3(4) + 7(6))x_1 + (3(2) + (7)5)x_2 \end{bmatrix}.$$

Since AB is the matrix of this composition map, we must have

$$AB = \begin{bmatrix} 9(4) + (-1)(6) & 9(2) + (-1)(5) \\ 3(4) + 7(6) & 3(2) + (7)5 \end{bmatrix}.$$

Comparing this to $A = \begin{bmatrix} 9 & -1 \\ 3 & 7 \end{bmatrix}$ and $B = \begin{bmatrix} 4 & 2 \\ 6 & 5 \end{bmatrix}$, we can see that the top left entry of AB is a sum of products of corresponding entries from the top row of A and the left column of B. This pattern of combining a row of A with a column of B continues throughout the rest of AB's entries.

To see that the pattern we observed in the previous example holds more generally, suppose

$$f\left(\begin{bmatrix} x_1 \\ x_2 \\ \vdots \\ x_k \end{bmatrix}\right) = \begin{bmatrix} a_{11}x_1 + a_{12}x_2 + \cdots + a_{1k}x_k \\ a_{21}x_1 + a_{22}x_2 + \cdots + a_{2k}x_k \\ \vdots \\ a_{m1}x_1 + a_{m2}x_2 + \cdots + a_{mk}x_k \end{bmatrix}$$

and

$$g\left(\begin{bmatrix} x_1 \\ x_2 \\ \vdots \\ x_n \end{bmatrix}\right) = \begin{bmatrix} b_{11}x_1 + b_{12}x_2 + \cdots + b_{1n}x_n \\ b_{21}x_1 + b_{22}x_2 + \cdots + b_{2n}x_n \\ \vdots \\ b_{k1}x_1 + b_{k2}x_2 + \cdots + b_{kn}x_n \end{bmatrix}.$$

To find the ith entry of $f(g(\vec{x}))$, we need to plug the entries of $g(\vec{x})$ into the ith entry of $f(\vec{x})$ as $x_1, \ldots, x_k$. The ith entry of $f(\vec{x})$ is $a_{i1}x_1 + a_{i2}x_2 + \cdots + a_{ik}x_k$ so plugging in the entries of $g(\vec{x})$ gives us

$$a_{i1}(b_{11}x_1 + b_{12}x_2 + \cdots + b_{1n}x_n) + \cdots + a_{ik}(b_{k1}x_1 + b_{k2}x_2 + \cdots + b_{kn}x_n).$$

This looks incredibly messy, but remember that to find the ijth entry of AB, we only care about the coefficient on x_j. The first term in the sum above contains $a_{i1}b_{1j}x_j$, the second term contains $a_{i2}b_{2j}x_j$, and so on until the last (kth) term which contains $a_{ik}b_{kj}x_j$. Thus the x_j term in the ith entry of $f(g(\vec{x}))$ is

$$(a_{i1}b_{1j} + a_{i2}b_{2j} + \cdots + a_{ik}b_{kj})x_j.$$

This allows us to make the following definition.

Definition. The product of the $m \times k$ matrix $A = \begin{bmatrix} a_{11} & a_{12} & \cdots & a_{1k} \\ a_{21} & a_{22} & \cdots & a_{2k} \\ \vdots & & \ddots & \\ a_{m1} & a_{m2} & \cdots & a_{mk} \end{bmatrix}$

and the $k \times n$ matrix $B = \begin{bmatrix} b_{11} & b_{12} & \cdots & b_{1n} \\ b_{21} & b_{22} & \cdots & b_{2n} \\ \vdots & & \ddots & \\ b_{k1} & b_{k2} & \cdots & b_{kn} \end{bmatrix}$ is the $m \times n$ matrix AB

whose ijth entry is $a_{i1}b_{1j} + a_{i2}b_{2j} + \cdots + a_{ik}b_{kj}$.

Notice that to find the ijth entry of AB, we add up the pairwise products along the ith row of A and down the jth column of B. This should feel similar to our method for multiplying a vector by a matrix. In fact, the jth column of AB is A times the vector which is the jth column of B.

Example 9. Compute AB where $A = \begin{bmatrix} -2 & -1 & 1 \\ 0 & 6 & -2 \end{bmatrix}$ and $B = \begin{bmatrix} 5 & 3 \\ 0 & 1 \\ -4 & 2 \end{bmatrix}$.

The first matrix A is 2×3 and the second matrix B is 3×2, so the product AB makes sense and is a 2×2 matrix. This means we need to compute each of the four entries of AB.

Let's start with the entry in the 1st row and 1st column of AB, so $i = 1$ and $j = 1$. (Here $k = 3$ since that is the number of columns of A and rows of B.) We want to compute the sum of pairwise products along the 1st row of A and 1st column of B as shown below.

$$\begin{bmatrix} \boxed{-2 \quad -1 \quad 1} \\ 0 \quad 6 \quad -2 \end{bmatrix} \begin{bmatrix} \boxed{\begin{matrix} 5 \\ 0 \\ -4 \end{matrix}} & \begin{matrix} 3 \\ 1 \\ 2 \end{matrix} \end{bmatrix}$$

Multiplying the corresponding entries in this row and column we get

$$(-2)(5) + (-1)(0) + (1)(-4) = -14$$

so $AB = \begin{bmatrix} -14 & - \\ - & - \end{bmatrix}$.

Moving to the entry in the 1st row and 2nd column of AB where $i = 1$ and $j = 2$, means we're looking at the 1st row of A and 2nd column of B as shown below.

$$\begin{bmatrix} \boxed{-2 \quad -1 \quad 1} \\ 0 \quad 6 \quad -2 \end{bmatrix} \begin{bmatrix} \begin{matrix} 5 \\ 0 \\ -4 \end{matrix} & \boxed{\begin{matrix} 3 \\ 1 \\ 2 \end{matrix}} \end{bmatrix}$$

Multiplying the corresponding entries in this row and column we get

$$(-2)(3) + (-1)(1) + (1)(2) = -5$$

so $AB = \begin{bmatrix} -14 & -5 \\ - & - \end{bmatrix}$.

The entry in the 2nd row and 1st column of AB has $i = 2$ and $j = 1$, which means we're looking at the 2nd row of A and 1st column of B as shown below.

$$\begin{bmatrix} -2 \quad -1 \quad 1 \\ \boxed{0 \quad 6 \quad -2} \end{bmatrix} \begin{bmatrix} \boxed{\begin{matrix} 5 \\ 0 \\ -4 \end{matrix}} & \begin{matrix} 3 \\ 1 \\ 2 \end{matrix} \end{bmatrix}$$

Multiplying the corresponding entries in this row and column we get

$$(0)(5) + (6)(0) + (-2)(-4) = 8$$

so $AB = \begin{bmatrix} -14 & -5 \\ 8 & \underline{} \end{bmatrix}$.

The entry in the 2nd row and 2nd column of AB has $i = 2$ and $j = 2$, which means we're looking at the 2nd row of A and 2nd column of B as shown below.

$$\begin{bmatrix} -2 & -1 & 1 \\ \boxed{0 \quad 6 \quad -2} \end{bmatrix} \begin{bmatrix} 5 & \boxed{3} \\ 0 & \boxed{1} \\ -4 & \boxed{2} \end{bmatrix}$$

Multiplying the corresponding entries in this row and column we get

$$(0)(3) + (6)(1) + (-2)(2) = 2$$

so

$$AB = \begin{bmatrix} -14 & -5 \\ 8 & 2 \end{bmatrix}.$$

The fact that this new operation is called matrix multiplication will tempt you to assume that it behaves like multiplication in the real numbers. However, that is a dangerous parallel to draw, because matrix multiplication isn't based on multiplication at all — it is based on function composition. With that in mind, let's explore two very important differences between matrix multiplication and our usual notion of multiplication.

Theorem 3. Matrix multiplication isn't commutative, i.e., $AB \neq BA$.

The easiest way we could have $AB \neq BA$ is if only one of these products can be computed.

Example 10. If $A = \begin{bmatrix} 1 & 3 \\ -2 & 4 \end{bmatrix}$ and $B = \begin{bmatrix} 7 & -4 & 0 \\ 3 & -1 & 9 \end{bmatrix}$, then the product AB makes sense but the product BA does not.

Here A is 2×2 and B is 2×3. Since A's number of columns equals B's number of rows, the we can compute the product AB. However, B's number of columns doesn't equal A's number of rows, so we can't compute BA.

The second obvious issue is that even when AB and BA both exist, they may have different dimensions.

Example 11. The two matrices A and B from Example 9 have both AB and BA defined, but they are different sizes.

The matrix A from Example 9 is 2×3 and the matrix B is 3×2. No matter which order we list them, the number of rows in the first matrix equals the number of columns in the second matrix. The product AB that we computed in Example 9 has as many rows as A and as many columns as B, so it is 2×2. However, the other product BA has as many rows as B and as many columns as A, so it is 3×3. Since AB and BA are different sizes, it is impossible for them to be equal.

Even if both of these matrix products make sense and are the same size, remember that AB is the matrix of $f \circ g$ while BA is the matrix of $g \circ f$. From working with function composition in calculus, you should be familiar with the fact that $f(g(\vec{x}))$ and $g(f(\vec{x}))$ may be very different functions. This means it makes sense that AB can easily be quite different from BA. (We will see some special cases where $AB = BA$, but they are the exceptions rather than the rule.)

Example 12. Check that $A = \begin{bmatrix} 1 & -2 \\ -5 & 1 \end{bmatrix}$ and $B = \begin{bmatrix} 3 & 1 \\ 0 & 4 \end{bmatrix}$ have $AB \neq BA$.

Both A and B are 2×2 matrices, so AB and BA are both defined and both have the same size (also 2×2). However

$$AB = \begin{bmatrix} 1 & -2 \\ -5 & 1 \end{bmatrix} \begin{bmatrix} 3 & 1 \\ 0 & 4 \end{bmatrix} = \begin{bmatrix} 1(3) - 2(0) & 1(1) - 2(4) \\ -5(3) + 1(0) & -5(1) + 1(4) \end{bmatrix} = \begin{bmatrix} 3 & -7 \\ -15 & -1 \end{bmatrix}$$

while

$$BA = \begin{bmatrix} 3 & 1 \\ 0 & 4 \end{bmatrix} \begin{bmatrix} 1 & -2 \\ -5 & 1 \end{bmatrix} = \begin{bmatrix} 3(1) + 1(-5) & 3(-2) + 1(1) \\ 0(1) + 4(-5) & 0(-2) + 4(1) \end{bmatrix} = \begin{bmatrix} -2 & -5 \\ -20 & 4 \end{bmatrix}$$

so $AB \neq BA$.

The moral of the story here is to be very careful about the order of matrix multiplication and fight against the urge to assume you can switch that order without changing the product.

Another oddity of matrix multiplication is given by the following theorem.

Theorem 4. It is possible to have $AB = 0$ with $A \neq 0$ and $B \neq 0$.

Note that the "0" in this theorem is the appropriately sized zero matrix.

Example 13. Check that $AB = 0$ where $A = \begin{bmatrix} 2 & 6 \\ 1 & 3 \end{bmatrix}$ and $B = \begin{bmatrix} 12 & 0 \\ -4 & 0 \end{bmatrix}$.

First of all, notice that neither A nor B are the zero matrix, since each of them contains at least one nonzero entry.

Since both A and B are 2×2 matrices, their product is defined and is also a 2×2 matrix. To compute the product, we follow the same process as in Example 9 and get

$$\begin{bmatrix} 2 & 6 \\ 1 & 3 \end{bmatrix} \begin{bmatrix} 12 & 0 \\ -4 & 0 \end{bmatrix} = \begin{bmatrix} 2(12)+6(-4) & 2(0)+6(0) \\ 1(12)+3(-4) & 1(0)+3(0) \end{bmatrix} = \begin{bmatrix} 0 & 0 \\ 0 & 0 \end{bmatrix}.$$

Since all entries of the product are zero, we do indeed have $AB = 0$.

(For further proof that $AB \neq BA$, you can check that BA is not the zero matrix!)

This contradicts our usual intuition from the real numbers where we often factor an equation of the form $f(x) = 0$ and then set each factor equal to zero to solve. This won't work to solve $AB = 0$ or even $A\vec{x} = \vec{0}$. We will have to develop new tools for this more complicated matrix situation.

One property matrix multiplication does share with multiplication of real numbers is the existence of an identity element which mimics the way 1 acts in $\mathbb{R}$. There $1r = r1 = r$ for any real number r. The identity element for matrix multiplication is the appropriately named *identity matrix* discussed in Example 4 of Section 2.2, which is the matrix of the identity map. Since there is an identity map from $\mathbb{R}^n$ to itself for every n, this is actually a collection of $n \times n$ identity matrices

$$I_n = \begin{bmatrix} 1 & 0 & 0 & \cdots & 0 \\ 0 & 1 & 0 & \cdots & 0 \\ \vdots & \vdots & \ddots & & \vdots \\ 0 & 0 & & 1 & 0 \\ 0 & 0 & \cdots & 0 & 1 \end{bmatrix}.$$

Any $m \times n$ matrix A corresponds to a map $f : \mathbb{R}^n \to \mathbb{R}^m$. The composition of f with the $n \times n$ identity map is $f \circ id_n$ which is just f, while the composition of the $m \times m$ identity map with f is $id_m \circ f$ which is again f. Translating this into matrix multiplication tells us that $AI_n = A$ and $I_m A = A$.

Example 14. Let $n = 2$. Show that $I_2 A = A$ for any 2×3 matrix A.

The 2×2 identity matrix is $I_2 = \begin{bmatrix} 1 & 0 \\ 0 & 1 \end{bmatrix}$ which corresponds to the map $id_2 : \mathbb{R}^2 \to \mathbb{R}^2$ with $id_2 \left(\begin{bmatrix} x_1 \\ x_2 \end{bmatrix} \right) = \begin{bmatrix} x_1 \\ x_2 \end{bmatrix}$.

This means

$$I_2 A = \begin{bmatrix} 1 & 0 \\ 0 & 1 \end{bmatrix} \begin{bmatrix} a & b & c \\ d & e & f \end{bmatrix} = \begin{bmatrix} 1(a)+0(d) & 1(b)+0(e) & 1(c)+0(f) \\ 0(a)+1(d) & 0(b)+1(e) & 0(c)+1(f) \end{bmatrix}$$

$$= \begin{bmatrix} a & b & c \\ d & e & f \end{bmatrix} = A.$$

You can check for yourself that $AI_3 = A$ where $I_3 = \begin{bmatrix} 1 & 0 & 0 \\ 0 & 1 & 0 \\ 0 & 0 & 1 \end{bmatrix}$.

Exercises 2.3.

1. Compute $3 \begin{bmatrix} 1 & -1 \\ 2 & 0 \\ 3 & -2 \end{bmatrix}$.

2. Compute $4 \begin{bmatrix} 10 & -5 \\ 5 & 2 \end{bmatrix}$.

3. Compute $\frac{1}{2} \begin{bmatrix} 2 & 4 \\ -6 & 0 \\ 1 & 8 \end{bmatrix}$.

4. Compute $3 \begin{bmatrix} -1 & 3 \\ 2 & 4 \\ -5 & 0 \end{bmatrix}$.

5. Compute $-2 \begin{bmatrix} 3 & 7 \\ 1 & -1 \end{bmatrix}$.

6. Compute $3 \begin{bmatrix} 1 & -2 \\ -1 & 4 \end{bmatrix}$.

7. Find $4 \begin{bmatrix} 2 & 2 \\ 1 & -4 \end{bmatrix}$.

8. Compute $\begin{bmatrix} -1 & 5 \\ 2 & 3 \end{bmatrix} + \begin{bmatrix} 1 & -2 & 5 \\ 2 & 0 & 7 \end{bmatrix}$ or explain why it isn't possible.

9. Compute $\begin{bmatrix} -1 & 5 \\ 2 & 3 \end{bmatrix} + \begin{bmatrix} 2 & -3 \\ 0 & 1 \end{bmatrix}$ or explain why it isn't possible.

10. Let $A = \begin{bmatrix} 1 & -1 \\ 2 & 0 \\ 3 & -2 \end{bmatrix}$, $B = \begin{bmatrix} 1 & 2 & 3 \\ 2 & 1 & -1 \\ 3 & 4 & 0 \end{bmatrix}$, $C = \begin{bmatrix} -4 & 2 \\ 1 & -1 \\ 2 & 0 \end{bmatrix}$.
    Compute whichever of $A + B$, $B + C$ and $A + C$ are possible.

11. Let $A = \begin{bmatrix} 3 & 1 \\ -2 & 2 \\ 0 & -1 \end{bmatrix}$, $B = \begin{bmatrix} 10 & -5 \\ 5 & 2 \end{bmatrix}$, $C = \begin{bmatrix} -1 & 3 \\ 0 & -2 \\ 4 & 9 \end{bmatrix}$.
    Compute whichever of $A + B$, $A + C$ and $B + C$ are possible.

12. Let $A = \begin{bmatrix} 2 & 4 \\ -6 & 0 \\ 1 & 8 \end{bmatrix}$, $B = \begin{bmatrix} 1 & 2 \\ 3 & 4 \end{bmatrix}$, $C = \begin{bmatrix} -1 & 3 \\ 0 & 10 \\ 2 & 9 \end{bmatrix}$.
    Compute whichever of $A + B$, $A + C$, and $B + C$ are possible.

13. Let $A = \begin{bmatrix} 1 & 2 & 0 \\ -2 & 3 & -2 \\ 0 & -1 & 4 \end{bmatrix}$, $B = \begin{bmatrix} -1 & 3 \\ 2 & 4 \\ -5 & 0 \end{bmatrix}$, $C = \begin{bmatrix} 3 & 1 & 7 \\ -2 & -5 & 10 \\ 0 & 3 & 4 \end{bmatrix}$.

    Compute whichever of $A + B$, $A + C$ and $B + C$ are possible.

14. Let $A = \begin{bmatrix} 1 & -4 & 2 \\ 0 & -1 & 3 \end{bmatrix}$, $B = \begin{bmatrix} 5 & 2 & -3 \\ 4 & 0 & 6 \end{bmatrix}$, $C = \begin{bmatrix} 3 & 7 \\ 1 & -1 \end{bmatrix}$.

    Compute whichever of $A + B$, $A + C$ and $B + C$ are possible.

15. Let $A = \begin{bmatrix} 1 & -2 & 0 \\ -4 & 2 & 5 \end{bmatrix}$, $B = \begin{bmatrix} 2 & 2 \\ 1 & -4 \end{bmatrix}$, $C = \begin{bmatrix} 1 & 0 & 7 \\ 3 & -2 & 0 \end{bmatrix}$.

    Compute whichever of $A + B$, $A + C$ and $B + C$ are possible.

16. Suppose A is a 10×6 matrix.

    (a) What size matrix B would make computing AB possible?
    (b) What size matrix B would make computing BA possible?

17. Suppose A is a 5×9 matrix.

    (a) What size matrix B would make computing AB possible?
    (b) What size matrix B would make computing BA possible?

18. Suppose A is a 7×3 matrix.

    (a) What value of n would make computing AI_n possible?
    (b) What value of m would make computing $I_m A$ possible?

19. Let $A = \begin{bmatrix} a & b \\ c & d \end{bmatrix}$. Use matrix multiplication to check that $I_2 A = A$ and $AI_2 = A$.

20. Compute $\begin{bmatrix} 1 & 2 & 5 \\ 3 & -1 & 1 \end{bmatrix} \cdot \begin{bmatrix} -2 & 2 \\ 4 & -1 \end{bmatrix}$ or explain why it isn't possible.

21. Compute $\begin{bmatrix} -2 & 2 \\ 4 & -1 \end{bmatrix} \cdot \begin{bmatrix} 1 & 2 & 5 \\ 3 & -1 & 1 \end{bmatrix}$ or explain why it isn't possible.

22. Let $A = \begin{bmatrix} 1 & -1 \\ 2 & 0 \\ 3 & -2 \end{bmatrix}$, $B = \begin{bmatrix} 1 & 2 & 3 \\ 2 & 1 & -1 \\ 3 & 4 & 0 \end{bmatrix}$, $C = \begin{bmatrix} -4 & 2 \\ 1 & -1 \\ 2 & 0 \end{bmatrix}$.

    Compute whichever of AB, BC and AC are possible.

23. Let $A = \begin{bmatrix} 3 & 1 \\ -2 & 2 \\ 0 & -1 \end{bmatrix}$, $B = \begin{bmatrix} 10 & -5 \\ 5 & 2 \end{bmatrix}$, $C = \begin{bmatrix} -1 & 3 \\ 0 & -2 \\ 4 & 9 \end{bmatrix}$.

    Compute whichever of AB, BC and CA are possible.

24. Let $A = \begin{bmatrix} 2 & 4 \\ -6 & 0 \\ 1 & 8 \end{bmatrix}$, $B = \begin{bmatrix} 1 & 2 \\ 3 & 4 \end{bmatrix}$, $C = \begin{bmatrix} -1 & 3 \\ 0 & 10 \\ 2 & 9 \end{bmatrix}$.

    Compute whichever of AB, AC, and BC are possible.

25. Let $A = \begin{bmatrix} 1 & 2 & 0 \\ -2 & 3 & -2 \\ 0 & -1 & 4 \end{bmatrix}$, $B = \begin{bmatrix} -1 & 3 \\ 2 & 4 \\ -5 & 0 \end{bmatrix}$, $C = \begin{bmatrix} 3 & 1 & 7 \\ -2 & -5 & 10 \\ 0 & 3 & 4 \end{bmatrix}$.

    Compute whichever of AB, BC, BA are possible.

26. Let $A = \begin{bmatrix} 1 & -4 & 2 \\ 0 & -1 & 3 \end{bmatrix}$, $B = \begin{bmatrix} 5 & 2 & -3 \\ 4 & 0 & 6 \end{bmatrix}$, $C = \begin{bmatrix} 3 & 7 \\ 1 & -1 \end{bmatrix}$.

    Compute whichever of AB, BC and AC are possible.

27. Let $A = \begin{bmatrix} 1 & -2 & 0 \\ -4 & 2 & 5 \end{bmatrix}$, $B = \begin{bmatrix} 2 & 2 \\ 1 & -4 \end{bmatrix}$, $C = \begin{bmatrix} 1 & 0 & 7 \\ 3 & -2 & 0 \end{bmatrix}$.

    Compute whichever of AB, BC and AC are possible.

28. Let $A = \begin{bmatrix} 7 & -3 & 5 \\ -2 & 0 & 8 \end{bmatrix}$ and $B = \begin{bmatrix} 2 & 4 & 1 \\ -4 & -2 & 1 \\ 0 & 3 & 6 \end{bmatrix}$. Explain why AB

    can be computed, but BA cannot.

29. Let $A = \begin{bmatrix} 3 & -2 \\ 6 & 1 \end{bmatrix}$ and $B = \begin{bmatrix} 4 & 1 \\ -1 & 2 \end{bmatrix}$. Show $AB \neq BA$ even though

    both AB and BA can be computed.

30. Let $A = \begin{bmatrix} 1 & -1 \\ 0 & 2 \end{bmatrix}$. Find a nonzero matrix B for which $AB = BA$.

31. Let $A = \begin{bmatrix} 1 & -1 \\ -2 & 2 \end{bmatrix}$. Find a nonzero matrix B for which $AB = 0$.

2.4   Matrix Vector Spaces

In the last section, we worked out the right way to add matrices and multiply matrices by scalars so that they correctly mirror adding functions and multiplying functions by scalars. Since we can only add matrices of the same size, let's pick a (generic) size for our matrices and consider only matrices of that fixed size $m \times n$. The set of all matrices of size $m \times n$ is called M_{mn}. As usual, we want to take stock of some nice properties of these two matrix operations. Stay alert and see if you can remember where we've discussed these properties before! First let's consider matrix addition.

If we add two $m \times n$ matrices, the result will be another $m \times n$ matrix. To see this, think of our sum as the addition of two functions f and g which both map from $\mathbb{R}^n$ to $\mathbb{R}^m$. The sum should be a new function which still maps from $\mathbb{R}^n$ to $\mathbb{R}^m$, which will also have an $m \times n$ matrix. This means M_{mn} is closed under matrix addition.

The order in which we add matrices doesn't matter. Suppose we have two matrices A and B, which correspond to linear functions f and g respectively. Since we saw in Chapter 1 that order of addition of vectors doesn't matter, it is clear that

$$f(\vec{x}) + g(\vec{x}) = g(\vec{x}) + f(\vec{x})$$

so we must have $f + g = g + f$ and hence

$$A + B = B + A.$$

This also jives with our intuition from calculus where $x^2 + 2x$ and $2x + x^2$ are the same function. This means matrix addition is commutative.

We can use a similar line of reasoning to see that if we're adding three matrices, we can start by adding any pair of them and still get the same result. This property is also inherited from vector addition, this time from the fact that $(\vec{v} + \vec{w}) + \vec{u} = \vec{v} + (\vec{w} + \vec{u})$. This means for any three functions from $\mathbb{R}^n$ to $\mathbb{R}^m$ we have

$$(f(\vec{x}) + g(\vec{x})) + h(\vec{x}) = f(\vec{x} + (g(\vec{x}) + h(\vec{x}))$$

so we also have $(f + g) + h = f + (g + h)$ and therefore

$$(A + B) + C = A + (B + C).$$

This means matrix addition is associative.

Matrix addition has an identity element, which is a matrix that we can add to any other matrix and get that other matrix back. Since the zero vector is the additive identity in $\mathbb{R}^n$, the function $f(\vec{x}) = \vec{0}$ is our additive identity in M_{mn} because adding it to any other function from $\mathbb{R}^n$ to $\mathbb{R}^m$ has no effect. This zero function's matrix is the $m \times n$ matrix with all entries equal to 0, which we'll call the zero matrix for short.

What about additive inverses? In other words, for each matrix A in M_{mn} we want a partner matrix $-A$ so that $-A + A$ is the zero matrix. Here it is easier to think computationally instead of in terms of functions. To get each entry of this sum to be zero, we need the entries of $-A$ to be the entries of A just with the opposite sign. This means that

$$-A = (-1) \cdot A.$$

Next let's consider multiplication of matrices by scalars.

If we multiply an $m \times n$ matrix by a scalar, the result will be another $m \times n$ matrix. As with addition, scaling a function f which maps from $\mathbb{R}^n$ to $\mathbb{R}^m$ gives another function from $\mathbb{R}^n$ to $\mathbb{R}^m$. In other words, M_{mn} is closed under scalar multiplication.

Multiplication by scalars is also associative in the sense that if we want to scale a matrix A by two scalars r and s, we can choose to either scale A by s and then scale sA by r, or just scale A by the product rs all at once. In other words,

$$r(sA) = (rs)A.$$

Thinking in terms of A's function f, this is like scaling the output vector of f by either s and then r or scaling by rs. Vectors have associative scalar multiplication, so f and A inherit it as well.

The scalar 1 leaves matrices unchanged, i.e., $1 \cdot A = A$. This is clear because multiplying each entry of A by 1 produces no change.

Finally, let's consider how these two operations interact with each other.

We saw that scalar multiplication of vectors distributes over scalar addition, i.e.,

$$(r + s) \cdot \vec{v} = r \cdot \vec{v} + s \cdot \vec{v}.$$

This means that for any linear function f from $\mathbb{R}^n$ to $\mathbb{R}^m$ we know

$$(r + s)f(\vec{x}) = rf(\vec{x}) + sf(\vec{x})$$

and therefore

$$(r + s)A = rA + sA$$

for the $m \times n$ matrix of f.

Similarly, we know scalar multiplication of vectors distributes over vector addition. In function terms, this means we know

$$rf(\vec{x}) + rg(\vec{x}) = r(f(\vec{x}) + g(\vec{x})).$$

Translating this to matrices, we see

$$rA + rB = r(A + B).$$

Did these properties look familiar? That's because they're the same as the good properties of vector addition and multiplication of vectors by scalars summarized in Theorem 1 of 1.1. One way to think of this is that M_{mn} is acting like $\mathbb{R}^n$ if we think of matrices in place of the vectors and use matrix addition and scalar multiplication instead of vector addition and scalar multiplication. This motivates the following definition.

Definition. A collection of mathematical objects, V, with two operations "+" and "·" is a *vector space* if for all $\vec{v}, \vec{u}, \vec{w}$ in V and r, s in $\mathbb{R}$ the following conditions are satisfied:

1. $\vec{v} + \vec{u}$ is in V (closure of addition)

2. $\vec{v} + \vec{u} = \vec{u} + \vec{v}$ (addition is commutative)

3. $(\vec{v} + \vec{u}) + \vec{w} = \vec{v} + (\vec{u} + \vec{w})$ (addition is associative)

4. There is a $\vec{0}$ in V so that $\vec{v} + \vec{0} = \vec{v}$ (additive identity)

5. For each $\vec{v}$, there is a $-\vec{v}$ so that $-\vec{v} + \vec{v} = \vec{0}$ (additive inverses)

6. $r \cdot \vec{v}$ is in V (closure of scalar multiplication)

7. $r \cdot (s \cdot \vec{v}) = (rs) \cdot \vec{v}$ (scalar multiplication is associative)

8. $1 \cdot \vec{v} = \vec{v}$ (scalar multiplication's identity is 1)

9. $(r + s) \cdot \vec{v} = r \cdot \vec{v} + s \cdot \vec{v}$ (scalar multiplication distributes over addition of scalars)

10. $r \cdot (\vec{v} + \vec{u}) = r \cdot \vec{v} + r \cdot \vec{u}$ (scalar multiplication distributes over addition of vectors)

Note that we will usually call the elements $\vec{v}$ of a vector space V vectors and put an arrow over their variables even though we know that they may not be vectors in $\mathbb{R}^n$ in the sense of our definition from 1.1. If this is confusing, you can feel free to mentally add air quotes around the word "vector" when thinking about this privately.

Example 1. The set $V = \mathbb{R}^n$ with vector addition as + and scalar multiplication of vectors as · is a vector space.

We saw in 1.1 that vector addition and scalar multiplication of vectors have these properties.

Example 2. The set $V = M_{mn}$ with matrix addition as + and scalar multiplication of matrices as · is a vector space.

We saw at the beginning of this section that matrix addition and scalar multiplication of matrices have these properties.

This gives us tons of vector spaces to work with: an $\mathbb{R}^n$ for every choice of n and an M_{mn} for every choice of m and n.

To get even more examples of vector spaces, remember that in 1.2 we discussed subsets of $\mathbb{R}^n$ which were self-contained subspaces under our vector operations. We can now reformulate our definition of a subspace using our definition of a vector space.

Definition. Let V with $+$ and $\cdot$ be a vector space. A subset W of V is a *subspace* if W is a vector space using the same operations $+$ and $\cdot$ as V.

Before we look at some examples, let's also generalize our subspace test from 1.2. Otherwise we'd be forced to recheck all ten properties from our vector space definition every time we wanted to say something was a subspace, which is much more work than we really need to do.

Theorem 1. Let V with $+$ and $\cdot$ be a vector space. $W \subseteq V$ is a subspace of V if it satisfies the following conditions:

1. $\vec{w}_1 + \vec{w}_2$ is in W for every $\vec{w}_1$ and $\vec{w}_2$ in W (closure of addition)

4. $\vec{0}_V$ is in W (zero vector of V is in W)

6. $r \cdot \vec{w}$ is in W for every scalar r and $\vec{w}$ in W (closure of scalar multiplication)

I've labeled each condition with the same number as its corresponding property from our definition of a vector space.

To see that this is true, let's convince ourselves that if we have these three properties we get the other seven from the definition of a vector space.

Notice that since V is a vector space and $W \subseteq V$, we automatically get properties 2, 3, 7, 8, 9, and 10. This means that we really just need to understand why we get property 5.

Pick any $\vec{w}$ in W. Since $\vec{w}$ is in V, we know that it has an additive inverse $-\vec{w}$ in V. If we can show that $-\vec{w}$ is also in W, we'll be done. To see that, consider

$$\vec{0}_V = 0 \cdot \vec{w} = (-1 + 1) \cdot \vec{w} = (-1) \cdot \vec{w} + 1 \cdot \vec{w} = (-1) \cdot \vec{w} + \vec{w}$$

which tells us that

$$(-1) \cdot \vec{w} = -\vec{w}.$$

Now we can invoke property 6 to say that since $\vec{w}$ is in W, $-\vec{w}$ is also in W.

Therefore W also satisfies property 5, and is a subspace of V. With this shortcut in hand, let's look at an example.

Example 3. Show $W = \left\{ \begin{bmatrix} a & 0 \\ 0 & b \end{bmatrix} \right\}$ with the usual addition and scalar multiplication of matrices is a subspace of M_{22}.

We know M_{22} is a vector space under matrix addition and scalar multiplication, and W is clearly a subset of M_{22}. That means we can use the subspace test to show W is a subspace. Since all three properties used in the subspace test ask us to show something is in W, we should first remind ourselves that to be in W a 2×2 matrix simply needs to have the top right and bottom left entries equal 0.

We'll start by checking that W is closed under addition. If we have two matrices $\begin{bmatrix} a & 0 \\ 0 & b \end{bmatrix}$ and $\begin{bmatrix} c & 0 \\ 0 & d \end{bmatrix}$ in W, their sum is

$$\begin{bmatrix} a & 0 \\ 0 & b \end{bmatrix} + \begin{bmatrix} c & 0 \\ 0 & d \end{bmatrix} = \begin{bmatrix} a+c & 0 \\ 0 & b+d \end{bmatrix}$$

which is clearly also in W. Next we need to show $\vec{0}_{M_{22}}$ is in W. Since

$$\vec{0}_{M_{22}} = \begin{bmatrix} 0 & 0 \\ 0 & 0 \end{bmatrix}$$

we can see that $\vec{0}_{M_{22}}$ is in W by letting $a = b = 0$. Finally, we need to show W is closed under scalar multiplication. If we have a matrix $\begin{bmatrix} a & 0 \\ 0 & b \end{bmatrix}$ in W and a scalar r, then their product is

$$r \begin{bmatrix} a & 0 \\ 0 & b \end{bmatrix} = \begin{bmatrix} ra & 0 \\ 0 & rb \end{bmatrix}.$$

Since this is in W, we've satisfied all three conditions of the subspace test. Therefore, W is a subspace of M_{22}. This also shows that W is a vector space.

We saw in 1.2 that the span of a set of vectors is automatically a subset of $\mathbb{R}^n$, which gave us another way to check that something was a subspace. We can generalize that idea to M_{mn}, but first we'll need to update our definition of a span in $\mathbb{R}^n$ to a general vector space.

Definition. Let V with $+$ and $\cdot$ be a vector space. The *span* of $\vec{v}_1, \ldots, \vec{v}_n$ in V is $\text{Span}\{\vec{v}_1, \ldots, \vec{v}_n\} = \{a_1\vec{v}_1 + \cdots + a_n\vec{v}_n\}$ where $a_1, \ldots, a_n$ are scalars.

Remember that the $+$ and $\cdot$ in this definition are the operations from V.

Example 4. Find the span of $\begin{bmatrix} 1 & 0 \\ 0 & 1 \end{bmatrix}$ and $\begin{bmatrix} 1 & 1 \\ 0 & 0 \end{bmatrix}$ in M_{22}.

From our definition, we have

$$\text{Span}\left\{ \begin{bmatrix} 1 & 0 \\ 0 & 1 \end{bmatrix}, \begin{bmatrix} 1 & 1 \\ 0 & 0 \end{bmatrix} \right\} = \left\{ a_1 \begin{bmatrix} 1 & 0 \\ 0 & 1 \end{bmatrix} + a_2 \begin{bmatrix} 1 & 1 \\ 0 & 0 \end{bmatrix} \right\}.$$

The right-hand side can be simplified to

$$\left\{ \begin{bmatrix} a_1 & 0 \\ 0 & a_1 \end{bmatrix} + \begin{bmatrix} a_2 & a_2 \\ 0 & 0 \end{bmatrix} \right\} = \left\{ \begin{bmatrix} a_1 + a_2 & a_2 \\ 0 & a_1 \end{bmatrix} \right\}$$

so

$$\text{Span}\left\{ \begin{bmatrix} 1 & 0 \\ 0 & 1 \end{bmatrix}, \begin{bmatrix} 1 & 1 \\ 0 & 0 \end{bmatrix} \right\} = \left\{ \begin{bmatrix} a_1 + a_2 & a_2 \\ 0 & a_1 \end{bmatrix} \right\}.$$

Another way to say this is that the span of these two matrices is the set of all matrices whose bottom left entry is 0 and whose top left entry is the sum of the top right and bottom right entries.

Theorem 2. Let $\vec{v}_1, \ldots, \vec{v}_k$ be in a vector space V. Then $\text{Span}\{\vec{v}_1, \ldots, \vec{v}_k\}$ is a subspace of V.

To check that any span is automatically a subspace of V, we can use the subspace test. Pick any two vectors $\vec{w}$ and $\vec{u}$ in $\text{Span}\{\vec{v}_1, \ldots, \vec{v}_n\}$. Since they are in the span of the $\vec{v}$s, we can rewrite them as

$$\vec{w} = a_1 \vec{v}_1 + \cdots + a_n \vec{v}_n$$

and

$$\vec{u} = b_1 \vec{v}_1 + \cdots + b_n \vec{v}_n$$

for some scalars $a_1, \ldots, a_n$ and $b_1, \ldots, b_n$. This means their sum is

$$\begin{aligned} \vec{w} + \vec{u} &= (a_1 \vec{v}_1 + \cdots + a_n \vec{v}_n) + (b_1 \vec{v}_1 + \cdots + b_n \vec{v}_n) \\ &= (a_1 \vec{v}_1 + b_1 \vec{v}_1) + \cdots + (a_n \vec{v}_n + b_n \vec{v}_n) \\ &= (a_1 + b_1)\vec{v}_1 + \cdots + (a_n + b_n)\vec{v}_n \end{aligned}$$

which is clearly in $\text{Span}\{\vec{v}_1, \ldots, \vec{v}_n\}$. If we let all our scalars equal zero, we have

$$0 \cdot \vec{v}_1 + \cdots + 0 \cdot \vec{v}_n = \vec{0}_V$$

in $\text{Span}\{\vec{v}_1, \ldots, \vec{v}_n\}$. If $\vec{w}$ is in $\text{Span}\{\vec{v}_1, \ldots, \vec{v}_n\}$ and r is any scalar, then

$$\begin{aligned} r \cdot \vec{w} &= r \cdot (a_1 \vec{v}_1 + \cdots + a_n \vec{v}_n) = r \cdot (a_1 \vec{v}_1) + \cdots + r \cdot (a_n \vec{v}_n) \\ &= (ra_1)\vec{v}_1 + \cdots + (ra_n)\vec{v}_n \end{aligned}$$

which is also in $\text{Span}\{\vec{v}_1, \ldots, \vec{v}_n\}$. Since $\text{Span}\{\vec{v}_1, \ldots, \vec{v}_n\}$ satisfies all three conditions of the subspace test, it is a subspace of V.

Example 5. Show $W = \left\{ \begin{bmatrix} a & 0 \\ 0 & b \end{bmatrix} \right\}$ is a subspace of M_{22}.

We showed this in Example 3 using the subspace test, but it can also be shown by writing W as a span.

To do this, we need to rewrite our general element of W as a linear combination of a set of other matrices. We'll start by splitting an element of W up into the sum of two parts: one containing terms with a and one containing terms with b. This gives us

$$W = \left\{ \begin{bmatrix} a & 0 \\ 0 & b \end{bmatrix} \right\} = \left\{ \begin{bmatrix} a & 0 \\ 0 & 0 \end{bmatrix} + \begin{bmatrix} 0 & 0 \\ 0 & b \end{bmatrix} \right\}.$$

Now we can factor an a out of the first matrix and a b out of the second to get.

$$W = \left\{ a \begin{bmatrix} 1 & 0 \\ 0 & 0 \end{bmatrix} + b \begin{bmatrix} 0 & 0 \\ 0 & 1 \end{bmatrix} \right\}.$$

We're more used to seeing x_1 and x_2 as our scalar coefficients, but there's no harm in calling them a and b instead. This means

$$W = \left\{ a \begin{bmatrix} 1 & 0 \\ 0 & 0 \end{bmatrix} + b \begin{bmatrix} 0 & 0 \\ 0 & 1 \end{bmatrix} \right\} = \text{Span} \left\{ \begin{bmatrix} 1 & 0 \\ 0 & 0 \end{bmatrix}, \begin{bmatrix} 0 & 0 \\ 0 & 1 \end{bmatrix} \right\}.$$

Since

$$W = \text{Span} \left\{ \begin{bmatrix} 1 & 0 \\ 0 & 0 \end{bmatrix}, \begin{bmatrix} 0 & 0 \\ 0 & 1 \end{bmatrix} \right\}$$

Theorem 2 tells us that W is a subspace of M_{22}.

Another important idea from $\mathbb{R}^n$ that we can generalize to any vector space is the idea of linear independence and linear dependence. In 1.3 we gave two equivalent definitions, so we have a choice here about which one to generalize. I'll generalize the vector equation version here, and you can work through generalizing the span version and arguing that the two are still equivalent in the exercises.

Definition. A set of elements $\vec{v}_1, \ldots, \vec{v}_k$ from a vector space V are *linearly dependent* if the equation $x_1 \cdot \vec{v}_1 + \cdots + x_k \cdot \vec{v}_k = \vec{0}_V$ has a solution where at least one of the x_is is nonzero. Otherwise $\vec{v}_1, \ldots, \vec{v}_k$ are *linearly independent*.

As with span, here $+$ and $\cdot$ are the addition and scalar multiplication of V and $\vec{0}_V$ is V's zero vector.

Example 6. Are $\begin{bmatrix} 1 & -1 \\ 0 & 2 \end{bmatrix}$, $\begin{bmatrix} -2 & 2 \\ 1 & 1 \end{bmatrix}$, and $\begin{bmatrix} -5 & 5 \\ 1 & -5 \end{bmatrix}$ linearly independent or linearly dependent?

Here $V = M_{22}$ with matrix addition and scalar multiplication as its operations and $\vec{0}_{M_{22}} = \begin{bmatrix} 0 & 0 \\ 0 & 0 \end{bmatrix}$. This means the equation from our generalized definition of linear independence and linear dependence is

$$x_1 \begin{bmatrix} 1 & -1 \\ 0 & 2 \end{bmatrix} + x_2 \begin{bmatrix} -2 & 2 \\ 1 & 1 \end{bmatrix} + x_3 \begin{bmatrix} -5 & 5 \\ 1 & -5 \end{bmatrix} = \begin{bmatrix} 0 & 0 \\ 0 & 0 \end{bmatrix}.$$

If we have no solutions to this equation apart from setting all the variables equal to zero, then our three matrices are linearly independent. If we can find a solution where any of the variables is nonzero, our matrices are linearly dependent.

Simplifying the left-hand side of this equation gives us

$$\begin{bmatrix} x_1 - 2x_2 - 5x_3 & -x_1 + 2x_2 + 5x_3 \\ x_2 + x_3 & 2x_1 + x_2 - 5x_3 \end{bmatrix} = \begin{bmatrix} 0 & 0 \\ 0 & 0 \end{bmatrix}.$$

Setting corresponding entries of these two matrices equal gives us the equations $x_1 - 2x_2 - 5x_3 = 0$, $-x_1 + 2x_2 + 5x_3 = 0$, $x_2 + x_3 = 0$, and $2x_1 + x_2 - 5x_3 = 0$. The third equation tells us that $x_2 = -x_3$. Plugging this into the first equation gives $x_1 + 2x_3 - 5x_3 = 0$ or $x_1 - 3x_3 = 0$, so $x_1 = 3x_3$. However, if we plug $x_1 = 3x_3$ and $x_2 = -x_3$ into any of our original equations they simplify to $0 = 0$. Therefore x_3 can equal anything and in particular can be nonzero. This tells us that our three matrices are linearly dependent.

Example 7. Are $\begin{bmatrix} 1 & -1 \\ 1 & -1 \end{bmatrix}$, $\begin{bmatrix} 1 & 0 \\ 0 & 1 \end{bmatrix}$, and $\begin{bmatrix} 0 & -1 \\ -1 & 0 \end{bmatrix}$ linearly independent or linearly dependent?

As in the previous example, $V = M_{22}$ and we can tackle this question of linear independence or dependence via the equation

$$x_1 \begin{bmatrix} 1 & -1 \\ 1 & -1 \end{bmatrix} + x_2 \begin{bmatrix} 1 & 0 \\ 0 & 1 \end{bmatrix} + x_3 \begin{bmatrix} 0 & -1 \\ -1 & 0 \end{bmatrix} = \begin{bmatrix} 0 & 0 \\ 0 & 0 \end{bmatrix}.$$

This simplifies to

$$\begin{bmatrix} x_1 + x_2 & -x_1 - x_3 \\ x_1 - x_3 & -x_1 + x_2 \end{bmatrix} = \begin{bmatrix} 0 & 0 \\ 0 & 0 \end{bmatrix}$$

which gives us the equations $x_1 + x_2 = 0$, $-x_1 - x_3 = 0$, $x_1 - x_3 = 0$, and $-x_1 + x_2 = 0$. The first equation gives us $x_1 = -x_2$, while the fourth equation gives us $x_1 = x_2$. This means $-x_2 = x_2$ which can only be satisfied if $x_2 = 0$. This also means $x_1 = 0$. The third equation tells us $x_1 = x_3$, so $x_3 = 0$ as well. Since our only solution is to have all three variables equal zero, our three matrices are linearly independent.

Just as we did with subspaces, spans, and linear independence, we can also generalize our idea of linear functions. We'll use the same properties as in 2.1, but now we'll allow the domain and codomain of our map to be general vector spaces instead of restricting them to be $\mathbb{R}^n$ and $\mathbb{R}^m$.

Definition. Let V and W be vector spaces. A function $f : V \to W$ is *linear* if $f(\vec{v} + \vec{u}) = f(\vec{v}) + f(\vec{u})$ and $f(r \cdot \vec{v}) = r \cdot f(\vec{v})$ for all vectors $\vec{v}$ and $\vec{u}$ in V and all scalars r.

Remember that the addition and scalar multiplication used on the left-hand side of each of this definition's examples are the operations from the vector space V, while the addition and scalar multiplication used on the right-hand sides are the operations from W. These may be different if V and W are different types of vector spaces!

Example 8. Show that $f : M_{23} \to \mathbb{R}^3$ by $f\left(\begin{bmatrix} a & b & c \\ d & e & f \end{bmatrix}\right) = \begin{bmatrix} a+d \\ 2b \\ c+e-f \end{bmatrix}$ is linear.

To show that f is a linear function, we need to show that it satisfies both conditions from the definition above. Let's start by checking addition.

$$f\left(\begin{bmatrix} a & b & c \\ d & e & f \end{bmatrix} + \begin{bmatrix} r & s & t \\ x & y & z \end{bmatrix}\right) = f\left(\begin{bmatrix} a+r & b+s & c+t \\ d+x & e+y & f+z \end{bmatrix}\right)$$

$$= \begin{bmatrix} (a+r)+(d+x) \\ 2(b+s) \\ (c+t)+(e+y)-(f+z) \end{bmatrix}$$

$$= \begin{bmatrix} (a+d)+(r+x) \\ 2b+2s \\ (c+e-f)+(t+y-z) \end{bmatrix}$$

$$= \begin{bmatrix} a+d \\ 2b \\ c+e-f \end{bmatrix} + \begin{bmatrix} r+x \\ 2s \\ t+y-z \end{bmatrix}$$

$$= f\left(\begin{bmatrix} a & b & c \\ d & e & f \end{bmatrix}\right) + f\left(\begin{bmatrix} r & s & t \\ x & y & z \end{bmatrix}\right).$$

This means f satisfies the first condition. Next we need to check scalar multiplication.

$$f\left(r\begin{bmatrix} a & b & c \\ d & e & f \end{bmatrix}\right) = f\left(\begin{bmatrix} ra & rb & rc \\ rd & re & rf \end{bmatrix}\right) = \begin{bmatrix} ra+rd \\ 2rb \\ rc+re-rf \end{bmatrix}$$

$$= \begin{bmatrix} r(a+d) \\ r(2b) \\ r(c+e-f) \end{bmatrix} = r\begin{bmatrix} a+d \\ 2b \\ c+e-f \end{bmatrix}.$$

Since f also satisfies the second condition, it is a linear function.

Example 9. Show that $f : \mathbb{R}^2 \to M_{22}$ by $f\left(\begin{bmatrix} x_1 \\ x_2 \end{bmatrix}\right) = \begin{bmatrix} x_1 & x_2 \\ x_1 + x_2 & x_1 x_2 \end{bmatrix}$ is not linear.

To show this function isn't linear, we only need to show that it fails one of our two conditions. This function actually fails both, but here I'll just show that it fails the condition for scalar multiplication.

$$f\left(r\begin{bmatrix} x_1 \\ x_2 \end{bmatrix}\right) = f\left(r\begin{bmatrix} rx_1 \\ rx_2 \end{bmatrix}\right) = \begin{bmatrix} rx_1 & rx_2 \\ rx_1 + rx_2 & rx_1 rx_2 \end{bmatrix} = \begin{bmatrix} rx_1 & rx_2 \\ rx_1 + rx_2 & r^2 x_1 x_2 \end{bmatrix}$$

but

$$rf\left(\begin{bmatrix} x_1 \\ x_2 \end{bmatrix}\right) = r\begin{bmatrix} x_1 & x_2 \\ x_1 + x_2 & x_1 x_2 \end{bmatrix} = \begin{bmatrix} rx_1 & rx_2 \\ r(x_1 + x_2) & rx_1 x_2 \end{bmatrix}.$$

Comparing the bottom right entries shows that these two options are not equal, so f is not linear.

Exercises 2.4.

1. What is the additive inverse of $\begin{bmatrix} 3 & 17 & 0 \\ -5 & 2 & -8 \end{bmatrix}$ in M_{23}?

2. What is the additive inverse of $\begin{bmatrix} -1 & 9 \\ 8 & 0 \end{bmatrix}$ in M_{22}?

3. Show $W = \left\{ \begin{bmatrix} a & -2a \\ 0 & a \end{bmatrix} \right\}$ with the usual matrix addition and scalar multiplication is a subspace of M_{22}.

4. Is $W = \left\{ \begin{bmatrix} a & b \\ -a & 2b \end{bmatrix} \right\}$ a subspace of M_{22}?

5. Let W be the set of all 2×2 matrices of the form $\begin{bmatrix} a & b \\ c & 0 \end{bmatrix}$ where $a + b + c = 1$. Is W a subspace of M_{22}?

6. The trace of an $n \times n$ matrix A, written $tr(A)$, is the sum of A's diagonal entries. Is $W = \{A \text{ in } M_{22} \mid tr(A) = 0\}$ a subspace of M_{22}?

7. Show that $V = \left\{ \begin{bmatrix} a & b \\ c & d \end{bmatrix} \mid a, b, c, d \geq 0 \right\}$ with the usual matrix addition and scalar multiplication is NOT a vector space.

8. Explain why we can't make the set of $n \times n$ matrices into a vector space where the "+" is defined as matrix multiplication, in other words, where $A + B = AB$.

9. Explain why we can't make $\mathbb{R}$ into a vector space where "+" is defined to be the usual multiplication of real numbers, i.e., $a + b = ab$.

10. Show that $D = \left\{ \begin{bmatrix} a & 0 \\ 0 & 2a \end{bmatrix} \right\}$ with a, b in $\mathbb{R}$ is a vector space under the usual operations of M_{22}.

11. Fix a particular 3×3 matrix A. Let W be the set of all 3×3 matrices which commute with A, i.e., $W = \{B \text{ in } M_{33} \mid AB = BA\}$. Show that W is a vector space.

12. Find the span of $\begin{bmatrix} 1 & 0 \\ 0 & 2 \end{bmatrix}$ and $\begin{bmatrix} 0 & 1 \\ 1 & 0 \end{bmatrix}$ in M_{22}.

13. Find the span of $\begin{bmatrix} -1 & 0 & 1 \\ 0 & 1 & 0 \end{bmatrix}$ and $\begin{bmatrix} 0 & 4 & 0 \\ 0 & -1 & 1 \end{bmatrix}$ in M_{23}.

14. Is $\begin{bmatrix} 1 & -2 \\ -1 & 0 \end{bmatrix}$ in the span of $\begin{bmatrix} 2 & -1 \\ 0 & 2 \end{bmatrix}$, $\begin{bmatrix} -1 & 0 \\ 3 & -1 \end{bmatrix}$, and $\begin{bmatrix} 1 & 1 \\ 1 & 1 \end{bmatrix}$?

15. Is $\begin{bmatrix} 2 & -2 & 10 \\ -1 & 1 & 8 \end{bmatrix}$ in the span of $\begin{bmatrix} 3 & -1 & 4 \\ 2 & 0 & 5 \end{bmatrix}$ and $\begin{bmatrix} 4 & 0 & -2 \\ 5 & -1 & 2 \end{bmatrix}$?

16. Are $\vec{v}_1 = \begin{bmatrix} 1 & 2 \\ 3 & 4 \end{bmatrix}$, $\vec{v}_2 = \begin{bmatrix} 4 & 3 \\ 2 & -1 \end{bmatrix}$ and $\vec{v}_3 = \begin{bmatrix} -2 & 1 \\ 4 & 9 \end{bmatrix}$ linearly independent or linearly dependent in M_{22}?

17. Are $\vec{v}_1 = \begin{bmatrix} 2 & 0 \\ 0 & 2 \end{bmatrix}$, $\vec{v}_2 = \begin{bmatrix} 1 & 1 \\ 0 & 1 \end{bmatrix}$, and $\vec{v}_3 = \begin{bmatrix} 0 & 1 \\ 1 & 0 \end{bmatrix}$ linearly independent or linearly dependent in M_{22}?

18. Are $\vec{v}_1 = \begin{bmatrix} 2 & 0 \\ 0 & 2 \end{bmatrix}$, $\vec{v}_2 = \begin{bmatrix} 1 & 1 \\ 0 & 1 \end{bmatrix}$, and $\vec{v}_3 = \begin{bmatrix} 0 & 1 \\ 1 & 0 \end{bmatrix}$ linearly independent or linearly dependent in M_{22}?

19. Are $\vec{v}_1 = \begin{bmatrix} 1 & -1 \\ 0 & -2 \end{bmatrix}$, $\vec{v}_2 = \begin{bmatrix} -3 & 2 \\ 1 & -1 \end{bmatrix}$, $\vec{v}_3 = \begin{bmatrix} 0 & -1 \\ 1 & -7 \end{bmatrix}$ linearly independent or linearly dependent in M_{22}?

20. In 1.3 we had two equivalent definitions for linear dependence, one in terms of an equation and one involving spans. In this section, we generalized the equation definition to get our definition of linear dependence in a vector space.

    (a) Generalize the definition of linear dependence involving spans to a vector space.

    (b) Generalize our explanation from 1.3 of why the two definitions were equivalent to a vector space.

21. Let $f\left(\begin{bmatrix} x_1 \\ x_2 \end{bmatrix} \right) = \begin{bmatrix} x_1 + 3x_2 & 0 \\ 0 & 2x_1 - x_2 \end{bmatrix}$.

    (a) What is the domain of f?

    (b) What is the codomain of f?

22. Let $f\left(\begin{bmatrix} a & b \\ c & d \end{bmatrix} \right) = \begin{bmatrix} a + b \\ b - c \\ 2d \end{bmatrix}$.

(a) What is the domain of f?

(b) What is the codomain of f?

23. Let $f\left(\begin{bmatrix} a & b \\ c & d \end{bmatrix}\right) = \begin{bmatrix} -a & a+b & b \\ a+c & 0 & c+d \end{bmatrix}$.

(a) What is the domain of f?

(b) What is the codomain of f?

24. Let $f\left(\begin{bmatrix} a & b & c \\ d & e & f \end{bmatrix}\right) = \begin{bmatrix} f & 0 & c \\ 0 & a+e & 0 \\ a & 0 & b \end{bmatrix}$.

(a) What is the domain of f?

(b) What is the codomain of f?

25. Let $f\left(\begin{bmatrix} x_1 \\ x_2 \end{bmatrix}\right) = \begin{bmatrix} x_1 + 3x_2 & 0 \\ 0 & 2x_1 - x_2 \end{bmatrix}$. Compute $f\left(\begin{bmatrix} 3 \\ -1 \end{bmatrix}\right)$.

26. Let $f\left(\begin{bmatrix} a & b \\ c & d \end{bmatrix}\right) = \begin{bmatrix} a+b \\ b-c \\ 2d \end{bmatrix}$. Compute $f\left(\begin{bmatrix} 4 & -1 \\ 2 & 5 \end{bmatrix}\right)$.

27. Let $f\left(\begin{bmatrix} a & b \\ c & d \end{bmatrix}\right) = \begin{bmatrix} -a & a+b & b \\ a+c & 0 & c+d \end{bmatrix}$. Compute $f\left(\begin{bmatrix} 1 & -3 \\ -2 & 6 \end{bmatrix}\right)$.

28. Let $f\left(\begin{bmatrix} a & b & c \\ d & e & f \end{bmatrix}\right) = \begin{bmatrix} f & 0 & c \\ 0 & a+e & 0 \\ a & 0 & b \end{bmatrix}$. Compute $f\left(\begin{bmatrix} 7 & -1 & 2 \\ 3 & 1 & 4 \end{bmatrix}\right)$.

29. Show that $f\left(\begin{bmatrix} a & b \\ c & d \end{bmatrix}\right) = \begin{bmatrix} a+b & 0 & 0 \\ 0 & c+d & 0 \\ 0 & 0 & a-d+1 \end{bmatrix}$ is not a linear function.

30. Show that the map from Example 9 doesn't satisfy the additive condition $f(\vec{x} + \vec{u}) = f(\vec{x}) + f(\vec{u})$ from the definition of a linear function.

31. Show that $f\left(\begin{bmatrix} x_1 \\ x_2 \end{bmatrix}\right) = \begin{bmatrix} x_1 + 3x_2 & 0 \\ 0 & 2x_1 - x_2 \end{bmatrix}$ is a linear function.

32. Show that $f\left(\begin{bmatrix} a & b \\ c & d \end{bmatrix}\right) = \begin{bmatrix} a+b \\ b-c \\ 2d \end{bmatrix}$ is a linear function.

2.5   Kernel and Range

In this section we'll explore how a linear function creates two special subspaces within its domain and codomain. We'll start in the domain, where we'll explore solutions to the equation $f(\vec{x}) = \vec{0}$. This is an important equation in many applications, for example, when finding maximums and minimums using the techniques of calculus. It also helps us figure out what is mapped to the origin in the codomain, which can be visualized geometrically as the subset of the domain which is collapsed to the point $\vec{0}$ by the map f.

Example 1. Let $f : \mathbb{R}^2 \to \mathbb{R}^3$ by $f\left(\begin{bmatrix} x_1 \\ x_2 \end{bmatrix}\right) = \begin{bmatrix} 0 \\ x_1 - 2x_2 \\ -3x_1 + 6x_2 \end{bmatrix}$. Which vectors from $\mathbb{R}^2$ are mapped to $\vec{0}$ in $\mathbb{R}^3$?

We can answer this question by solving $f(\vec{x}) = \vec{0}$. In our case this is

$$\begin{bmatrix} 0 \\ x_1 - 2x_2 \\ -3x_1 + 6x_2 \end{bmatrix} = \begin{bmatrix} 0 \\ 0 \\ 0 \end{bmatrix}.$$

This means we need $x_1 - 2x_2 = 0$ and $-3x_1 + 6x_2 = 0$. Both of these equations simplify to $x_1 = 2x_2$, so a vector from $\mathbb{R}^2$ maps to $\vec{0}$ in $\mathbb{R}^3$ exactly when it has the form $\begin{bmatrix} x_1 \\ 2x_1 \end{bmatrix}$.

We can see this subset of $\mathbb{R}^2$ geometrically by relabeling x_1 and x_2 as x and y, so our requirement to map to $\vec{0}$ becomes $x = 2y$. Now solving for y allows us to view this set in a familiar format as the line $y = \frac{1}{2}x$ pictured below.

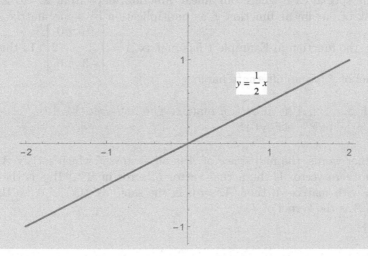

Since this subset of the domain is clearly important, we make the following definition.

Definition.   Suppose $f : V \to W$ is a linear function between two vector spaces. The *kernel* of f is $\ker f = \{\vec{v} \text{ in } V \mid f(\vec{v}) = \vec{0}_W\}$.

In other words, the kernel of a linear function is the set of all vectors in its domain which map to the zero vector of the codomain. (Keep in mind that we're using "zero vector of the codomain" in the context of our vector space definition, so if $W = M_{mn}$ our zero vector would be the zero matrix.)

Example 2.   Find the kernel of the function $f : \mathbb{R}^3 \to M_{22}$ where

$$f\left(\begin{bmatrix} x_1 \\ x_2 \\ x_3 \end{bmatrix}\right) = \begin{bmatrix} x_2 & x_1 - x_3 \\ 0 & x_2 \end{bmatrix}.$$

The domain of f is $\mathbb{R}^3$ and the codomain is M_{22}, so the kernel of f is the set of 3-vectors which map to the 2×2 zero matrix. We can solve for those vectors using the equation $f(\vec{x}) = \vec{0}_{M_{22}}$ which can be rewritten as

$$\begin{bmatrix} x_2 & x_1 - x_3 \\ 0 & x_2 \end{bmatrix} = \begin{bmatrix} 0 & 0 \\ 0 & 0 \end{bmatrix}.$$

Setting corresponding matrix entries equal, we get $x_2 = 0$ and $x_1 - x_3 = 0$. This means for $\vec{x}$ to be in the kernel we need $x_2 = 0$ and $x_1 = x_3$. Thus

$$\ker(f) = \left\{ \begin{bmatrix} x_1 \\ 0 \\ x_1 \end{bmatrix} \right\}.$$

In the special case where our linear function maps from $\mathbb{R}^n$ to $\mathbb{R}^m$, we can think of our linear function f as multiplication by some matrix A. For example, the function in Example 1 has matrix $A = \begin{bmatrix} 0 & 0 \\ 1 & -2 \\ -3 & 6 \end{bmatrix}$. In this case, the kernel of f has an alternate name.

Definition.   Let A be an $m \times n$ matrix. The *null space* of A is $Nul(A) = \{\vec{x} \text{ in } \mathbb{R}^n \mid A\vec{x} = \vec{0}\}$.

In other words, the null space of A is all vectors $\vec{x}$ which satisfy $A\vec{x} = \vec{0}$. (Here our zero vector is the actual vector of zeros in $\mathbb{R}^m$.) If f is the linear function with matrix A, then $A\vec{x} = \vec{0}$ is the same as $f(\vec{x}) = \vec{0}$, so the null space of A is the kernel of f.

Example 3. Find the null space of $A = \begin{bmatrix} 1 & -2 \\ -\frac{1}{2} & 1 \end{bmatrix}$.

The null space of A is all 2-vectors $\vec{x}$ where $A\vec{x} = \vec{0}$. We can rewrite this as

$$\begin{bmatrix} 1 & -2 \\ -\frac{1}{2} & 1 \end{bmatrix} \begin{bmatrix} x_1 \\ x_2 \end{bmatrix} = \begin{bmatrix} 0 \\ 0 \end{bmatrix}.$$

Multiplying out the left-hand side gives us

$$\begin{bmatrix} x_1 - 2x_2 \\ -\frac{1}{2}x_1 + x_2 \end{bmatrix} = \begin{bmatrix} 0 \\ 0 \end{bmatrix}.$$

Setting corresponding entries equal gives us the equations $x_1 - 2x_2 = 0$ and $-\frac{1}{2}x_1 + x_2 = 0$. Solving for x_1 in either equation gives us $x_1 = 2x_2$, so

$$Nul(A) = \left\{ \begin{bmatrix} 2x_2 \\ x_2 \end{bmatrix} \right\}.$$

So far we've seen the kernel (or null space) as a subset of the domain of our function. However, the situation is more special than that.

Theorem 1. If $f : V \to W$ is a linear function, then $\ker(f)$ is a subspace of V.

Another way to say this (while practicing our new vocabulary) is that the kernel of a linear function is a subspace of its domain, which we can check using our subspace test. This means we need to verify that the kernel contains the zero vector of the domain and is closed under addition and scalar multiplication.

To show that $\vec{0}_V$ (the zero vector of the domain V) is in the kernel of f, we need to show $f(\vec{0}_V) = \vec{0}_W$. To do this we'll use the fact that f is linear along with a trick: because $(-1) \cdot \vec{x}$ is the additive inverse, $-\vec{x}$, of $\vec{x}$, we get $\vec{0} = \vec{x} + (-1) \cdot \vec{x}$ for any vector $\vec{x}$ in any vector space V. This means

$$f(\vec{0}_V) = f(\vec{v} + (-1) \cdot \vec{v}) = f(\vec{v}) + (-1) \cdot f(\vec{v}) = \vec{0}_W$$

so $\vec{0}_V$ is in $\ker(f)$.

To show that the kernel of f is closed under addition, suppose we have two vectors $\vec{v}_1$ and $\vec{v}_2$ in $\ker(f)$. We need to show $\vec{v}_1 + \vec{v}_2$ is also in $\ker(f)$. Since $\vec{v}_1$ and $\vec{v}_2$ are in $\ker(f)$ we know $f(\vec{v}_1) = \vec{0}_W$ and $f(\vec{v}_2) = \vec{0}_W$. Therefore we have

$$f(\vec{v}_1 + \vec{v}_2) = f(\vec{v}_1) + f(\vec{v}_2) = \vec{0}_W + \vec{0}_W = \vec{0}_W$$

so $\vec{v}_1 + \vec{v}_2$ is in $\ker(f)$.

To show that the kernel of f is closed under scalar multiplication, suppose we have a vector $\vec{v}_1$ in $\ker(f)$ and a scalar r. We need to show $r \cdot \vec{v}_1$ is also in $\ker(f)$. As above, we have $f(\vec{v}_1) = \vec{0}_W$, so

$$f(r \cdot \vec{v}_1) = r \cdot f(\vec{v}_1) = r \cdot \vec{0}_W = \vec{0}_W$$

which means $r \cdot \vec{v}_1$ is in $\ker(f)$.

Since the kernel satisfies all three conditions of the subspace test, we can see that the kernel of any linear map is a subspace of its domain.

This means that we automatically know that the sets we found in Examples 1 and 3 are subspaces of $\mathbb{R}^2$ and the set we found in Example 2 is a subspace of $\mathbb{R}^3$.

During our subspace check above, we showed that the kernel of any linear map contains the zero vector of its domain. One interesting question we can ask about a function is whether its kernel contains anything besides the zero vector. This is equivalent to asking when the only vector that maps to the zero vector of the codomain is the zero vector of the domain. Geometrically, this would mean that there is no collapsing as we map to $\vec{0}$ since only a single unique vector is being mapped there. We can expand this idea of a function having no collapsing as we map to anything in the codomain, which motivates the following definition.

Definition.   A function f is *1-1* if $\vec{x} \neq \vec{y}$ guarantees that $f(\vec{x}) \neq f(\vec{y})$.

If we have $\vec{x} \neq \vec{y}$ with $f(\vec{x}) = f(\vec{y})$, then f has collapsed $\vec{x}$ and $\vec{y}$ to the same point in its codomain. This means a function is 1-1 if it doesn't allow any collapsing to occur.

Example 4.   The function $f : \mathbb{R}^2 \to \mathbb{R}^2$ by $f\left(\begin{bmatrix} x \\ y \end{bmatrix}\right) = \begin{bmatrix} y \\ x \end{bmatrix}$ is 1-1.

This function switches the two components of our vector, which means f is reflection about the line $y = x$. It is impossible to map two different points in the plane to the same point with this reflection, so f is 1-1.

Example 5.   The function $f : \mathbb{R}^3 \to \mathbb{R}^2$ by $f\left(\begin{bmatrix} x \\ y \\ z \end{bmatrix}\right) = \begin{bmatrix} x \\ y \end{bmatrix}$ is not 1-1.

Geometrically, this function can be thought of as a projection of $\mathbb{R}^3$ into the xy-plane. This projection definitely collapses multiple different vectors to the same output vector as each line in the z direction is identified with a single vector in $\mathbb{R}^2$. Algebraically, f identifies all vectors in $\mathbb{R}^3$ which differ only in the z component. From either perspective, f is not 1-1.

Since not every function is as easy to visualize as those in our previous two examples, it can be useful to have a more computational way to check whether or not a function is 1-1. It turns out that if a function maps two different vectors to the same output, then it must also map multiple different vectors to the zero vector, which is stated more precisely in the Theorem below.

Theorem 2. A function $f : V \to W$ is 1-1 if and only if $\ker(f) = \{\vec{0}_V\}$.

This means another way to view 1-1 functions is that they are the functions with the smallest possible kernels.

To see why this theorem is true, suppose we have a linear map $f : V \to W$ and two vectors $\vec{x}$ and $\vec{y}$ in V for which $f(\vec{x}) = f(\vec{y})$. This can be rewritten as

$$f(\vec{x}) - f(\vec{y}) = \vec{0}_W$$

and since f is linear, we can rewrite the left-hand side to get

$$f(\vec{x} - \vec{y}) = \vec{0}_W.$$

This means that $\vec{x} - \vec{y}$ is an element of $\ker(f)$ whenever $f(\vec{x}) = f(\vec{y})$. Having $\vec{x} = \vec{y}$ is equivalent to having $\vec{x} - \vec{y} = \vec{0}_V$. Therefore the only way to have $\ker(f) = \{\vec{0}_V\}$ is to have $\vec{x} = \vec{y}$, and the only way to have $\vec{x} = \vec{y}$ whenever $f(\vec{x}) = f(\vec{y})$ is to have $\ker(f) = \{\vec{0}_V\}$.

Example 6. Use Theorem 2 to show that $f\left(\begin{bmatrix} a & b \\ c & d \end{bmatrix}\right) = \begin{bmatrix} 4d \\ 3c \\ a-b \\ a+b \end{bmatrix}$ is 1-1.

Since we have $f : M_{22} \to \mathbb{R}^4$, to see that this map is 1-1 we need to show that $\ker(f)$ contains only the zero matrix (which is the zero vector of M_{22}). We can do this using the equation $f\left(\begin{bmatrix} a & b \\ c & d \end{bmatrix}\right) = \vec{0}$ which in this case is

$$\begin{bmatrix} 4d \\ 3c \\ a-b \\ a+b \end{bmatrix} = \begin{bmatrix} 0 \\ 0 \\ 0 \\ 0 \end{bmatrix}.$$

Setting corresponding entries equal, we get $4d = 0$, $3c = 0$, $a - b = 0$, and $a+b = 0$. The first equation gives us $d = 0$, and the second gives us $c = 0$. The third equation tells us $a = b$, which can be plugged into the fourth equation to give $b + b = 2b = 0$, so $a = b = 0$. This means the kernel of f is the 2×2 zero matrix, so f is 1-1.

Example 7. Use Theorem 2 to show that $f\left(\begin{bmatrix} x_1 \\ x_2 \\ x_3 \end{bmatrix}\right) = \begin{bmatrix} x_2 & x_1 - x_3 \\ 0 & x_2 \end{bmatrix}$ is not 1-1.

As we saw in Example 2, the kernel of this map contains vectors other than the zero vector. Thus f is not 1-1.

Suppose for a moment that we are in the special case $f : \mathbb{R}^n \to \mathbb{R}^m$, where the kernel of f can be thought of as the null space of its $m \times n$ matrix A. If f is 1-1, this means $Nul(A) = \{\vec{0}\}$. This is the same as saying $A\vec{x} = \vec{0}$ is only satisfied by $\vec{x} = \vec{0}$. If we mentally transform this matrix equation into a vector equation, this is equivalent to saying A's columns are linearly independent. Since there are n columns in our matrix, this means the span of the columns has dimension n. However, A's columns are m-vectors, so their span is a subset of $\mathbb{R}^m$, which means its dimension is at most m. Therefore, we get the following.

Theorem 3. It is only possible for f to be 1-1 if $n \leq m$. Thus if $n > m$, we know that f cannot be 1-1.

Note that this theorem can't be reversed, i.e., having $n \leq m$ doesn't guarantee f is 1-1.

Example 8. Use Theorem 3 to check whether or not the function f with matrix $\begin{bmatrix} -3 & 0 & 1 & 2 \\ 4 & 7 & -6 & 1 \end{bmatrix}$ is 1-1.

This matrix is 2×4, so we know $f : \mathbb{R}^4 \to \mathbb{R}^2$. Since $4 > 2$, Theorem 3 tells us the function corresponding to this matrix is not 1-1.

Example 9. Use Theorem 3 to check whether or not the function f with matrix $\begin{bmatrix} -1 & 7 \\ 0 & 2 \\ 4 & -5 \end{bmatrix}$ is 1-1.

This matrix is 3×2, so we know $f : \mathbb{R}^2 \to \mathbb{R}^3$. Since $2 \not> 3$, Theorem 3 can't tell us anything about whether the function corresponding to this matrix is 1-1 or not.

Now that we've discussed the kernel within our domain, let's switch gears to look at a special subset of our codomain. Since the codomain was defined as the vector space where the outputs of the function live, it's very natural to ask what those outputs are.

Example 10. What are the outputs of the function $f : \mathbb{R}^2 \to \mathbb{R}^2$ by $f\left(\begin{bmatrix} x \\ y \end{bmatrix}\right) = \begin{bmatrix} x - y \\ -x + y \end{bmatrix}$?

The outputs of this function are all vectors in the codomain, $\mathbb{R}^2$, which are $f(\vec{x})$ for some vector $\vec{x}$ in the domain. Looking at the formula for our function, this means our outputs are all 2-vectors of the form $\begin{bmatrix} x - y \\ -x + y \end{bmatrix}$ for some scalars x and y. We can rewrite this output vector as $\begin{bmatrix} x \quad y \\ -(x - y) \end{bmatrix}$, which shows that our outputs can be thought of as any 2-vector whose second entry is the negative of the first entry. To summarize, f's outputs are all 2-vectors of the form $\begin{bmatrix} z \\ -z \end{bmatrix}$ for some scalar z. We can visualize this set as the line $y = -x$ in the picture below.

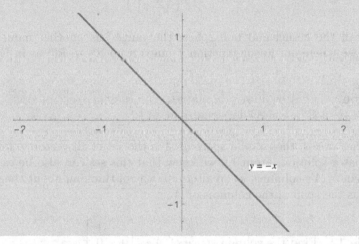

Definition. Suppose $f : V \to W$ is a linear function between two vector spaces. The *range* of f is $range(f) = \{\vec{w} \text{ in } W \mid \vec{w} = f(\vec{v}) \text{ for some } \vec{v} \text{ in } V\}$.

In other words, the range of a function is the set of all its outputs. This is clearly a subset of its codomain.

Example 11. Find the range of $f : M_{22} \to \mathbb{R}^3$ by $f\left(\begin{bmatrix} a & b \\ c & d \end{bmatrix}\right) = \begin{bmatrix} a + c \\ b \\ 2b \end{bmatrix}$.

The range of f is the set of all 3-vectors which are $f(A)$ for some 2×2 matrix A. We can express this as the set of $\vec{x}$ where $\vec{x} = f\left(\begin{bmatrix} a & b \\ c & d \end{bmatrix}\right)$, i.e., we need some a, b, c, and d so that

$$\begin{bmatrix} x_1 \\ x_2 \\ x_3 \end{bmatrix} = \begin{bmatrix} a + c \\ b \\ 2b \end{bmatrix}.$$

(Remember that we are assuming we know x_1, x_2, and x_3 and are looking for the matrix $\begin{bmatrix} a & b \\ c & d \end{bmatrix}$.) Setting corresponding entries equal gives us $a + c = x_1$, $b = x_2$, and $2b = x_3$. We can solve this for any value of x_1 and x_2 by letting $a = x_1$, $b = x_2$, $c = 0$, and d any number we like. However, this only works if $2b = 2x_2 = x_3$. Therefore we have

$$\text{range}(f) = \left\{ \begin{bmatrix} x_1 \\ x_2 \\ 2x_2 \end{bmatrix} \right\}.$$

As with the kernel and null space, the range has another name in the special case where our linear function f maps from $\mathbb{R}^n$ to $\mathbb{R}^m$ as in Example 10.

Definition. Let A be an $m \times n$ matrix. The *column space* of A is $Col(A) = \{\vec{b} \text{ in } \mathbb{R}^m \mid \vec{b} = A\vec{x} \text{ for some } \vec{x} \text{ in } \mathbb{R}^n\}$.

In other words, the column space of A is the set of all vectors $\vec{b}$ for which $A\vec{x} = \vec{b}$ has a solution. From 1.2 we know that this set can also be expressed as the span of A's columns, so an alternate way of thinking about the column space is as the span of the columns of A.

Example 12. Find the column space of $A = \begin{bmatrix} 1 & 1 \\ 0 & 0 \\ 0 & -2 \end{bmatrix}$.

Our matrix is 3×2, so its column space is the set of all 3-vectors $\vec{b}$ where $A\vec{x} = \vec{b}$ for some 2-vector $\vec{x}$. We can rewrite this equation as

$$\begin{bmatrix} 1 & 1 \\ 0 & 0 \\ 0 & -2 \end{bmatrix} \begin{bmatrix} x_1 \\ x_2 \end{bmatrix} = \begin{bmatrix} b_1 \\ b_2 \\ b_3 \end{bmatrix}$$

which simplifies to

$$\begin{bmatrix} x_1 + x_2 \\ 0 \\ -2x_2 \end{bmatrix} = \begin{bmatrix} b_1 \\ b_2 \\ b_3 \end{bmatrix}.$$

(As in the previous example, we view b_1, b_2, and b_3 as known quantities and solve for x_1 and x_2.) Setting corresponding entries equal gives us $x_1 + x_2 = b_1$,

$0 = b_2$, and $-2x_2 = b_3$. Solving for x_2 in the third equation gives $x_2 = -\frac{1}{2}b_3$. Plugging that into the first equation and solving for x_1 gives $x_1 = b_1 + \frac{1}{2}b_3$. This means as long as $b_2 = 0$ we can find a $\vec{x}$ for which $A\vec{x} = \vec{b}$, so

$$Col(A) = \left\{ \begin{bmatrix} b_1 \\ 0 \\ b_3 \end{bmatrix} \right\}.$$

As with the kernel and domain, the range (or column space) is more than just a subset of the codomain.

Theorem 4. If $f : V \to W$ is a linear function, then $range(f)$ is a subspace of W.

As with the kernel, we'll use the subspace test by checking that the range contains the zero vector of the codomain and is closed under addition and scalar multiplication.

To show that $\vec{0}_W$ (the zero vector of the codomain W) is in the range, we need to find some $\vec{v}$ in V with $f(\vec{v}) = \vec{0}_W$. However, we've already shown that $f(\vec{0}_V) = \vec{0}_W$, so $\vec{0}_W$ is in the range of f.

To show that the range of f is closed under addition, suppose we have two vectors $\vec{w}_1$ and $\vec{w}_2$ in the range. We need to show $\vec{w}_1 + \vec{w}_2$ is also in the range. Since $\vec{w}_1$ and $\vec{w}_2$ are in the range, we must have $\vec{v}_1$ and $\vec{v}_2$ in V with $f(\vec{v}_1) = \vec{w}_1$ and $f(\vec{v}_2) = \vec{w}_2$. Since f is linear, we get

$$\vec{w}_1 + \vec{w}_2 = f(\vec{v}_1) + f(\vec{v}_2) = f(\vec{v}_1 + \vec{v}_2)$$

and $\vec{w}_1 + \vec{w}_2$ is in the range of f.

To show that the range of f is closed under scalar multiplication, suppose we have $\vec{w}_1$ in the range and a scalar r. We need to show $r \cdot \vec{w}_1$ is also in the range. As above, we have $\vec{v}_1$ with $f(\vec{v}_1) = \vec{w}_1$, so

$$r \cdot \vec{w}_1 = r \cdot f(\vec{v}_1) = f(r \cdot \vec{v}_1)$$

so $r \cdot \vec{w}_1$ is in the range of f.

Therefore the range is a subspace of the codomain.

This means we get that the set from Example 10 is a subspace of $\mathbb{R}^2$ and the sets from Examples 11 and 12 are subspaces of $\mathbb{R}^3$.

Unlike in our discussion of 1-1 where we asked which functions had the smallest possible kernel, we'll ask which functions have the largest possible range. Since the range is a subset of the codomain, the largest the range could be is the whole codomain. Geometrically, this means that the function's outputs fill up the entire codomain, which prompts the following definition.

Definition.   A function $f : V \to W$ is *onto* if its range is W.

In other words, a function is onto if its range is its entire codomain.

Example 13. The function $f : M_{22} \to M_{22}$ by $f\left(\begin{bmatrix} a & b \\ c & d \end{bmatrix}\right) = \begin{bmatrix} d & -b \\ -c & a \end{bmatrix}$ is onto.

For any 2×2 matrix $A = \begin{bmatrix} x & y \\ z & w \end{bmatrix}$, we have $f\left(\begin{bmatrix} w & -y \\ -z & x \end{bmatrix}\right) = A$. This means the range of f is all of M_{22}, so f is onto.

Example 14. The function $f : M_{22} \to \mathbb{R}^3$ by $f\left(\begin{bmatrix} a & b \\ c & d \end{bmatrix}\right) = \begin{bmatrix} a+c \\ b \\ 2b \end{bmatrix}$ is not onto.

We saw in Example 11 that $\text{range}(f) = \left\{ \begin{bmatrix} x_1 \\ x_2 \\ 2x_2 \end{bmatrix} \right\}$. Since not all 3-vectors have $x_3 = 2x_2$, this is strictly smaller than f's codomain $\mathbb{R}^3$. Therefore f is not onto.

As we did with the kernel, let's explore onto functions in the case where $f : \mathbb{R}^n \to \mathbb{R}^m$. Here the range of f can also be thought of as the column space of f's $m \times n$ matrix A. If f is onto, then $Col(A) = \mathbb{R}^m$. This means $Col(A)$ has dimension m. Since the column space of A is the span of A's columns, this means that A must have m linearly independent columns. Since A has n columns in total, A has at most n linearly independent columns. Therefore we get a parallel to Theorem 2.

Theorem 5. It is only possible for f to be onto if $m \leq n$. Thus if $m > n$, we know that f cannot be onto.

This theorem also can't be reversed, i.e., having $m \leq n$ doesn't guarantee f is onto.

Example 15. Use Theorem 5 to check whether or not the function f with matrix $\begin{bmatrix} 7 & -2 & 1 \\ 3 & 4 & 0 \end{bmatrix}$ is onto.

This matrix is 2×3, so we know $f : \mathbb{R}^3 \to \mathbb{R}^2$. Since $2 \not> 3$, Theorem 5 can't tell us anything about whether or not the function corresponding to this matrix is onto.

Example 16. Use Theorem 5 to check whether or not the function f with matrix $\begin{bmatrix} 4 & 9 \\ -1 & 0 \\ 2 & 1 \end{bmatrix}$ is onto.

This matrix is 3×2, so we know $f : \mathbb{R}^2 \to \mathbb{R}^3$. Since $3 > 2$ we know the function corresponding to this matrix is not onto.

From a geometric perspective Example 16's conclusion should make sense, because it says that our linear function f cannot fill up three-dimensional space ($\mathbb{R}^3$) with its outputs if its inputs are only two-dimensional ($\mathbb{R}^2$).

Exercises 2.5.

1. Is $\begin{bmatrix} 0 & 1 & 0 \\ -1 & 0 & 1 \\ 2 & 1 & -2 \end{bmatrix}$ in the kernel of $f\left(\begin{bmatrix} a & b & c \\ d & e & f \\ g & h & k \end{bmatrix}\right) = \begin{bmatrix} a & c \\ 2a & e+k \end{bmatrix}$?

2. Is $\begin{bmatrix} 3 & -2 \\ 6 & 2 \end{bmatrix}$ in the kernel of $f\left(\begin{bmatrix} a & b \\ c & d \end{bmatrix}\right) = \begin{bmatrix} 4a - 6d \\ 3b + d \end{bmatrix}$?

3. Find the kernel of $f\left(\begin{bmatrix} x_1 \\ x_2 \\ x_3 \\ x_4 \end{bmatrix}\right) = \begin{bmatrix} 4x_1 + 8x_3 \\ x_2 - x_3 + 2x_4 \end{bmatrix}$.

4. Find the kernel of $f\left(\begin{bmatrix} x_1 \\ x_2 \\ x_3 \end{bmatrix}\right) = \begin{bmatrix} x_1 - 4x_2 + 2x_3 \\ -x_2 + 3x_3 \\ x_1 - 3x_2 - x_3 \end{bmatrix}$.

5. Find the kernel of $f\left(\begin{bmatrix} a & b \\ c & d \end{bmatrix}\right) = \begin{bmatrix} 2a - b \\ a + c - d \\ -3c \end{bmatrix}$.

6. Find the kernel of $f\left(\begin{bmatrix} x_1 \\ x_2 \\ x_3 \end{bmatrix}\right) = \begin{bmatrix} x_1 + x_2 & 0 \\ x_2 + x_3 & -x_1 - x_2 \end{bmatrix}$.

7. Find the kernel of $f\left(\begin{bmatrix} a & b \\ c & d \end{bmatrix}\right) = \begin{bmatrix} d & a - b \\ a + c & 5d \end{bmatrix}$.

8. Find the kernel of $f\left(\begin{bmatrix} a & b & c \\ d & e & f \end{bmatrix}\right) = \begin{bmatrix} 2f + 4b & c - 3a \\ a - 2b & -e \end{bmatrix}$.

9. Find the null space of $A = \begin{bmatrix} 1 & 3 & 0 \\ -2 & 2 & -8 \\ 0 & -1 & 1 \end{bmatrix}$.

10. Find the null space of $A = \begin{bmatrix} 2 & 0 & 4 & -2 \\ 0 & 1 & -3 & 1 \end{bmatrix}$.

11. Find the null space of $A = \begin{bmatrix} 1 & -1 & 0 & 2 & 0 \\ 0 & 0 & 1 & -3 & 0 \\ 0 & 0 & 0 & 0 & 1 \end{bmatrix}$.

12. Find the null space of $A = \begin{bmatrix} 1 & -4 & 0 & 1 & 0 & 0 \\ 0 & 0 & 1 & -2 & 0 & -1 \\ 0 & 0 & 0 & 0 & 1 & 5 \end{bmatrix}$.

13. Explain, without doing any computations, why the linear map whose matrix is $A = \begin{bmatrix} 1 & 0 & 0 & -2 \\ 0 & 0 & 1 & 3 \end{bmatrix}$ cannot be 1-1.

14. Explain, without doing any computations, why the linear map whose matrix is $A = \begin{bmatrix} 1 & -5 & 0 & 0 \\ 0 & 0 & 1 & -2 \\ 0 & 0 & -4 & 8 \end{bmatrix}$ cannot be 1-1.

15. Is $f\left(\begin{bmatrix} x_1 \\ x_2 \end{bmatrix} \right) = \begin{bmatrix} x_1 + x_2 \\ 0 \\ 2x_1 + x_2 \end{bmatrix}$ a 1-1 function?

16. Is $f\left(\begin{bmatrix} a & b \\ c & d \end{bmatrix} \right) = \begin{bmatrix} a - d & c \\ d & b + c \end{bmatrix}$ a 1-1 function?

17. Is $\begin{bmatrix} 1 & 2 \\ 3 & 4 \end{bmatrix}$ in the range of $f\left(\begin{bmatrix} a & b & c \\ d & e & f \\ g & h & k \end{bmatrix} \right) = \begin{bmatrix} a & c \\ 2a & e + k \end{bmatrix}$?

18. Is $\begin{bmatrix} -1 \\ 7 \end{bmatrix}$ in the range of the function with matrix $\begin{bmatrix} 1 & 0 & 0 & -2 \\ 0 & 0 & 1 & 3 \end{bmatrix}$?

19. Find the range of the linear map f whose matrix is $\begin{bmatrix} 1 & -1 \\ 4 & -3 \\ 0 & 3 \end{bmatrix}$.

20. Find the range of the linear map f whose matrix is
$$A = \begin{bmatrix} 1 & -3 & 0 \\ 0 & 0 & 1 \\ 0 & 0 & 0 \\ 0 & 0 & 0 \end{bmatrix}.$$

21. Find the range of $f\left(\begin{bmatrix} x_1 \\ x_2 \\ x_3 \end{bmatrix} \right) = \begin{bmatrix} 2x_1 & x_2 \\ x_2 + x_3 & x_3 \end{bmatrix}$.

22. Find the range of $f\left(\begin{bmatrix} a & b & c \\ d & e & f \end{bmatrix} \right) = \begin{bmatrix} a + 2e + f \\ d + c \\ -4b \end{bmatrix}$.

23. Find the range of $f\left(\begin{bmatrix} a & b & c \\ d & e & f \end{bmatrix} \right) = \begin{bmatrix} a + b & c \\ d & e + f \end{bmatrix}$.

24. Find the range of $f\left(\begin{bmatrix} a & b \\ c & d \end{bmatrix}\right) = \begin{bmatrix} a+c & 0 & 0 \\ 0 & b+d & 0 \\ 0 & 0 & a+b+c+d \end{bmatrix}$.

25. Is $\vec{v} = \begin{bmatrix} -1 \\ 2 \\ 1 \end{bmatrix}$ in the column space of $A = \begin{bmatrix} 1 & -4 & 2 \\ 0 & -1 & 3 \\ 1 & -3 & -1 \end{bmatrix}$?

26. Is $\vec{v} = \begin{bmatrix} 4 \\ 10 \\ 2 \end{bmatrix}$ in the column space of $A = \begin{bmatrix} 2 & -1 & 5 \\ 10 & 3 & -1 \\ 0 & -1 & 4 \end{bmatrix}$?

27. Find the column space of $A = \begin{bmatrix} 1 & 1 \\ 0 & 0 \\ 2 & 5 \end{bmatrix}$.

28. Find the column space of $A = \begin{bmatrix} 2 & 5 \\ 4 & 10 \end{bmatrix}$.

29. Explain without doing any computations why the linear map whose matrix is $A = \begin{bmatrix} 9 & -2 \\ 3 & 1 \\ -4 & 6 \end{bmatrix}$ cannot be onto.

30. Explain without doing any computations why the linear map whose matrix is $A = \begin{bmatrix} 2 & 0 & 4 \\ -3 & 1 & 1 \\ 6 & -6 & 2 \\ 0 & -1 & 1 \end{bmatrix}$ cannot be onto.

31. Is $f\left(\begin{bmatrix} x_1 \\ x_2 \\ x_3 \end{bmatrix}\right) = \begin{bmatrix} 2x_1 \\ -x_2 \end{bmatrix}$ an onto function?

32. Is $f\left(\begin{bmatrix} x_1 \\ x_2 \\ x_3 \end{bmatrix}\right) = \begin{bmatrix} x_1 & x_2 \\ x_3 & x_1 + x_2 \end{bmatrix}$ an onto function?

33. Let $f : \mathbb{R}^3 \to \mathbb{R}^2$ be a linear map whose matrix is A.

    (a) Give an example of an A for which f is an onto map.
    (b) Give an example of an A for which f is not an onto map.
    (c) Briefly explain why we can't give an A for which f is 1-1.

34. Let $f : \mathbb{R}^2 \to \mathbb{R}^3$ be a linear map whose matrix is A.

    (a) Give an example of an A for which f is a 1-1 map.
    (b) Give an example of an A for which f is not a 1-1 map.
    (c) Briefly explain why we can't give an A for which f is onto.

2.6   Row Reduction

In many of our previous sections, we've worked on problems that eventually reduced to solving $f(\vec{x}) = \vec{b}$ for $\vec{x}$ or solving $x_1\vec{v}_1 + \cdots x_n\vec{v}_n = \vec{b}$ for $x_1, \ldots, x_n$. In Section 2.2, we learned how to find a matrix A so that both types of problems could be restated as solving $A\vec{x} = \vec{b}$ for $\vec{x}$. In this section, we'll develop an algorithm to systematically solve these matrix equations.

Before we start building our algorithm, we need to introduce two more ways of writing the equation $A\vec{x} = \vec{b}$. The first way doesn't contain any new mathematical insights, but allows us to write down the minimum information needed to specify the matrix equation we're working with. Since our algorithm will often be done on a computer or calculator, the efficiency of this format will be very helpful.

If you consider a matrix equation $A\vec{x} = \vec{b}$, many of the symbols used can be inferred from the rest of the setup. For example, from the number of columns of the matrix A we can already tell the number of entries in $\vec{x}$. This means we really just need to tell our algorithm the entries of A and $\vec{b}$.

Definition.   The *augmented coefficient matrix* of $A\vec{x} = \vec{b}$ is the matrix $\left[A \mid \vec{b}\right]$ formed by joining $\vec{b}$ onto the right-hand side of A.

When writing augmented coefficient matrices, I will always include a vertical line separating the part of the matrix coming from A from the last column, which gives $\vec{b}$. (This line occurs where the equals sign is in $A\vec{x} = \vec{b}$.) If you don't see that line, I'm not describing an augmented coefficient matrix. Some people call A the coefficient matrix, but I'll avoid using that terminology to avoid confusion.

Note that if A is $m \times n$ (making $\vec{b}$ an m-vector) then the augmented coefficient matrix of $A\vec{x} = \vec{b}$ will be $m \times (n+1)$.

Example 1.   Find the augmented coefficient matrix of $A\vec{x} = \vec{b}$ where
$A = \begin{bmatrix} -3 & 4 & -1 \\ 0 & 2 & 6 \end{bmatrix}$ and $\vec{b} = \begin{bmatrix} 5 \\ -2 \end{bmatrix}$.

The augmented coefficient matrix is the matrix A with $\vec{b}$ added as its new fourth column to give us

$$\left[\begin{array}{ccc|c} -3 & 4 & -1 & 5 \\ 0 & 2 & 6 & -2 \end{array}\right].$$

(Note the vertical line where we joined $\vec{b}$ onto A, which tells us this is an augmented coefficient matrix.)

Since writing down augmented coefficient matrices is easier than writing down matrix equations and our algorithm to solve $A\vec{x} = \vec{b}$ will involve writing down many matrix equations, I'll write the algorithm in terms of augmented coefficient matrices.

Our second new way to rewrite $A\vec{x} = \vec{b}$ is a more familiar format from algebra class: as a list of equations to be solved simultaneously. To see this, we can use the matrix-vector multiplication we developed in 2.2 to compute

$$A\vec{x} = \begin{bmatrix} a_{11}x_1 + a_{12}x_2 + \cdots + a_{1n}x_n \\ a_{21}x_1 + a_{22}x_2 + \cdots + a_{2n}x_n \\ \vdots \\ a_{m1}x_1 + a_{m2}x_2 + \cdots + a_{mn}x_n \end{bmatrix}.$$

Setting this vector equal to $\vec{b} = \begin{bmatrix} b_1 \\ \vdots \\ b_m \end{bmatrix}$, we get

$$\begin{bmatrix} a_{11}x_1 + a_{12}x_2 + \cdots + a_{1n}x_n \\ a_{21}x_1 + a_{22}x_2 + \cdots + a_{2n}x_n \\ \vdots \\ a_{m1}x_1 + a_{m2}x_2 + \cdots + a_{mn}x_n \end{bmatrix} = \begin{bmatrix} b_1 \\ \vdots \\ b_m \end{bmatrix}.$$

From 1.1 we know two vectors are equal precisely when each pair of their corresponding entries is equal, so the equation above can be broken up into m separate equations which must all hold at once. These equations are

$$a_{11}x_1 + a_{12}x_2 + \cdots + a_{1n}x_n = b_1$$

$$a_{21}x_1 + a_{22}x_2 + \cdots + a_{2n}x_n = b_2$$

$$\vdots$$

$$a_{m1}x_1 + a_{m2}x_2 + \cdots + a_{mn}x_n = b_m.$$

Notice that each equation corresponds to a row of our augmented coefficient matrix $\begin{bmatrix} A \mid \vec{b} \end{bmatrix}$, since the coefficients on the x_i's in one equation all come from the entries of a single row of A and the equation equals the entry of $\vec{b}$ from that row. Thus a matrix equation involving an $m \times n$ matrix A gives us an $m \times (n+1)$ augmented coefficient matrix which corresponds to a list of m linear equations. Each variable, on the other hand, corresponds to one of the first n columns, since the entries down the jth column of A appear as the coefficients on x_j in each equation. The entries of $\vec{b}$ form the added $n + 1$st column, which is sometimes called the augmentation column.

The left-hand side of each of these equations is a linear combination of the variables $x_1, \ldots, x_n$. For this reason, many people call each of these equations a linear equation and the whole list of linear equations a linear system.

Example 2. Write the matrix equation $A\vec{x} = \vec{b}$ as list of linear equations, where $A = \begin{bmatrix} -3 & 4 & -1 \\ 0 & 2 & 6 \end{bmatrix}$ and $\vec{b} = \begin{bmatrix} 5 \\ -2 \end{bmatrix}$.

Each row of A gives us the coefficients for one equation, which will equal the corresponding entry in $\vec{b}$. Since A has 2 rows and $\vec{b}$ has 2 entries, this means we'll have 2 linear equations. Each column of A corresponds to a variable, so A's 3 columns mean we have 3 variables which I'll call x_1, x_2, and x_3. Putting this all together, the first row of A and first entry of $\vec{b}$ give us the equation $-3x_1 + 4x_2 - x_3 = 5$. The second row of A and second entry of $\vec{b}$ give us $2x_2 + 6x_3 = -2$. (Note that there is no x_1 term in the second equation, because the first entry in A's second row is 0, which means x_1's coefficient in the second equation is 0.) This means we can write $A\vec{x} = \vec{b}$ as

$$-3x_1 + 4x_2 - x_3 = 5$$
$$2x_2 + 6x_3 = -2.$$

The benefit of rewriting a matrix equation as a list of linear equations is that we can then use several tools from algebra to transform our linear equations or augmented coefficient matrix. Our goal is to change our original formulation of the problem into a version where the solutions are easy to find. Of course, we want this transformation to occur without changing our solutions.

Consider the linear equations $3x_1 + x_2 = 14$ and $x_1 + x_2 = 4$ compared to the linear equations $x_1 = 5$ and $x_2 = -1$. The second pair of equations is much easier to use, even though both pairs of equations have the same solutions.

What properties made the second pair of equations so much easier? To start with, each equation only contained one of our variables. Additionally, those variables had coefficient 1. We also listed our equations so that we started with the equation for x_1, then proceeded to the equation for x_2.

Even when we aren't lucky enough to get such a clean final format, we can still look for a simpler format. For example, we can consider the linear equations $2x_1 - 4x_2 + x_3 = 12$, $x_1 - 2x_2 + 3x_3 = 11$, and $-x_1 + 2x_2 = -5$ compared to the linear equations $x_1 - 2x_2 = 5$, $x_3 = 2$, and $0 = 0$. This second pair of equations is simpler, and again has the same solutions as the first pair.

What properties made our second set of equations simpler? We don't have each variable in its own equation, but we still have a coefficient of 1 on the first variable to appear in each equation. Additionally, the first variable in each equation doesn't appear in any other equation. This set of equations had a redundant equation which we've replaced with $0 = 0$. Finally, we've ordered our list of equations based on their starting variables and placed $0 = 0$ at the end of our list since we can basically ignore it.

We're planning to write down our algorithm using augmented coefficient matrices, so let's try to translate the nice properties of our simpler sets of linear equations into properties of an augmented coefficient matrix. The augmented coefficient matrix of $x_1 = 5$ and $x_2 = -1$ is

$$\begin{bmatrix} 1 & 0 & | & 5 \\ 0 & 1 & | & -1 \end{bmatrix}$$

and the augmented coefficient matrix of $x_1 - 2x_2 = 5$, $x_3 = 2$, and $0 = 0$ is

$$\begin{bmatrix} 1 & -2 & 0 & | & 5 \\ 0 & 0 & 1 & | & 2 \\ 0 & 0 & 0 & | & 0 \end{bmatrix}.$$

The fact that the first variable in each equation had a coefficient of 1 means that the first nonzero entry in each row of our matrices is 1. Those variables didn't appear in any other equation, so if a column contains one of these 1s then the rest of that column's entries are zero. The equations were ordered by their first variables, so these special 1s appear in lower rows as we move from left to right across the columns of our augmented coefficient matrices. Finally, we put $0 = 0$ equations last in the list, so any rows that are all zeros appear at the bottom of our matrices. These properties are summed up in the following definition.

Definition. A matrix is in *reduced echelon form* if it satisfies the following conditions:

- Any rows whose entries are all 0 occur at the bottom.

- The leftmost nonzero entry in each row is a 1, called a *leading 1*.

- If a column contains a leading 1, then all its other entries are 0s.

- If a leading 1 appears in a row above another leading 1, then it must also appear in a column to the left of the other leading 1.

Example 3. Circle all leading 1s of $A = \begin{bmatrix} 1 & 0 & | & -3 \\ 0 & 1 & | & 8 \end{bmatrix}$. Is A in reduced echelon form?

To check whether A is in reduced echelon form, we need to check each of the conditions in the definition above. Our matrix A doesn't have any rows of zeros, and it does have a 1 as the leftmost nonzero entry in each row. These are our leading 1s, which are circled below.

$$\begin{bmatrix} ① & 0 & | & -3 \\ 0 & ① & | & 8 \end{bmatrix}$$

In the first two columns, which have leading 1s, all other entries are 0. The leading 1 in the first row is in a row above the leading 1 in the second row, and it is also to the left of the second row's leading 1. Since A satisfies all the definition's conditions, it is in reduced echelon form.

Example 4. Circle all leading 1s of $A = \begin{bmatrix} 1 & 0 & 0 & -1 \\ 0 & 1 & 0 & 2 \\ 0 & 0 & 1 & 5 \\ 0 & 0 & 0 & 0 \end{bmatrix}$. Is A in reduced

echelon form?

This matrix has a row of zeros, which is at the bottom. The first nonzero entry in each of the other rows is a 1. These leading 1s are circled below.

$$\begin{bmatrix} \circled{1} & 0 & 0 & -1 \\ 0 & \circled{1} & 0 & 2 \\ 0 & 0 & \circled{1} & 5 \\ 0 & 0 & 0 & 0 \end{bmatrix}$$

The first three columns, which contain the leading 1s, have all other entries 0. The leading 1 in the top row is above those in the second and third rows, and it is also to their left. Similarly, the leading 1 in the second row is above that in the third row, and it is also to its left. Thus A is in reduced echelon form.

Note that this matrix isn't in our ideal format because of the bottom row of zeros. However, this row corresponds to the equation $0 = 0$, so isn't really a problem.

Example 5. Circle all leading 1s of $A = \begin{bmatrix} 1 & 0 & 3 & 0 & -1 \\ 0 & 1 & 1 & 0 & 2 \\ 0 & 0 & 0 & 1 & 9 \end{bmatrix}$. Is A in reduced

echelon form?

Our matrix doesn't have any rows of zeros, and the leftmost nonzero entry of each row is a 1. See the picture below where these leading 1s are circled.

$$\begin{bmatrix} \circled{1} & 0 & 3 & 0 & -1 \\ 0 & \circled{1} & 1 & 0 & 2 \\ 0 & 0 & 0 & \circled{1} & 9 \end{bmatrix}$$

(Note that the 1 in the second row and third column is not a leading 1, because it is not the leftmost nonzero entry.)

The first, second, and fourth columns with the leading 1s have all other entries 0. The leading 1 in the top row is above those in the second and third rows, and it is also to their left. Similarly, the leading 1 in the second row is above that in the third row, and it is also to its left. Thus A is in reduced echelon form.

This matrix also isn't in our ideal format, because it doesn't have an equation of the form $x_3 = s$. This is because there is no leading 1 in the third column. We also have x_3 terms in the equations defining x_1 and x_2. This does happen sometimes, and we'll discuss how to deal with this type of outcome in 2.7 and 2.8.

Example 6. Why isn't $A = \begin{bmatrix} 1 & 0 & | & 1 \\ 0 & -2 & | & 3 \end{bmatrix}$ in reduced echelon form?

The leftmost nonzero entry in the second row is -2 instead of a leading 1, so A isn't in reduced echelon form.

Example 7. Why isn't $A = \begin{bmatrix} 1 & 0 & 0 \\ 0 & 0 & 1 \\ 0 & 1 & 0 \end{bmatrix}$ in reduced echelon form?

Note that this isn't an augmented coefficient matrix. That's okay, we can still think about whether or not it is in reduced echelon form.

The leading 1 in the second row of A is above the leading 1 in the third row, but it is not to the left of the third row's leading 1. This is why A isn't in reduced echelon form.

Example 8. Why isn't $A = \begin{bmatrix} 0 & 1 & | & 3 \\ 0 & 0 & | & 1 \end{bmatrix}$ in reduced echelon form?

The third column has a leading 1, but its other entry isn't 0 (instead it is 3). This means A isn't in reduced echelon form. (The fact that the first column of the matrix doesn't have a leading 1 isn't a problem.)

Now that we understand reduced echelon form, we can start developing the tools needed to create an algorithm which transforms an augmented coefficient matrix into an augmented coefficient matrix in reduced echelon form while at every step keeping the solutions to our matrix equation the same. These tools are based on methods from algebra used to manipulate sets of equations, and since those equations correspond to rows of our augmented coefficient matrix our algorithm's tools are called row operations.

The first of our three row operations is to swap the order of the rows in our augmented coefficient matrix, which corresponds to swapping the order of equations in our list. The order in which the equations appear in a list doesn't matter, so it is clear that reordering the rows of our augmented coefficient matrix doesn't change the solutions. We'll usually do this reordering by swapping one pair of rows at a time. I'll use the notation $r_i \leftrightarrow r_j$ to indicate that we're swapping the ith and jth rows of our matrix.

Example 9. The solutions to $\begin{bmatrix} 1 & 1 & | & 2 \\ 1 & -1 & | & 1 \end{bmatrix}$ and $\begin{bmatrix} 1 & -1 & | & 1 \\ 1 & 1 & | & 2 \end{bmatrix}$ are the same where $\begin{bmatrix} 1 & 1 & | & 2 \\ 1 & -1 & | & 1 \end{bmatrix} \rightarrow_{r_1 \leftrightarrow r_2} \begin{bmatrix} 1 & -1 & | & 1 \\ 1 & 1 & | & 2 \end{bmatrix}$.

The first augmented coefficient matrix corresponds to the pair of equations $x_1 + x_2 = 2$ and $x_1 - x_2 = 1$, while the second corresponds to the equations $x_1 - x_2 = 1$ and $x_1 + x_2 = 2$. Adding these equations together (in either order) gives $2x_1 = 3$ or $x_1 = \frac{3}{2}$. Plugging this back into one of the original equations then gives $x_2 = \frac{1}{2}$. The order in which we originally listed the equations doesn't affect their solutions.

Our next row operation is to multiply one row of our augmented coefficient matrix by a nonzero scalar. This corresponds to multiplying an entire equation by a nonzero scalar, i.e., multiplying all coefficients and the value the equation equals by that scalar. Since our scalar isn't zero, this doesn't change the solutions of that equation. I'll use the notation $c\,r_i$ to indicate that we're multiplying the ith row of the augmented coefficient matrix by c.

Example 10. The solutions to $\begin{bmatrix} 2 & 1 & | & 6 \\ 0 & 4 & | & 8 \end{bmatrix}$ and $\begin{bmatrix} 2 & 1 & | & 6 \\ 0 & 1 & | & 4 \end{bmatrix}$ are the same where $\begin{bmatrix} 2 & 1 & | & 6 \\ 0 & 4 & | & 8 \end{bmatrix} \to_{\frac{1}{4}r_2} \begin{bmatrix} 2 & 1 & | & 6 \\ 0 & 1 & | & 2 \end{bmatrix}$.

The first matrix corresponds to the equations $2x_1 + x_2 = 6$ and $4x_2 = 8$ while the second corresponds to the equations $2x_1 + x_2 = 6$ and $x_2 = 4$. When solving the first pair of equations, we'd automatically divide the second equation by 4 (i.e., multiply by $\frac{1}{4}$) to get $x_2 = 2$. This is the second equation from our second augmented coefficient matrix, so this row operation doesn't change the fact that $x_2 = 2$. The first rows of our two matrices are the same, so we'd be plugging $x_2 = 2$ back into the same equation $2x_1 + x_2 = 6$ to get $x_1 = 3$ in both cases. Therefore multiplying our second row by $\frac{1}{4}$ didn't change the solution.

Our final row operation is to add a multiple of one row of our augmented coefficient matrix to another row. This corresponds to adding a multiple of one equation to another equation, and is commonly done to use part of one equation to cancel out one of the variables from another equation (see the example below). It also doesn't change the solutions of our list of equations. I'll use the notation $r_i + c\,r_j$ to indicate that we're replacing the ith row of our augmented coefficient matrix by the ith row plus c times the jth row. (Note that the notation for this row operation is not symmetric, since the row operation itself isn't symmetric. The row listed first is the only row of the matrix being changed.)

Example 11. The solutions to $\begin{bmatrix} -1 & 2 & | & 3 \\ 1 & -1 & | & 5 \end{bmatrix}$ and $\begin{bmatrix} -1 & 2 & | & 3 \\ 0 & 1 & | & 8 \end{bmatrix}$ are the same where $\begin{bmatrix} -1 & 2 & | & 3 \\ 1 & -1 & | & 5 \end{bmatrix} \to_{r_2+r_1} \begin{bmatrix} -1 & 2 & | & 3 \\ 0 & 1 & | & 8 \end{bmatrix}$.

Here our first matrix corresponds to the equations $-x_1 + 2x_2 = 3$ and $x_1 - x_2 = 5$. When solving this, a common strategy is to add two equations to cancel out one of the variables so we can solve for the other variable. In this case, if we simply add the equations together, we'll cancel out x_1 to get $x_2 = 8$. Plugging this back into one of the original equations then gives us $x_1 = 13$. The second matrix corresponds to the equations $-x_1 + 2x_2 = 3$ and $x_2 = 8$. Essentially, here we've already done the addition of equations. In any case, you can see that the solution is still $x_1 = 13$ and $x_2 = 8$.

Now that we understand the three building blocks of our algorithm, it's time to create the algorithm itself. Our goal is to use these three row operations to transform our augmented coefficient matrix into its reduced echelon form. Since row operations don't change the solutions, we'll be able to solve our original matrix equation by reading off the solutions of the reduced echelon form's matrix equation. The row reduction algorithm has two parts. In the first half, we start at the top left corner of our matrix and work our way downward and to the right, creating the leading 1s and the zeros beneath them. In the second half, we start at the bottom right corner of our matrix and work our way up and to the left, creating the zeros above each leading 1. Since our matrices have finitely many rows and columns, this means our algorithm will always terminate in finitely many steps. Formally, the row reduction algorithm's instructions are as follows:

Row Reduction Algorithm:

Part 1:

- If possible, swap rows to put a nonzero entry in the top left corner of the matrix. If not possible, skip to the last step in this part.

- Multiply the top row by a nonzero constant to make its first nonzero entry into a leading 1.

- Add a multiple of the top row to each lower row to get zero entries below the top row's leading 1.

- Ignore the top row and leftmost column of the matrix. If there are any rows remaining, go back to the top of Part 1, and repeat the process on the remaining entries of the matrix. If there are no rows remaining after you ignore the top row, go to Part 2.

Part 2:

- If the bottom row has a leading 1, add a multiple of the bottom row to each higher row to get zero entries above that leading 1. If the bottom row doesn't have a leading 1, skip to the next step.

- Ignore the bottom row of the matrix. If there are two or more rows remaining, go back to the top of Part 2, and repeat the process on the remaining entries of the matrix. If you have reached the top row after you ignore the bottom row, you're done with the algorithm.

Let's see how this works on an example.

Example 12. Find the reduced echelon form of $\begin{bmatrix} 0 & 3 & 3 & | & 0 \\ 2 & 0 & 0 & | & -4 \\ -1 & 0 & 1 & | & 6 \end{bmatrix}$.

We have a zero in the top left corner of our matrix, so we'll start by swapping the order of rows to put something nonzero up there. Since both other rows have nonzero first entries, we have a choice of which row to swap with row 1. It doesn't matter, so I'll start by swapping rows 1 and 2 which gives us

$$\begin{bmatrix} 0 & 3 & 3 & | & 0 \\ 2 & 0 & 0 & | & -4 \\ -1 & 0 & 1 & | & 6 \end{bmatrix} \rightarrow_{r_1 \leftrightarrow r_2} \begin{bmatrix} 2 & 0 & 0 & | & -4 \\ 0 & 3 & 3 & | & 0 \\ -1 & 0 & 1 & | & 6 \end{bmatrix}.$$

Next I have to multiply the top row by a constant to create our first leading 1. Since the first nonzero entry in the top row is 2, I'll multiply the top row by $\frac{1}{2}$ which is

$$\begin{bmatrix} 2 & 0 & 0 & | & -4 \\ 0 & 3 & 3 & | & 0 \\ -1 & 0 & 1 & | & 6 \end{bmatrix} \rightarrow_{\frac{1}{2}r_1} \begin{bmatrix} 1 & 0 & 0 & | & -2 \\ 0 & 3 & 3 & | & 0 \\ -1 & 0 & 1 & | & 6 \end{bmatrix}.$$

(Note that we multiplied the whole row by $\frac{1}{2}$, not just the first entry!) Next we need to add multiples of the top row to the lower rows to get zeros below our leading 1. The second row already has a zero as its first entry, so we just need to tackle the third row. Its first entry is -1, so we can simply add row 1 to row 3 which gives us

$$\begin{bmatrix} 1 & 0 & 0 & | & -2 \\ 0 & 3 & 3 & | & 0 \\ -1 & 0 & 1 & | & 6 \end{bmatrix} \rightarrow_{r_3 + r_1} \begin{bmatrix} 1 & 0 & 0 & | & -2 \\ 0 & 3 & 3 & | & 0 \\ 0 & 0 & 1 & | & 4 \end{bmatrix}.$$

We've reached the end of our first repetition of Part 1 of the algorithm, so we'll ignore the top row and left column of our matrix and repeat Part 1 on the remaining matrix entries shown in the picture below.

$$\begin{bmatrix} 1 & 0 & 0 & | & 2 \\ 0 & 3 & 3 & | & 0 \\ 0 & 0 & 1 & | & 4 \end{bmatrix}$$

The top left corner of this remaining section is nonzero, so we can skip swapping rows and go directly to creating our leading 1. Since the current entry is 3, we'll multiply the second row of the matrix by $\frac{1}{3}$ and get

$$\begin{bmatrix} 1 & 0 & 0 & -2 \\ 0 & 3 & 3 & 0 \\ 0 & 0 & 1 & 4 \end{bmatrix} \rightarrow_{\frac{1}{3}r_2} \begin{bmatrix} 1 & 0 & 0 & -2 \\ 0 & 1 & 1 & 0 \\ 0 & 0 & 1 & 4 \end{bmatrix}.$$

We already have zero underneath our new leading 1, so we can again drop the top row and left column of our block as shown below.

$$\begin{bmatrix} 1 & 0 & 0 & -2 \\ 0 & 1 & 1 & 0 \\ 0 & 0 & 1 & 4 \end{bmatrix}$$

Our remaining section's top left corner is not only nonzero, it is already a leading 1! There are no matrix entries below this leading 1 and we've reached the bottom row of our matrix, so we're done with Part 1 of our algorithm and ready to start Part 2.

The bottom row of our matrix has a leading 1, so we need to add multiples of this third row to the rows above to get zeros above that leading 1. The top row already has a zero there, but the second row doesn't. Since the second row's entry is 1, we'll need to add -1 times the third row to the second row to cancel out that entry. This looks like

$$\begin{bmatrix} 1 & 0 & 0 & -2 \\ 0 & 1 & 1 & 0 \\ 0 & 0 & 1 & 4 \end{bmatrix} \rightarrow_{r_2 - r_3} \begin{bmatrix} 1 & 0 & 0 & -2 \\ 0 & 1 & 0 & -4 \\ 0 & 0 & 1 & 4 \end{bmatrix}.$$

We've reached the end of our first repetition of Part 2, so we'll ignore the bottom row and repeat Part 2 on the top two rows of our matrix. This makes our new "bottom row" the second row, which has a leading 1. Since the top row has a zero above this leading 1, we don't have to do anything. Ignoring the second row leaves us at the top row of our matrix, so we're done with the algorithm and can see that

$$\begin{bmatrix} 0 & 3 & 3 & 0 \\ 2 & 0 & 0 & -4 \\ -1 & 0 & 1 & 6 \end{bmatrix}$$

has reduced echelon form

$$\begin{bmatrix} 1 & 0 & 0 & -2 \\ 0 & 1 & 0 & -4 \\ 0 & 0 & 1 & 4 \end{bmatrix}.$$

Now that we understand how to find the reduced echelon form of a matrix, let's practice using that to solve a matrix equation.

Example 13. Solve the matrix equation $A\vec{x} = \vec{b}$ where $A = \begin{bmatrix} 0 & 3 & 3 \\ 2 & 0 & 0 \\ -1 & 0 & 1 \end{bmatrix}$

and $\vec{b} = \begin{bmatrix} 0 \\ -4 \\ 6 \end{bmatrix}$.

In Example 12, we saw that the augmented coefficient matrix of this equation has reduced echelon form

$$\begin{bmatrix} 1 & 0 & 0 & | & -2 \\ 0 & 1 & 0 & | & -4 \\ 0 & 0 & 1 & | & 4 \end{bmatrix}.$$

Since the original augmented coefficient matrix and the reduced echelon form have the same solutions, we can read off the solutions from the reduced echelon form. This gives us $x_1 = -2$, $x_2 = -4$, and $x_3 = 4$ or $\vec{x} = \begin{bmatrix} -2 \\ -4 \\ 4 \end{bmatrix}$. If you'd like to check your work, you can multiply this vector by our original matrix A and see that your answer is $\vec{b}$.

Most calculators and computer mathematics packages have commands which implement this row reduction algorithm, and most of the time working mathematicians use technology to compute the reduced echelon form of a matrix. (You can read about how to do row reduction using Mathematica in Appendix A.2.) However, it is important to go through enough examples of this algorithm by hand that you understand how it works, since we will rely on some of its theoretical properties in later sections.

Exercises 2.6.

1. Find the augmented coefficient matrix of the matrix equation $\begin{bmatrix} 3 & -4 & 0 \\ 4 & 1 & 17 \end{bmatrix} \vec{x} = \begin{bmatrix} -5 \\ 8 \end{bmatrix}$.

2. Find the augmented coefficient matrix of the matrix equation $\begin{bmatrix} 3 & 4 \\ 6 & 2 \\ 0 & 1 \end{bmatrix} \vec{x} = \begin{bmatrix} 4 \\ -2 \\ 0 \end{bmatrix}$.

3. Find the augmented coefficient matrix of the matrix equation $\begin{bmatrix} -1 & 3 & 5 \\ 2 & 4 & 6 \\ 0 & 9 & -2 \end{bmatrix} \vec{x} = \begin{bmatrix} 0 \\ -6 \\ 10 \end{bmatrix}$.

4. Find the augmented coefficient matrix of the matrix equation $\begin{bmatrix} 9 & 2 & 0 & -1 \\ 5 & -8 & 3 & 4 \end{bmatrix} \vec{x} = \begin{bmatrix} 1 \\ 2 \end{bmatrix}$.

5. Find the augmented coefficient matrix of the following set of equations: $2x_1 + x_2 - 11x_3 = 5$, $-5x_1 + 3x_3 = -1/2$, and $10x_1 - 9x_2 + 2x_3 = 0$.

6. Find the augmented coefficient matrix of the following set of equations: $8x_1 - 3x_2 + x_3 - x_4 = 14$, $x_1 + 6x_2 + 2x_3 + x_4 = -4$, and $-2x_1 - 4x_2 + 10x_3 + 8x_4 = 2$.

7. Find the augmented coefficient matrix of the following set of equations: $-4x_1 - x_3 = 7$, $x_2 + x_3 + 2x_4 = 9$, and $x_1 + 5x_4 = 13$.

8. Find the augmented coefficient matrix of the following set of equations: $x_1 + 2x_2 + 3x_3 = 15$, $x_1 - 5x_3 = 6$, $-2x_1 + 9x_2 = 13$, and $4x_2 - x_3 = -1$.

9. Find the augmented coefficient matrix of the vector equation
$$x_1 \begin{bmatrix} 2 \\ 1 \end{bmatrix} + x_2 \begin{bmatrix} 4 \\ 2 \end{bmatrix} + x_3 \begin{bmatrix} -2 \\ -1 \end{bmatrix} + x_4 \begin{bmatrix} 4 \\ 0 \end{bmatrix} = \begin{bmatrix} -1 \\ 9 \end{bmatrix}.$$

10. Find the augmented coefficient matrix of the vector equation
$$x_1 \begin{bmatrix} -1 \\ 0 \\ 4 \end{bmatrix} + x_2 \begin{bmatrix} 6 \\ 2 \\ 0 \end{bmatrix} + x_3 \begin{bmatrix} -9 \\ 1 \\ -2 \end{bmatrix} = \begin{bmatrix} 0 \\ 0 \\ 1 \end{bmatrix}.$$

11. Find the augmented coefficient matrix of the vector equation
$$x_1 \begin{bmatrix} 3 \\ 0 \\ -6 \\ 7 \end{bmatrix} + x_2 \begin{bmatrix} -2 \\ 1 \\ 8 \\ 4 \end{bmatrix} = \begin{bmatrix} 0 \\ -5 \\ 1 \\ 2 \end{bmatrix}.$$

12. Find the augmented coefficient matrix of the vector equation
$$x_1 \begin{bmatrix} 3 \\ -5 \end{bmatrix} + x_2 \begin{bmatrix} 0 \\ 12 \end{bmatrix} + x_3 \begin{bmatrix} -5 \\ 1 \end{bmatrix} = \begin{bmatrix} 19 \\ -11 \end{bmatrix}.$$

13. Consider the augmented coefficient matrix $\left[\begin{array}{cc|c} 0 & -1 & 4 & 2 \\ 6 & 0 & -5 & 11 \end{array} \right]$.

    (a) Rewrite this augmented coefficient matrix as a matrix equation.
    (b) Rewrite this augmented coefficient matrix as a list of equations.
    (c) Rewrite this augmented coefficient matrix as a vector equation.

14. Consider the augmented coefficient matrix $\left[\begin{array}{ccc|c} -7 & 2 & 0 & 1 \\ 1 & 0 & -4 & 3 \\ 2 & -9 & 1 & 1 \\ 6 & -6 & 1 & 0 \end{array} \right]$.

    (a) Rewrite this augmented coefficient matrix as a matrix equation.
    (b) Rewrite this augmented coefficient matrix as a list of equations.
    (c) Rewrite this augmented coefficient matrix as a vector equation.

15. Decide whether or not the following matrix is in reduced echelon form. If it is, circle all its leading 1s. $\begin{bmatrix} 1 & -3 & 0 & 0 \\ 0 & 0 & 1 & 0 \\ 0 & 0 & 0 & 1 \end{bmatrix}$

16. Decide whether or not the following matrix is in reduced echelon form. If it is, circle all its leading 1s.
$$\begin{bmatrix} 1 & 0 & 2 \\ 0 & 1 & 0 \\ 0 & 0 & -1 \end{bmatrix}$$

17. Decide whether or not the following matrix is in reduced echelon form. If it is, circle all its leading 1s.
$$\begin{bmatrix} 1 & 0 & 1 & 4 \\ 0 & 1 & 0 & -3 \end{bmatrix}$$

18. Decide whether or not the following matrix is in reduced echelon form. If it is, circle all its leading 1s.
$$\begin{bmatrix} 1 & -3 & 0 & 0 \\ 0 & 0 & 1 & 0 \\ 0 & 1 & 0 & 0 \end{bmatrix}$$

19. Find the reduced echelon form of $A = \begin{bmatrix} 1 & -2 & 0 & 4 & 5 \\ 0 & 0 & 2 & 1 & 3 \\ 0 & 0 & 0 & -1 & -1 \\ 0 & 0 & 1 & 2 & 3 \end{bmatrix}$.

20. Find the reduced echelon form of $A = \begin{bmatrix} 4 & 8 & 0 & -4 \\ 0 & -1 & 2 & 1 \\ 3 & 6 & 0 & -3 \\ 0 & 2 & -4 & 6 \end{bmatrix}$.

21. Find the reduced echelon form of $A = \begin{bmatrix} 0 & 1 & -5 & 0 \\ -1 & 2 & 0 & -2 \\ 0 & -4 & 12 & 4 \end{bmatrix}$.

22. Find the reduced echelon form of $A = \left[\begin{array}{ccc|c} -5 & 0 & -10 & 5 \\ 6 & 1 & 12 & 8 \\ 2 & 0 & 3 & -1 \end{array}\right]$.

23. Find the reduced echelon form of $A = \left[\begin{array}{cc|c} 4 & 0 & 8 \\ 2 & 1 & 4 \\ -3 & 0 & -7 \end{array}\right]$.

24. Find the reduced echelon form of $A = \left[\begin{array}{ccc|c} 4 & -2 & 4 & 14 \\ 7 & 0 & 7 & 14 \end{array}\right]$.

25. The reduced echelon form of $A\vec{x} = \vec{b}$'s augmentation matrix is
$$\left[\begin{array}{ccc|c} 1 & 0 & 0 & -7 \\ 0 & 1 & 0 & 2 \\ 0 & 0 & 1 & 6 \end{array}\right]$$. Use this to solve for $\vec{x}$.

26. The reduced echelon form of $A\vec{x} = \vec{b}$'s augmentation matrix is
$$\left[\begin{array}{ccc|c} 1 & 0 & 0 & 4 \\ 0 & 1 & 0 & -3 \\ 0 & 0 & 1 & 12 \end{array}\right]$$. Use this to solve for $\vec{x}$.

27. Use row reduction to solve $\begin{bmatrix} 2 & 0 & 2 \\ 3 & 1 & -1 \\ 1 & -3 & 1 \end{bmatrix} \vec{x} = \begin{bmatrix} 14 \\ 0 \\ 10 \end{bmatrix}$.

28. Use row reduction to solve $\begin{bmatrix} 2 & 4 & 0 \\ 5 & 1 & -3 \\ 1 & 0 & 1 \end{bmatrix} \vec{x} = \begin{bmatrix} 2 \\ -1 \\ -7 \end{bmatrix}$.

29. Use row reduction to solve $x_1 \begin{bmatrix} -2 \\ 0 \\ -1 \end{bmatrix} + x_2 \begin{bmatrix} 0 \\ 3 \\ 0 \end{bmatrix} + x_3 \begin{bmatrix} 4 \\ 2 \\ 4 \end{bmatrix} = \begin{bmatrix} -14 \\ 7 \\ -9 \end{bmatrix}$.

30. Use row reduction to solve $x_1 \begin{bmatrix} 6 \\ 10 \end{bmatrix} + x_2 \begin{bmatrix} -3 \\ 4 \end{bmatrix} = \begin{bmatrix} -12 \\ 7 \end{bmatrix}$.

31. Use row reduction to solve the following equations simultaneously: $-3x_1 + x_3 = -1$, $x_1 + x_2 + x_3 = 0$, and $2x_1 - 4x_2 + x_3 = -5$.

32. Use row reduction to solve the following equations simultaneously: $x_1 + x_2 + x_3 = 6$, $-3x_2 + 2x_3 = 9$, and $-5x_1 + 4x_2 + x_3 = 15$.

2.7   Applications of Row Reduction

Now that we understand how to use row reduction to solve matrix equations, we can apply it to many of the computational aspects of ideas discussed in previous sections. (At this point, I'll assume you've practiced doing row reduction by hand enough to understand its mechanics and are now row reducing matrices either on your calculator or a computer. See Appendix A.2 for help using *Mathematica*.) Let's start by applying row reduction to solving $f(\vec{x}) = \vec{b}$ using f's matrix.

Example 1. Let $f : \mathbb{R}^2 \to \mathbb{R}^2$ by $f\left(\begin{bmatrix} x_1 \\ x_2 \end{bmatrix}\right) = \begin{bmatrix} -\frac{1}{2}x_1 + \frac{\sqrt{3}}{2}x_2 \\ \frac{\sqrt{3}}{2}x_1 + \frac{1}{2}x_2 \end{bmatrix}$. Find $\vec{x}$ for which $f(\vec{x}) = \begin{bmatrix} -1 \\ 1 \end{bmatrix}$.

This may look like a function nobody could love, but in fact it is reflection about the line with slope $\sqrt{3}$ as pictured below. (See 2.2's Exercise 36 for the general formula of the matrix which reflects $\mathbb{R}^2$ about $y = mx$.)

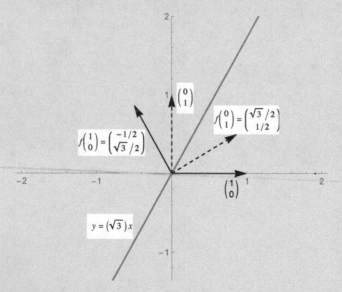

Essentially we're asking which vector reflects across the line $y = (\sqrt{3})x$ to land on $\vec{b} = \begin{bmatrix} -1 \\ 1 \end{bmatrix}$. This could perhaps be worked out geometrically with 30-60 right triangles, but I think it is much easier to solve via row reduction. Our function has matrix $\begin{bmatrix} -\frac{1}{2} & \frac{\sqrt{3}}{2} \\ \frac{\sqrt{3}}{2} & \frac{1}{2} \end{bmatrix}$, so the augmented coefficient matrix of our equation is

$$\left[\begin{array}{cc|c} -\frac{1}{2} & \frac{\sqrt{3}}{2} & -1 \\ \frac{\sqrt{3}}{2} & \frac{1}{2} & 1 \end{array}\right].$$

This augmented coefficient matrix has reduced echelon form

$$\left[\begin{array}{cc|c} 1 & 0 & \frac{1}{2} + \frac{\sqrt{3}}{2} \\ 0 & 1 & \frac{1}{2} - \frac{\sqrt{3}}{2} \end{array}\right]$$

so our solution is $x_1 = \frac{1}{2} + \frac{\sqrt{3}}{2}$ and $x_2 = \frac{1}{2} - \frac{\sqrt{3}}{2}$ or

$$\vec{x} = \left[\begin{array}{c} \frac{1}{2} + \frac{\sqrt{3}}{2} \\ \frac{1}{2} - \frac{\sqrt{3}}{2} \end{array}\right] \approx \left[\begin{array}{c} 1.366 \\ -0.366 \end{array}\right].$$

We can check this answer geometrically by plotting $\vec{x}$ and $\vec{b}$ to verify that $\vec{b}$ is $\vec{x}$'s reflection across $y = (\sqrt{3})x$.

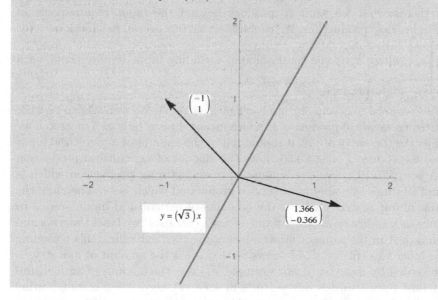

In economics, matrix equations are used to model the interplay between industries using the Leontif input-output model. Here we set up an input-output matrix that describes the inputs and outputs of each industry and use that information to solve for each industry's production needed to meet a specified level of demand.

Example 2. Let's look at three industrial sectors: agriculture, manufacturing, and service. To produce goods or services in one industry typically requires input from some of the others, including that same industry (think of the seed corn required to produce more corn). Suppose we know that for each dollar of agricultural output we need \$0.30 of agricultural input, \$0.10 of manufacturing input, and \$0.20 of service input. For each dollar of manufacturing output we need \$0.10 of agricultural input, \$0.50 of manufacturing input, and \$0.20 of service input. For each dollar of service output we need \$0 of agricultural input, \$0.10 of manufacturing input, and \$0.40 of service input. If we want to produce \$150,000 of agricultural output, \$100,000 of manufacturing output, and \$225,000 of service output above what is needed in the production process, how much does each industry need to produce?

Let's start by setting up the vectors that will model demand and production. We have three industries to keep track of, so we'll use 3-vectors to model both the production and demand. I'll let the first entry track agriculture, the second track manufacturing, and the third track services. This means that the net output or external demand vector is $\vec{b} = \begin{bmatrix} 150,000 \\ 100,000 \\ 225,000 \end{bmatrix}$ since this is what we want to produce beyond the input requirements of these industries' production. We can also set up our overall production vector $\vec{x} = \begin{bmatrix} x_1 \\ x_2 \\ x_3 \end{bmatrix}$ where x_i is the total amount, including input requirements, each industry should produce.

Next we need to set up our input-output matrix A. We want $A\vec{x}$ to give the input requirements of producing $\vec{x}$, which means A must be a 3×3 matrix. If we consider the top entry of $A\vec{x}$, it should tell us the amount of agricultural input needed to produce $\vec{x}$, i.e., \$0.30 times the amount of agricultural production which is x_1, \$0.10 times the amount of manufacturing production which is x_2, and \$0 times the amount of service production which is x_3. Therefore the top row of our matrix contains the amounts of agricultural input needed to produce each industry in order of our vector's entries. If we label the rows and columns of A in the same order as our vectors, then extending this reasoning to the other two entries of $A\vec{x}$ shows us that a_{ij} is the amount of industry i's input needed by industry j. For example, a_{12} was the amount of agricultural input (and agriculture is our vector's first entry) needed for manufacturing (which is our vector's second entry). This means our input-output matrix is

$$A = \begin{bmatrix} 0.3 & 0.1 & 0 \\ 0.1 & 0.5 & 0.1 \\ 0.2 & 0.2 & 0.4 \end{bmatrix}.$$

Now that we have A, $\vec{x}$, and $\vec{b}$, we need to relate them in a matrix equation. The total output we want to produce is the sum of the outputs needed as inputs for our various industries, $A\vec{x}$, plus the external demand $\vec{b}$, so our total output is $A\vec{x} + \vec{b}$. However, we plan to produce $\vec{x}$, so our total output is also $\vec{x}$. Setting these two versions of total output equal gives us $\vec{x} = A\vec{x} + \vec{b}$.

This is a matrix equation, but it doesn't have the right format for row reduction so we'll need to do some algebra to get it there. We can start by subtracting $A\vec{x}$ from both sides to get $\vec{x} - A\vec{x} = \vec{b}$. We can't factor $\vec{x}$ out of the left-hand side, because we'd be left with $1 - A$, which doesn't make sense. However, recall from Example 4 of 2.2 that the $n \times n$ identity matrix, I_n, has $I_n\vec{x} = \vec{x}$. Here $n = 3$, and if we substitute in $I_3\vec{x}$ for $\vec{x}$ we get $I_3\vec{x} - A\vec{x} = \vec{b}$. Now we can factor out $\vec{x}$ to get $(I_3 - A)\vec{x} = \vec{b}$ so our matrix equation is in a format where we can use row reduction to solve for $\vec{x}$.

Since

$$I_3 - A = \begin{bmatrix} 1 & 0 & 0 \\ 0 & 1 & 0 \\ 0 & 0 & 1 \end{bmatrix} - \begin{bmatrix} 0.3 & 0.1 & 0 \\ 0.1 & 0.5 & 0.1 \\ 0.2 & 0.2 & 0.4 \end{bmatrix} = \begin{bmatrix} 0.7 & -0.1 & 0 \\ -0.1 & 0.5 & -0.1 \\ -0.2 & -0.2 & 0.6 \end{bmatrix}$$

our augmented coefficient matrix is

$$\left[I_3 - A \mid \vec{b} \right] = \left[\begin{array}{ccc|c} 0.7 & -0.1 & 0 & 150,000 \\ -0.1 & 0.5 & -0.1 & 100,000 \\ -0.2 & -0.2 & 0.6 & 225,000 \end{array} \right]$$

which row reduces to

$$\left[\begin{array}{ccc|c} 1 & 0 & 0 & 267,287 \\ 0 & 1 & 0 & 371,011 \\ 0 & 0 & 1 & 587,766 \end{array} \right]$$

(with answers rounded to the nearest dollar).

This means that in order to end up with a net output of \$150,000 in agriculture, \$100,000 in manufacturing, and \$225,000 in service, we actually need to produce \$267,287 in agriculture, \$371,011 in manufacturing, and \$587,766 in service.

In the previous examples, we ended up with the nicest possible reduced echelon form, which gave us one unique solution to our matrix equation. However, we saw in Examples 4 and 5 of the last section that it's possible to have reduced echelon forms which are more complicated. We'll spend the rest of this section exploring what to do in such cases while simultaneously applying row reduction to previous ideas like kernel, range, and linear independence. Let's start by using our ability to quickly and systematically solve $f(\vec{x}) = \vec{b}$ to compute the kernel of a map.

Example 3. Compute the kernel of the function $f : \mathbb{R}^3 \to \mathbb{R}^2$ where
$$f\left(\begin{bmatrix} x_1 \\ x_2 \\ x_3 \end{bmatrix} \right) = \begin{bmatrix} 4x_1 + x_2 + 3x_3 \\ 2x_1 - 4x_3 \end{bmatrix}.$$

The kernel is all 3-vectors $\vec{x}$ with $f(\vec{x}) = \vec{0}$. This function has matrix

$$A = \begin{bmatrix} 4 & 1 & 3 \\ 2 & 0 & -4 \end{bmatrix}$$

so finding the kernel of f is equivalent to solving $A\vec{x} = \vec{0}$. This equation has the augmented coefficient matrix

$$\begin{bmatrix} 4 & 1 & 3 & | & 0 \\ 2 & 0 & -4 & | & 0 \end{bmatrix}$$

whose reduced echelon form is

$$\begin{bmatrix} 1 & 0 & -2 & | & 0 \\ 0 & 1 & 11 & | & 0 \end{bmatrix}.$$

This isn't the simplest kind of reduced echelon form we'd hoped for, but we can still read off the equations given by its rows. The top row gives us $x_1 - 2x_3 = 0$ and the bottom row gives us $x_2 + 11x_3 = 0$. Even though we don't have a unique one-number answer for each variable, these two equations are still enough for us to find the kernel of f. The first equation $x_1 - 2x_3 = 0$ can be solved for x_1 to give $x_1 = 2x_3$, while the second equation $x_2 + 11x_3 = 0$ can be solved for x_2 to give $x_2 = -11x_3$. Once we pick a value for x_3, we'll get values for x_1 and x_2, for example, if $x_3 = 2$ then $x_1 = 4$ and $x_2 = -22$. One way to write our overall answer is

$$\ker(f) = \left\{ \begin{bmatrix} 2x_3 \\ -11x_3 \\ x_3 \end{bmatrix} \right\}.$$

Notice that in the example above, our kernel contained more than just $\vec{0}$, so our map wasn't 1-1. We can tell that because we ended up with a variable, x_3, which didn't have its own equation with a leading 1 as its coefficient. That happened because x_3's column didn't have a leading 1 in A's reduced echelon form. This gives us an easy-to-check criterion for a function to be 1-1.

Theorem 1. A function $f : \mathbb{R}^n \to \mathbb{R}^m$ is 1-1 if and only if the reduced echelon form of its matrix A has a leading 1 in every column.

If we shift our focus from solving equations of the form $f(\vec{x}) = \vec{b}$ to solving vector equations, then we can apply row reduction to answering the question of whether $\vec{v}_1, \ldots, \vec{v}_k$ are linearly independent or linearly dependent. Remember from 1.3 that $\vec{v}_1, \ldots, \vec{v}_k$ are linearly dependent if we can find a solution to $x_1\vec{v}_1 + \cdots + x_k\vec{v}_k = \vec{0}$ where some x_i is nonzero, otherwise $\vec{v}_1, \ldots, \vec{v}_k$ are linearly independent.

Example 4. Are $\begin{bmatrix} 2 \\ -1 \\ 0 \\ 3 \end{bmatrix}$, $\begin{bmatrix} 0 \\ 0 \\ 4 \\ 0 \end{bmatrix}$, $\begin{bmatrix} -1 \\ -2 \\ 0 \\ 4 \end{bmatrix}$, and $\begin{bmatrix} 5 \\ 0 \\ 0 \\ 2 \end{bmatrix}$ linearly independent or linearly dependent?

We can check this by solving

$$x_1 \begin{bmatrix} 2 \\ -1 \\ 0 \\ 3 \end{bmatrix} + x_2 \begin{bmatrix} 0 \\ 0 \\ 4 \\ 0 \end{bmatrix} + x_3 \begin{bmatrix} -1 \\ -2 \\ 0 \\ 4 \end{bmatrix} + x_4 \begin{bmatrix} 5 \\ 0 \\ 0 \\ 2 \end{bmatrix} = \begin{bmatrix} 0 \\ 0 \\ 0 \\ 0 \end{bmatrix}$$

to see whether or not we have solutions where some of the variables are nonzero. From 2.2 we know that the vector equation $x_1 \vec{v}_1 + \cdots + x_k \vec{v}_k = \vec{b}$ is equivalent to the matrix equation $A\vec{x} = \vec{b}$ where A is the matrix whose columns are $\vec{v}_1, \ldots, \vec{v}_k$. This means we want to solve $A\vec{x} = \vec{b}$ where

$$A = \begin{bmatrix} 2 & 0 & -1 & 5 \\ -1 & 0 & -2 & 0 \\ 0 & 4 & 0 & 0 \\ 3 & 0 & 4 & 2 \end{bmatrix} \text{ and } \vec{b} = \begin{bmatrix} 0 \\ 0 \\ 0 \\ 0 \end{bmatrix}.$$

This matrix equation has the augmented coefficient matrix

$$\begin{bmatrix} 2 & 0 & -1 & 5 & | & 0 \\ -1 & 0 & -2 & 0 & | & 0 \\ 0 & 4 & 0 & 0 & | & 0 \\ 3 & 0 & 4 & 2 & | & 0 \end{bmatrix}$$

whose reduced echelon form is

$$\begin{bmatrix} 1 & 0 & 0 & 2 & | & 0 \\ 0 & 1 & 0 & 0 & | & 0 \\ 0 & 0 & 1 & -1 & | & 0 \\ 0 & 0 & 0 & 0 & | & 0 \end{bmatrix}.$$

From here we can see $x_1 = -2x_4$, $x_2 = 0$, $x_3 = x_4$, and x_4 is free. Since we can choose a nonzero value for x_4, our vectors are linearly dependent.

Notice that our vector equation above had a variable, x_4, which wasn't necessarily zero precisely because that variable's column didn't have a leading 1 in the reduced echelon form of our augmented coefficient matrix. This means that we have the following check for linear independence and linear dependence.

Theorem 2. A set of vectors $\vec{v}_1, \ldots, \vec{v}_k$ in $\mathbb{R}^n$ are linearly independent if the reduced echelon form of the matrix with columns $\vec{v}_1, \ldots, \vec{v}_k$ has a leading 1 in every column. If the reduced echelon form has a column with no leading 1, then $\vec{v}_1, \ldots, \vec{v}_k$ are linearly dependent.

So far, we've always been in the situation where $A\vec{x} = \vec{b}$ had a solution. However, we've seen that this isn't always the case. How can we tell that our matrix equation has no solution from the reduced echelon form of its augmented coefficient matrix? The trick is to look at where our leading 1s occur. If we have a leading 1 in the rightmost (augmentation) column, our matrix equation has no solutions. To understand why this is true, think back to our interpretation of a matrix equation as a list of linear equations. Since a leading 1 is the first nonzero entry in its row, having a leading 1 in the rightmost column means that its row looks like $0 \cdots 0 \,|\, 1$. This corresponds to the equation $0x_1 + \cdots + 0x_n = 1$, i.e., $0 = 1$, which is clearly impossible. Happily such a leading 1 will be formed during Part 1 of our row reduction algorithm, which saves us from having to do Part 2. (In fact you can see your matrix equation has no solution as soon as you see that the first nonzero entry in any row occurs in the last column.) It's always important to understand when a problem cannot be solved, because then you don't waste time continuing to work on it.

Example 5. The equation $A\vec{x} = \vec{b}$ with $A = \begin{bmatrix} 1 & -5 & 4 \\ 2 & -7 & 3 \\ -2 & 1 & 7 \end{bmatrix}$ and $\vec{b} = \begin{bmatrix} -3 \\ -2 \\ -1 \end{bmatrix}$ has no solution.

The augmented coefficient matrix of this equation is

$$\begin{bmatrix} 1 & -5 & 4 & -3 \\ 2 & -7 & 3 & -2 \\ -2 & 1 & 7 & -1 \end{bmatrix}$$

which has reduced echelon form

$$\begin{bmatrix} 1 & 0 & -\frac{13}{3} & 0 \\ 0 & 1 & \frac{-5}{3} & 0 \\ 0 & 0 & 0 & 1 \end{bmatrix}.$$

If we look at the bottom row of the matrix, we get the equation $0 = 1$. Since this is clearly a contradiction (there are no values of x_1, x_2, and x_3 which will make zero equal one), our original equation doesn't have any solutions.

A slightly different perspective on the question of whether or not a matrix equation has solutions is to ask which vectors $\vec{b}$ make $A\vec{x} = \vec{b}$ have solutions.

If our matrix A corresponds to a linear function f, this is the same as asking for the range of f.

Example 6. Compute the range of the function $f : \mathbb{R}^3 \to \mathbb{R}^3$ where f has

matrix $A = \begin{bmatrix} 1 & -5 & 4 \\ 2 & -7 & 3 \\ -2 & 1 & 7 \end{bmatrix}$.

Asking for the range of f is asking for every $\vec{b}$ where we have a solution

to $A\vec{x} = \vec{b}$. To figure this out, we'll pick a generic $\vec{b} = \begin{bmatrix} b_1 \\ b_2 \\ b_3 \end{bmatrix}$ and try to solve

$A\vec{x} = \vec{b}$. The augmented coefficient matrix of this equation is now

$$\begin{bmatrix} 1 & -5 & 4 & \bigm| & b_1 \\ 2 & -7 & 3 & \bigm| & b_2 \\ -2 & 1 & 7 & \bigm| & b_3 \end{bmatrix}.$$

Mathematica isn't as good about row reducing a matrix which has variable entries, so we'll do this one by hand (and get in a little extra row reduction practice along the way).

We already have a 1 in the top left corner, so we can move on to adding multiples of row 1 to rows 2 and 3 to create zeros under that leading 1. In our row reduction notation, this is

$$\begin{bmatrix} 1 & -5 & 4 & \bigm| & b_1 \\ 2 & -7 & 3 & \bigm| & b_2 \\ -2 & 1 & 7 & \bigm| & b_3 \end{bmatrix} \to_{r_2 - 2r_1} \begin{bmatrix} 1 & -5 & 4 & \bigm| & b_1 \\ 0 & 3 & -5 & \bigm| & b_2 - 2b_1 \\ -2 & 1 & 7 & \bigm| & b_3 \end{bmatrix}$$

$$\to_{r_3 + 2r_1} \begin{bmatrix} 1 & -5 & 4 & \bigm| & b_1 \\ 0 & 3 & -5 & \bigm| & b_2 - 2b_1 \\ 0 & -9 & 15 & \bigm| & b_3 + 2b_1 \end{bmatrix}.$$

Next we ignore the top row and first column, and we multiply the second row by $\frac{1}{3}$ to create our next leading 1. This gives us

$$\begin{bmatrix} 1 & -5 & 4 & \bigm| & b_1 \\ 0 & 3 & -5 & \bigm| & b_2 - 2b_1 \\ 0 & -9 & 15 & \bigm| & b_3 + 2b_1 \end{bmatrix} \to_{\frac{1}{3}r_2} \begin{bmatrix} 1 & -5 & 4 & \bigm| & b_1 \\ 0 & 1 & \frac{-5}{3} & \bigm| & \frac{1}{3}b_2 - \frac{2}{3}b_1 \\ 0 & -9 & 15 & \bigm| & b_3 + 2b_1 \end{bmatrix}.$$

We add a multiple of row 2 to row 3 to get a zero below our new leading 1, which looks like

$$\begin{bmatrix} 1 & -5 & 4 & \bigm| & b_1 \\ 0 & 1 & \frac{-5}{3} & \bigm| & \frac{1}{3}b_2 - \frac{2}{3}b_1 \\ 0 & -9 & 15 & \bigm| & b_3 + 2b_1 \end{bmatrix} \to_{r_3 + 9r_2} \begin{bmatrix} 1 & -5 & 4 & \bigm| & b_1 \\ 0 & 1 & \frac{-5}{3} & \bigm| & \frac{1}{3}b_2 - \frac{2}{3}b_1 \\ 0 & 0 & 0 & \bigm| & (b_3 + 2b_1) + (3b_2 - 6b_1) \end{bmatrix}.$$

Since the first three entries of the bottom row are zero, the bottom row will prevent us from having any solutions (as in Example 5) unless the bottom entry of the right column is also 0. This can be expressed as

$$(b_3 + 2b_1) + (3b_2 - 6b_1) = 0$$

which simplifies to

$$-4b_1 + 3b_2 + b_3 = 0.$$

As long as $\vec{b}$ satisfies this condition, we'll have a solution to $f(\vec{x}) = \vec{b}$ and $\vec{b}$ will be in the range of f. Another way to write this is

$$\text{range}(f) = \left\{ \begin{bmatrix} b_1 \\ b_2 \\ b_3 \end{bmatrix} \middle| -4b_1 + 3b_2 + b_3 = 0 \right\}.$$

Notice that the range of the example function above is smaller than its codomain, i.e., f isn't onto. This happened because there were some vectors $\vec{b}$ where $f(\vec{x}) = \vec{b}$ didn't have a solution. Thinking in terms of row reduction, this happened because there was a row in the reduced echelon form of f's matrix which was all zeros. This created a row in the augmented coefficient matrix of the form $0 \cdots 0 \mid *$ leaving the possibility of an unsolvable equation. As with 1-1, this gives us an easy-to-check criterion for when a function is onto.

Theorem 3. A function f is onto if and only if the reduced echelon form of its matrix A has a leading 1 in every row.

Note that Theorems 1 and 3 are similar, so it's important to remember that for onto we're checking rows and for 1-1 we're checking columns! This should be much easier since you've seen why each of these criteria hold, which is why it's important to understand the reasons behind our mathematical rules rather than just blindly memorizing them.

Finally, let's apply row reduction to the question of whether or not a vector is in the span of a set of vectors. We know from 1.2 that $\vec{b}$ is in the span of $\vec{v}_1, \ldots, \vec{v}_k$ exactly when we can solve $x_1\vec{v}_1 + \cdots + x_k\vec{v}_k = \vec{b}$. This means we can check whether or not $\vec{b}$ is in the span of $\vec{v}_1, \ldots, \vec{v}_k$ by using row reduction to check whether or not our vector equation has solutions.

Example 7. Is $\vec{b} = \begin{bmatrix} 1 \\ 2 \\ 3 \end{bmatrix}$ in the span of $\begin{bmatrix} 2 \\ -1 \\ -2 \end{bmatrix}, \begin{bmatrix} 1 \\ 0 \\ 1 \end{bmatrix}$, and $\begin{bmatrix} -4 \\ 5 \\ 16 \end{bmatrix}$?

Asking this question is equivalent to asking whether we can solve the vector equation

$$x_1 \begin{bmatrix} 2 \\ -1 \\ -2 \end{bmatrix} + x_2 \begin{bmatrix} 1 \\ 0 \\ 1 \end{bmatrix} + x_3 \begin{bmatrix} -4 \\ 5 \\ 16 \end{bmatrix} = \begin{bmatrix} 1 \\ 2 \\ 3 \end{bmatrix}.$$

This vector equation has augmented coefficient matrix

$$\left[\begin{array}{ccc|c} 2 & 1 & -4 & 1 \\ -1 & 0 & 5 & 2 \\ -2 & 1 & 16 & 3 \end{array}\right]$$

whose reduced echelon form is

$$\left[\begin{array}{ccc|c} 1 & 0 & -5 & 0 \\ 0 & 1 & 0 & 0 \\ 0 & 0 & 0 & 1 \end{array}\right].$$

Our augmented coefficient matrix has a leading 1 in the right column of its reduced echelon form. This means our vector equation has no solution, and therefore $\vec{b} = \begin{bmatrix} 1 \\ 2 \\ 3 \end{bmatrix}$ is not in the span of the other vectors.

Example 8. Is $\vec{b} = \begin{bmatrix} 0 \\ 4 \\ 16 \end{bmatrix}$ in the span of $\begin{bmatrix} 2 \\ -1 \\ -2 \end{bmatrix}$, $\begin{bmatrix} 1 \\ 0 \\ 1 \end{bmatrix}$, and $\begin{bmatrix} -4 \\ 5 \\ 16 \end{bmatrix}$?

As in the previous example, this is equivalent to asking whether we can solve

$$x_1 \begin{bmatrix} 2 \\ -1 \\ -2 \end{bmatrix} + x_2 \begin{bmatrix} 1 \\ 0 \\ 1 \end{bmatrix} + x_3 \begin{bmatrix} -4 \\ 5 \\ 16 \end{bmatrix} = \begin{bmatrix} 0 \\ 4 \\ 16 \end{bmatrix}.$$

This vector equation has augmented coefficient matrix

$$\left[\begin{array}{ccc|c} 2 & 1 & -4 & 0 \\ -1 & 0 & 5 & 5 \\ -2 & 1 & 16 & 16 \end{array}\right]$$

whose reduced echelon form is

$$\left[\begin{array}{ccc|c} 1 & 0 & -5 & -4 \\ 0 & 1 & 6 & 8 \\ 0 & 0 & 0 & 0 \end{array}\right].$$

Since our augmented coefficient matrix doesn't have a leading 1 in the right column of its reduced echelon form, our vector equation has a solution. Therefore $\vec{b} = \begin{bmatrix} 0 \\ 4 \\ 16 \end{bmatrix}$ is in the span of the other vectors.

Exercises 2.7.

1. Use row reduction to solve $f(\vec{x}) = \begin{bmatrix} 1 \\ 9 \\ -1 \end{bmatrix}$ where f has matrix

$$A = \begin{bmatrix} -1 & -2 & -1 & 3 \\ 0 & -3 & -6 & 4 \\ -2 & -3 & 0 & 3 \end{bmatrix}.$$

2. Use row reduction to solve $f(\vec{x}) = \begin{bmatrix} 18 \\ -4 \\ 17 \end{bmatrix}$ where f has matrix

$$A = \begin{bmatrix} 2 & -4 & 6 \\ -1 & 3 & 0 \\ 2 & -5 & 3 \end{bmatrix}.$$

3. Use row reduction to solve $f(\vec{x}) = \begin{bmatrix} 2 \\ 5 \\ 1 \end{bmatrix}$ where f has matrix

$$A = \begin{bmatrix} 2 & 8 & 4 \\ 2 & 5 & 1 \\ 4 & 10 & -1 \end{bmatrix}.$$

4. Use row reduction to solve $f(\vec{x}) = \begin{bmatrix} 3 \\ 1 \\ 8 \end{bmatrix}$ where f has matrix

$$A = \begin{bmatrix} 1 & 1 & -2 \\ 3 & -2 & 4 \\ 2 & -3 & 6 \end{bmatrix}.$$

5. Find the kernel of $f\left(\begin{bmatrix} x_1 \\ x_2 \\ x_3 \end{bmatrix} \right) = \begin{bmatrix} -x_1 - x_2 + 2x_3 \\ -9x_1 + 4x_2 + 5x_3 \\ 6x_1 - 2x_3 - 4x_3 \end{bmatrix}.$

6. Find the kernel of $f\left(\begin{bmatrix} x_1 \\ x_2 \\ x_3 \end{bmatrix} \right) = \begin{bmatrix} 2x_1 + 6x_2 - 5x_3 \\ x_1 + 3x_2 + x_3 \end{bmatrix}.$

7. Find the kernel of $f\left(\begin{bmatrix} x_1 \\ x_2 \\ x_3 \\ x_4 \end{bmatrix} \right) = \begin{bmatrix} -x_1 - 2x_2 - x_3 + 3x_4 \\ -3x_2 - 6x_3 + 4x_4 \\ -2x_1 - 3x_2 + 3x_4 \end{bmatrix}.$

8. Find the kernel of $f\left(\begin{bmatrix} x_1 \\ x_2 \\ x_3 \end{bmatrix} \right) = \begin{bmatrix} 5x_1 - 2x_2 + x_3 \\ 3x_2 - 3x_3 \\ 5x_1 - 5x_2 + 4x_3 \\ -x_2 + x_3 \end{bmatrix}.$

9. Is the linear map f with matrix $A = \begin{bmatrix} 1 & 2 & -3 & 0 \\ -1 & 2 & 4 & 1 \\ 0 & 4 & 1 & -2 \end{bmatrix}$ 1-1?

10. Is the linear map f with matrix $A = \begin{bmatrix} 1 & 1 & -2 \\ 3 & -2 & 4 \\ 2 & -3 & 6 \end{bmatrix}$ 1-1?

11. Is $f\left(\begin{bmatrix} x_1 \\ x_2 \\ x_3 \end{bmatrix}\right) = \begin{bmatrix} -4x_1 - 4x_3 \\ 3x_1 + x_2 - x_3 \\ -x_1 + 3x_2 - x_3 \end{bmatrix}$ 1-1?

12. Is $f\left(\begin{bmatrix} x_1 \\ x_2 \\ x_3 \end{bmatrix}\right) = \begin{bmatrix} 5 & -10 & 9 \\ 0 & -2 & 1 \\ 0 & 6 & -3 \end{bmatrix}$ 1-1?

13. Are $\begin{bmatrix} 3 \\ 1 \\ 0 \\ -1 \end{bmatrix}, \begin{bmatrix} 5 \\ -2 \\ 1 \\ 2 \end{bmatrix}, \begin{bmatrix} 4 \\ 5 \\ -1 \\ 2 \end{bmatrix}$ linearly dependent or linearly independent?

14. Are $\begin{bmatrix} 1 \\ 1 \\ 1 \end{bmatrix}, \begin{bmatrix} 2 \\ 3 \\ 1 \end{bmatrix}, \begin{bmatrix} 5 \\ 6 \\ 4 \end{bmatrix}$ linearly independent or linearly dependent?

15. Are $\begin{bmatrix} 2 \\ 1 \\ 0 \\ 1 \end{bmatrix}, \begin{bmatrix} 3 \\ 0 \\ 2 \\ 1 \end{bmatrix}, \begin{bmatrix} 1 \\ 0 \\ 2 \\ 0 \end{bmatrix}$ linearly independent or linearly dependent?

16. Let $\vec{v}_1 = \begin{bmatrix} -1 \\ 0 \\ -2 \end{bmatrix}, \vec{v}_2 = \begin{bmatrix} 3 \\ -2 \\ 2 \end{bmatrix}$, and $\vec{v}_3 = \begin{bmatrix} 5 \\ -2 \\ 6 \end{bmatrix}$

    (a) Are these vectors linearly independent or linearly dependent?
    (b) What is the dimension of the space spanned by $\vec{v}_1$, $\vec{v}_2$, and $\vec{v}_3$?

17. Let $\vec{v}_1 = \begin{bmatrix} 2 \\ -1 \\ 3 \end{bmatrix}, \vec{v}_2 = \begin{bmatrix} 0 \\ 9 \\ 5 \end{bmatrix}$, and $\vec{v}_3 = \begin{bmatrix} -3 \\ 6 \\ -2 \end{bmatrix}$.

    (a) Are these vectors linearly independent or linearly dependent?
    (b) What is the dimension of the space spanned by $\vec{v}_1$, $\vec{v}_2$, and $\vec{v}_3$?

18. Let $\vec{v}_1 = \begin{bmatrix} 1 \\ -1 \\ 0 \\ 0 \end{bmatrix}, \vec{v}_2 = \begin{bmatrix} 2 \\ -1 \\ 1 \\ 0 \end{bmatrix}$, and $\vec{v}_3 = \begin{bmatrix} 0 \\ -1 \\ 1 \\ 0 \end{bmatrix}$.

    (a) Are these vectors linearly independent or linearly dependent?
    (b) What is the dimension of the space spanned by $\vec{v}_1$, $\vec{v}_2$, and $\vec{v}_3$?

19. Let $\vec{v}_1 = \begin{bmatrix} 1 \\ 0 \\ -1 \\ 1 \end{bmatrix}, \vec{v}_2 = \begin{bmatrix} 0 \\ 2 \\ 1 \\ 3 \end{bmatrix}$, and $\vec{v}_3 = \begin{bmatrix} -4 \\ 4 \\ 6 \\ 2 \end{bmatrix}$.

(a) Show that $\vec{v}_1$, $\vec{v}_2$, $\vec{v}_3$ are linearly dependent.

(b) Write one of these vectors as a linear combination of the other two.

(c) What can you say about the dimension of $\text{Span}\{\vec{v}_1, \vec{v}_2, \vec{v}_3\}$?

20. Find the range of the linear map f with matrix $A = \begin{bmatrix} -1 & 4 \\ 0 & 1 \\ -2 & 3 \end{bmatrix}$.

21. Find the range of the linear map f with matrix $A = \begin{bmatrix} 1 & 2 \\ 2 & 5 \\ 0 & 4 \end{bmatrix}$.

22. Find the range of $f\left(\begin{bmatrix} x_1 \\ x_2 \\ x_3 \end{bmatrix}\right) = \begin{bmatrix} x_1 - 4x_3 \\ -2x_1 + x_2 + 9x_3 \\ 3x_2 + 3x_3 \end{bmatrix}$.

23. Find the range of $f\left(\begin{bmatrix} x_1 \\ x_2 \end{bmatrix}\right) = \begin{bmatrix} 2x_1 - 4x_2 \\ -x_1 + 3x_2 \end{bmatrix}$.

24. Is the linear map f with matrix $A = \begin{bmatrix} 1 & 2 & -3 & 0 \\ -1 & 2 & 4 & 1 \\ 0 & 4 & 1 & -2 \end{bmatrix}$ onto?

25. Is the linear map f with matrix $A = \begin{bmatrix} 1 & 1 & -2 \\ 3 & -2 & 4 \\ 2 & -3 & 6 \end{bmatrix}$ onto?

26. Is $f\left(\begin{bmatrix} x_1 \\ x_2 \\ x_3 \end{bmatrix}\right) = \begin{bmatrix} 5 & -10 & 9 \\ 0 & -2 & 1 \\ 0 & 6 & -3 \end{bmatrix}$ onto?

27. Is $f\left(\begin{bmatrix} x_1 \\ x_2 \\ x_3 \end{bmatrix}\right) = \begin{bmatrix} -4x_1 - 4x_3 \\ 3x_1 + x_2 - x_3 \\ -x_1 + 3x_2 - x_3 \end{bmatrix}$ onto?

28. Let $f : \mathbb{R}^4 \to \mathbb{R}^3$ by $f(\vec{x}) = A\vec{x}$ where $A = \begin{bmatrix} 1 & -5 & 0 & 0 \\ 0 & 0 & 1 & -2 \\ 0 & 0 & -4 & 8 \end{bmatrix}$.

(a) Show f is not onto.

(b) Write down an easy-to-check condition we can use to tell if a vector $\begin{bmatrix} x_1 \\ x_2 \\ x_3 \end{bmatrix}$ is in the range of f.

29. Write down a matrix A so the linear map $f(\vec{x}) = A\vec{x}$ is 1-1 but not onto.

30. Write down the matrix of a linear map which satisfies each of the given conditions. (Your matrices need not be square.)

(a) 1-1 and onto

(b) 1-1 but not onto

(c) onto but not 1-1

31. Is $\vec{b} = \begin{bmatrix} 1 \\ 1 \\ 1 \\ 1 \end{bmatrix}$ in Span $\left\{ \begin{bmatrix} 1 \\ 0 \\ -1 \\ 2 \end{bmatrix}, \begin{bmatrix} 2 \\ -3 \\ 0 \\ 1 \end{bmatrix}, \begin{bmatrix} -3 \\ 0 \\ 6 \\ -2 \end{bmatrix} \right\}$?

32. Is $\vec{u} = \begin{bmatrix} 2 \\ 10 \\ 7 \end{bmatrix}$ in the span of $\vec{v}_1 = \begin{bmatrix} 2 \\ 0 \\ 1 \end{bmatrix}, \vec{v}_1 = \begin{bmatrix} 1 \\ 1 \\ 1 \end{bmatrix}, \vec{v}_1 = \begin{bmatrix} 0 \\ 14 \\ 8 \end{bmatrix}$?

33. Is $\vec{u} = \begin{bmatrix} 2 \\ 1 \\ 1 \end{bmatrix}$ in the span of $\vec{v}_1 = \begin{bmatrix} 2 \\ 0 \\ -1 \end{bmatrix}, \vec{v}_2 = \begin{bmatrix} 0 \\ 1 \\ 0 \end{bmatrix}$, and $\vec{v}_3 = \begin{bmatrix} -4 \\ 1 \\ 4 \end{bmatrix}$?

34. Do $\begin{bmatrix} 2 \\ 0 \\ -1 \end{bmatrix}, \begin{bmatrix} 0 \\ 1 \\ 0 \end{bmatrix}$, and $\begin{bmatrix} -4 \\ 1 \\ 4 \end{bmatrix}$ span all of $\mathbb{R}^3$?

35. Do $\begin{bmatrix} 1 \\ 0 \\ -1 \end{bmatrix}, \begin{bmatrix} 0 \\ 1 \\ 2 \end{bmatrix}$, and $\begin{bmatrix} -4 \\ 0 \\ 6 \end{bmatrix}$ span all of $\mathbb{R}^3$?

36. Do $\begin{bmatrix} 1 \\ 0 \\ 1 \end{bmatrix}, \begin{bmatrix} 2 \\ 0 \\ 2 \end{bmatrix}$, and $\begin{bmatrix} 0 \\ 1 \\ 1 \end{bmatrix}$ span all of $\mathbb{R}^3$?

37. Briefly explain how you would figure out whether or not a set of 4-vectors spans all of $\mathbb{R}^4$.

38. Is it possible to span all of $\mathbb{R}^6$ with five 6-vectors, i.e., five vectors from $\mathbb{R}^6$? Explain why or why not.

39. Redo Example 2 if for each dollar of agricultural output we need $0.30 of agricultural input, $0.10 of manufacturing input, and $0.40 of service input, for each dollar of manufacturing output we need $0.10 of agricultural input, $0.50 of manufacturing input, and $0.40 of service input, and for each dollar of service output we need $0 of agricultural input, $0.10 of manufacturing input, and $0.40 of service input.

40. Suppose we divide industry into two categories: raw materials and manufacturing. For each dollar of raw materials produced, we need $0.10 of raw materials and $0.35 of manufactured goods while for each dollar of manufacturing we need $0.50 of raw materials and $0.20 of manufactured goods. If we want to end up with a net output of $750 of raw materials and $1000 of manufacturing, what should our production level be for each industry?

2.8   Solution Sets

Whether we are finding the range of a linear function from $\mathbb{R}^n$ to $\mathbb{R}^m$, solving a vector equation, or figuring out which points $\vec{x}$ are mapped to $\vec{b}$ by a geometric transformation, we often want to describe the set of vectors $\vec{x}$ which satisfy $A\vec{x} = \vec{b}$ for some matrix A. In this section, we'll explore the possible options for such sets and how to easily describe them.

> **Definition.**   A vector $\vec{x}$ which satisfies $A\vec{x} = \vec{b}$ is called a *solution*, and the set of all such vectors is called the *solution set*.

We saw in the last section that we may not have any solutions to $A\vec{x} = \vec{b}$, in which case we can say that there are no solutions or that our solution set is empty. We also saw that if there are solutions it's possible to have either a single solution or a set of infinitely many different solutions determined by our choice of values for some of the variables. Are no solutions, one solution, or infinitely many solutions our only options? Let's explore this question visually. For simplicity's sake, we'll do our exploration in 2D using a 2×2 matrix A.

In the 2×2 case, $A\vec{x} = \vec{b}$ is really

$$\begin{bmatrix} a & b \\ c & d \end{bmatrix} \begin{bmatrix} x_1 \\ x_2 \end{bmatrix} = \begin{bmatrix} b_1 \\ b_2 \end{bmatrix}$$

which we can write as the equations $ax_1 + bx_2 = b_1$ and $cx_1 + dx_2 = b_2$. For familiarity's sake, let's use x for x_1 and y for x_2, so our equations are $ax + by = b_1$ and $cx + dy = b_2$. Geometrically, this means we're looking at two lines in the plane. A solution to $A\vec{x} = \vec{b}$ is a pair of values for x and y which satisfy both of these equations, which geometrically means a point in the plane (x, y) which lies on both lines. This means our solution set is the set of intersection points of our two lines.

If our two lines are in a fairly standard configuration, they'll intersect in precisely one point as shown in Figure 2.2.

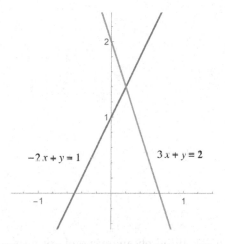

Figure 2.2: A single point intersection

Having only one point of intersection means we have one uniquely determined solution to $A\vec{x} = \vec{b}$, as in 2.7's Example 1.

If our two lines are parallel, they won't intersect at all as in Figure 2.3.

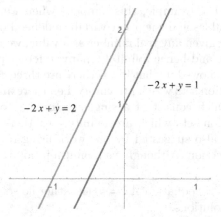

Figure 2.3: No points of intersection

Since there are no points of intersection, $A\vec{x} = \vec{b}$ has no solutions, as in 2.7's Example 5.

The last way to draw two lines in the plane may sound silly, but it is still a possibility. Here we draw two lines which are the same line, i.e., draw the second line directly on top of the first line as in Figure 2.4.

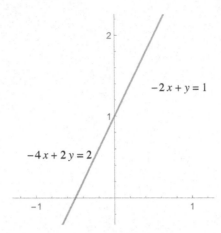

Figure 2.4: Infinitely many points of intersection

(Note that the equations of these two lines look different even though they actually describe the same line.) In this case, any point on our doubled line is a point of intersection, so there are infinitely many different points of intersection. This means $A\vec{x} = \vec{b}$ has infinitely many solutions as in 2.7's Example 3. As in that example, this happens when we can choose a value for some of the variables and the other variables depend on that choice. Since each variable can be given any real number as a value, we have infinitely many choices for its value, and hence infinitely many different possible solutions.

Our exploration showed us that in 2D there are three categories of solution sets: There are solution sets which are empty, i.e., there are no solutions, there are solution sets which consist of a single vector, i.e., one unique solution, and there are solution sets which are infinite, i.e., there are infinitely many solutions. This was also suggested by our work using row reduction to solve $A\vec{x} = \vec{b}$ in the last section. Although we'll omit a proof, this is true in general.

Theorem 1. A matrix equation $A\vec{x} = \vec{b}$ has either no solutions, one solution, or infinitely many solutions.

Let's do one more example of each type. We'll start with the case where we have no solutions.

Example 1. Find the solution set of $A\vec{x} = \vec{b}$ where $A = \begin{bmatrix} -3 & -1 & 5 \\ 1 & 2 & -2 \\ 0 & 5 & -1 \end{bmatrix}$ and $\vec{b} = \begin{bmatrix} 2 \\ 6 \\ 15 \end{bmatrix}$.

The augmented coefficient matrix of this equation is

$$\begin{bmatrix} -3 & -1 & 5 & | & 2 \\ 1 & 2 & -2 & | & 6 \\ 0 & 5 & -1 & | & 15 \end{bmatrix}$$

which has reduced echelon form

$$\begin{bmatrix} 1 & 0 & -\frac{8}{5} & | & 0 \\ 0 & 1 & -\frac{1}{5} & | & 0 \\ 0 & 0 & 0 & | & 1 \end{bmatrix}.$$

The last row of this matrix corresponds to the equation $0 = 1$, so our equation has no solutions.

Geometrically, our solution set lives in $\mathbb{R}^3$ because $\vec{x}$ is a 3-vector. The three rows of our original matrix correspond to three planes in $\mathbb{R}^3$ with equations: $-3x - y + 5z = 2$, $x + 2y - 2z = 6$, and $5y - z = 15$. (In general, a single linear equation in $\mathbb{R}^n$ describes an $(n-1)$-dimensional object.) Since we have no solutions, we know our three planes don't all intersect at the same point. Feel free to use Mathematica or another software package to plot these three planes together to verify this visually.

Again, notice that when we have no solutions we'll always end up with an impossible equation of the form $0 = b$ where b is nonzero. (If we're in reduced echelon form, we'll have $b = 1$.) This is easily spotted in the reduced echelon form, because we only have an equation of this format if there is a leading 1 in the rightmost column.

Next let's look at the case where we have one unique solution.

Example 2. Find the solution set of $f(\vec{x}) = \vec{b}$ where $\vec{b} = \begin{bmatrix} 3 \\ 4 \\ 1 \end{bmatrix}$ and

$$f\left(\begin{bmatrix} x_1 \\ x_2 \\ x_3 \end{bmatrix}\right) = \begin{bmatrix} -3x_1 + x_3 \\ x_1 - 2x_3 \\ -x_1 + x_2 + 3x_3 \end{bmatrix}.$$

This function has matrix $A = \begin{bmatrix} -3 & 0 & 1 \\ 1 & 0 & -2 \\ -1 & 1 & 3 \end{bmatrix}$, so we're looking for the solution set of $A\vec{x} = \vec{b}$. This equation has augmented coefficient matrix

$$\begin{bmatrix} -3 & 0 & 1 & | & 3 \\ 1 & 0 & -2 & | & 4 \\ -1 & 1 & 3 & | & 1 \end{bmatrix}$$

which has reduced echelon form

$$\left[\begin{array}{ccc|c} 1 & 0 & 0 & -2 \\ 0 & 1 & 0 & 8 \\ 0 & 0 & 1 & -3 \end{array}\right].$$

This means $x_1 = -2$, $x_2 = 8$, and $x_3 = -3$, so we have the unique solution

$$\vec{x} = \begin{bmatrix} -2 \\ 8 \\ -3 \end{bmatrix}.$$

Again, if we think about this geometrically, we know the three planes in $\mathbb{R}^3$ represented by the three rows of our augmented coefficient matrix all intersect

in exactly one point: $\begin{bmatrix} -2 \\ 8 \\ -3 \end{bmatrix}$.

Notice that this case can only occur if we have no impossible equations (so we have a solution) and each variable has its own equation where it has a leading 1 as its coefficient. This means we need a reduced echelon form which has a leading 1 in every column except the rightmost one.

Finally, let's look at the case where we have infinitely many solutions.

Example 3. Find the solution set of $A\vec{x} = \vec{b}$ where $A = \begin{bmatrix} 2 & -4 & 1 \\ 1 & 0 & 1 \\ 1 & -8 & -1 \end{bmatrix}$ and

$\vec{b} = \begin{bmatrix} 6 \\ 3 \\ 3 \end{bmatrix}$.

This equation has augmented coefficient matrix

$$\left[\begin{array}{ccc|c} 2 & -4 & 1 & 6 \\ 1 & 0 & 1 & 3 \\ 1 & -8 & -1 & 3 \end{array}\right]$$

which has reduced echelon form

$$\left[\begin{array}{ccc|c} 1 & 0 & 1 & 3 \\ 0 & 1 & \frac{1}{4} & 0 \\ 0 & 0 & 0 & 0 \end{array}\right].$$

This gives us the equations $x_1 + x_3 = 3$, $x_2 + \frac{1}{4}x_3 = 0$, and $0 = 0$. The last of these gives us no information, but at least it isn't an impossible equation of the form $0 = 1$. Solving for x_1 and x_2 in the first two equations gives us $x_1 = 3 - x_3$ and $x_2 = -\frac{1}{4}x_3$. There is no equation to determine the value of x_3, so we can pick any real number we like. This is often expressed by saying x_3 is a *free variable*. This means we have infinitely many different solutions depending on our infinite choices for x_3. We can write this as $x_1 = 3 - x_3$, $x_2 = -\frac{1}{4}x_3$, and x_3 is free, or as

$$\vec{x} = \begin{bmatrix} 3 - x_3 \\ -\frac{1}{4}x_3 \\ x_3 \end{bmatrix}.$$

Geometrically, we can think of our real number line of choices for x_3 as the line of triple intersection between the three planes corresponding to the rows of our augmented coefficient matrix. (If we had 2 free variables, we would have two real number lines worth of choices, i.e., a plane of intersection, and so on.)

To get infinitely many solutions we need to avoid impossible equations, but we also need at least one variable whose value we get to choose. Those variables are the ones which don't have an equation where their coefficient is a leading 1. In other words, we're in this case if we have no leading 1 in the rightmost column and no leading 1 in at least one other column.

Now that we understand how to check the reduced echelon form to see whether or not we're in each individual case of our solution set's format, let's combine them into one coherent whole. Suppose we're solving $A\vec{x} = \vec{b}$. No matter which type of solution set we end up with, we always start by doing our row reduction algorithm on the augmented coefficient matrix $[A \,|\, \vec{b}]$. In the interests of efficiency, we'd like to recognize as early as possible in that process which type of solution set $A\vec{x} = \vec{b}$ has. The earliest criteria to show up is whether or not we have a leading 1 in the rightmost column, which indicates no solutions. If we fail to see that, then our equation has at least one solution. To see whether it has more than one solution, we complete our row reduction to see whether or not all the rest of our columns contain leading 1s. If they do, we have only one solution. If they don't, we have infinitely many solutions.

We can visualize this process as a flow chart:

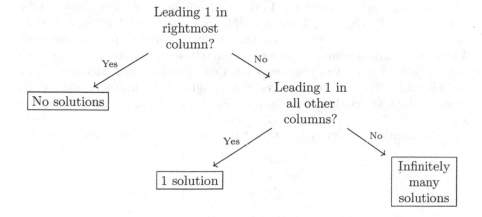

Now that we've figured out the possible options for the solution set of $A\vec{x} = \vec{b}$, we'll move on to figuring out how to concisely describe the solutions to a given matrix equation. We'll start with a special case of a matrix equation where $\vec{b} = \vec{0}$, so we're solving $A\vec{x} = \vec{0}$. (We've seen this before in 2.5 when we looked at the null space of a matrix.) This case is often called the homogeneous case (having $\vec{b} \neq \vec{0}$ is then called the nonhomogeneous case). Although it isn't immediately obvious, using linear algebra to balance chemical equations always results in matrix equations with $\vec{b} = \vec{0}$. To see how this works, let's reexamine our example from Chapter 0.

Example 4. When propane is burned, it combines with oxygen to form carbon dioxide and water. Propane molecules are composed of three carbon atoms and eight hydrogen atoms, so they are written as C_3H_8. Oxygen molecules each contain two atoms of oxygen, so they are written as O_2. Carbon dioxide molecules each contain one carbon atom and two oxygen atoms, so they are written CO_2. Water molecules are made up of two hydrogen atoms and one oxygen atom (which is why some aspiring chemistry comedians call water dihydrogen monoxide), so they are written H_2O. Using this notation, our chemical reaction can be written $C_3H_8 + O_2 \rightarrow CO_2 + H_2O$. Since atoms are neither created nor destroyed in a chemical reaction, how many of each kind of molecule must be included in this reaction?

This may look nothing like a linear algebra problem, let alone one of the form $A\vec{x} = \vec{b}$, but in fact row reduction is one of the best ways to tackle it. Let's suppose we have x_1 molecules of propane, x_2 molecules of oxygen, x_3 molecules of carbon dioxide, and x_4 molecules of water in our reaction. In this notation, our chemical reaction looks like

$$x_1(C_3H_8) + x_2(O_2) \rightarrow x_3(CO_2) + x_4(H_2O).$$

Since the number of a particular type of atom is the same before and after the reaction, this actually gives us a linear equation for each type of atom present in our various molecules. In this reaction, our molecules contain three types of atoms: carbon (C), hydrogen (H), and oxygen (O). Looking at carbon atoms, we have 3 from each propane molecule, 0 from each oxygen molecule, 1 from each carbon monoxide molecule, and 0 from each water molecule. This means we have $3x_1 + 0x_2$ before our reaction and $1x_3 + 0x_4$ afterward. Thus we must have $3x_1 = x_3$ or $3x_1 - x_3 = 0$. Looking at hydrogen atoms, by the same method we get that $8x_1 = 2x_4$ or $8x_1 - 2x_4 = 0$. Similarly, from oxygen we get $2x_2 = 2x_3 + x_4$ or $2x_2 - 2x_3 - x_4 = 0$.

This means we're really just trying to solve the equation $A\vec{x} = \vec{0}$ where

$$A = \begin{bmatrix} 3 & 0 & -1 & 0 \\ 8 & 0 & 0 & -2 \\ 0 & 2 & -2 & -1 \end{bmatrix}.$$

(This is a homogeneous equation since we're setting $A\vec{x}$ equal to $\vec{0}$.) In fact balancing a chemical reaction always gives a homogeneous equation, since we never have a constant term on either side of our chemical reaction.

The reduced echelon form of our augmented coefficient matrix is

$$\left[\begin{array}{cccc|c} 1 & 0 & 0 & -\frac{1}{4} & 0 \\ 0 & 1 & 0 & -\frac{5}{4} & 0 \\ 0 & 0 & 1 & -\frac{3}{4} & 0 \end{array}\right]$$

so we get $x_1 = \frac{1}{4}x_4$, $x_2 = \frac{5}{4}x_4$, $x_3 = \frac{3}{4}x_4$, and x_4 free. Since we have a free variable, our equation has infinitely many solutions, which shouldn't surprise you since we don't usually worry about only being allowed to burn a fixed number of propane molecules at one time.

This is perfectly fine as a linear algebra answer, but chemists don't want to think about fractions or negative quantities of a molecule. They want positive integer solutions rather than real number solutions. We can easily fix this problem by choosing a value of x_4 that is a positive integer multiple of 4. If $x_4 = 4k$ for some positive integer k, then our solution becomes $x_1 = k$, $x_2 = 5k$, $x_3 = 3k$, $x_4 = 4k$.

There are many similar problems where, for practical reasons, we want our solutions to have integer values. In fact, there are people who specifically study integer valued linear algebra, but that is outside the scope of this book.

The nice thing about solving homogeneous equations is that we always have at least one solution, namely $\vec{x} = \vec{0}$. This means we only have two possible options for our solution set instead of the usual three. Additionally, if there is one unique solution, we already know it is $\vec{x} = \vec{0}$. This means we can focus our attention on how to describe our solution set when we have infinitely many solutions. Since listing all possible solutions would literally take forever, we'll instead focus on finding a finite set of vectors which span our solution set. This is best illustrated by working through an example, which we'll do below.

Example 5. Find a finite set of vectors which span the solution set of $A\vec{x} = \vec{0}$ where $A = \begin{bmatrix} 3 & 2 & 0 & 4 & -1 \\ 0 & 2 & 6 & 0 & -8 \\ 1 & 1 & 1 & 4 & 1 \end{bmatrix}$.

The reduced echelon form of A is

$$\begin{bmatrix} 1 & 0 & -2 & 0 & 1 \\ 0 & 1 & 3 & 0 & -4 \\ 0 & 0 & 0 & 1 & 1 \end{bmatrix}$$

which gives us the equations $x_1 - 2x_3 + x_5 = 0$, $x_2 + 3x_3 - 4x_5 = 0$, and $x_4 + x_5 = 0$. (I didn't bother to write down the augmentation column here, because a column of zeros is unchanged by any row operations and so will remain a column of zeros in the reduced echelon form.) The third and fifth columns of A's reduced echelon form had no leading 1s, so x_3 and x_5 are free. Solving for x_1, x_2, and x_4 gives us $x_1 = 2x_3 - x_5$, $x_2 = -3x_3 + 4x_5$, and $x_4 = -x_5$. So in vector format, our solution set is

$$\left\{ \begin{bmatrix} 2x_3 - x_5 \\ -3x_3 + 4x_5 \\ x_3 \\ -x_5 \\ x_5 \end{bmatrix} \right\}.$$

A span is a set of linear combinations of vectors, i.e., a sum of vectors which each have their own scalar coefficient. There isn't a single scalar coefficient we can factor out of each entry of our solution vector, but we can split each entry up as a sum of two terms; one with x_3 and one with x_5. This gives us

$$\begin{bmatrix} 2x_3 - x_5 \\ -3x_3 + 4x_5 \\ x_3 \\ -x_5 \\ x_5 \end{bmatrix} = \begin{bmatrix} 2x_3 \\ -3x_3 \\ x_3 \\ 0 \\ 0 \end{bmatrix} + \begin{bmatrix} -x_5 \\ 4x_5 \\ 0 \\ -x_5 \\ x_5 \end{bmatrix}.$$

Now we can pull out our scalars, since every entry of the first vector in our sum is a multiple of x_3 and every entry in the second vector is a multiple of x_5. This means we have

$$\begin{bmatrix} 2x_3 \\ -3x_3 \\ x_3 \\ 0 \\ 0 \end{bmatrix} + \begin{bmatrix} -x_5 \\ 4x_5 \\ 0 \\ -x_5 \\ x_5 \end{bmatrix} = x_3 \begin{bmatrix} 2 \\ -3 \\ 1 \\ 0 \\ 0 \end{bmatrix} + x_5 \begin{bmatrix} -1 \\ 4 \\ 0 \\ -1 \\ 1 \end{bmatrix}.$$

This newest format is precisely the span of the two vectors being multiplied by x_3 and x_5! Therefore our solution set can be written as

$$\text{Span} \left\{ \begin{bmatrix} 2 \\ -3 \\ 1 \\ 0 \\ 0 \end{bmatrix}, \begin{bmatrix} -1 \\ 4 \\ 0 \\ -1 \\ 1 \end{bmatrix} \right\}.$$

(Notice that there is a shortcut through this process: We get one spanning vector per free variable, and that vector is precisely the vector of coefficients on that variable in our general solution set vector.)

If we wanted to describe the solutions to this equation to someone else, the most efficient way would be to tell them our spanning vectors. This same procedure works for the solution set of any homogeneous equation.

As a bonus, it turns out that the spanning set we constructed in this way conveys even more information. The spanning vectors form a linearly independent set! To see this, focus on the entries of the vectors which correspond to the free variables. If x_i is a free variable, the spanning vectors will all have entry 0 in the ith spot except for the spanning vector formed from x_i's coefficients. This means that no linear combination of these spanning vectors equals $\vec{0}$ unless all coefficients in the combination are zero, i.e., our spanning vectors are linearly independent. As discussed in 1.3, this means the dimension equals the number of spanning vectors. Therefore the dimension of the null space of a matrix A is the same as the number of free variables in the solution set of $A\vec{x} = \vec{0}$. In the example above, this means A has a two-dimensional null space.

Now that we've successfully figured out a good way to describe the solution set of $A\vec{x} = \vec{0}$, let's turn our attention to the situation where $A\vec{x} = \vec{b}$ and $\vec{b} \neq \vec{0}$. (Remember that $\vec{b} \neq \vec{0}$ doesn't mean all its entries are nonzero, but rather that $\vec{b}$ has at least one nonzero entry.) We have no problem describing the solution set when $A\vec{x} = \vec{b}$ has no solutions. Similarly if $A\vec{x} = \vec{b}$ has one unique solution, we can easily describe the solution set by simply giving that solution. As with $A\vec{x} = \vec{0}$, this leaves us with the task of concisely describing our solution set when we have infinitely many solutions.

Let's start with a geometric exploration. As at the beginning of this section we'll work in $\mathbb{R}^2$ for ease of drawing pictures, but here let's look at a single line. This means $A\vec{x} = \vec{b}$ looks like $\begin{bmatrix} a & c \end{bmatrix} \begin{bmatrix} x \\ y \end{bmatrix} = \begin{bmatrix} b \end{bmatrix}$ or $ax + cy = b$ which we can rewrite more familiarly as $y = \frac{a}{c}x + b$. If we're in the homogeneous case then $b = 0$, which means we're talking about a line through the origin as in Figure 2.5.

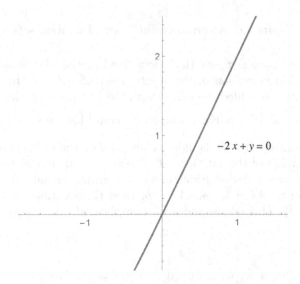

Figure 2.5: A homogeneous solution set

When we move back to the case where $b \neq 0$, we're adding some nonzero number b to our equation. Specifically, this is the difference between the lines $y = \frac{a}{c}x$ and $y = \frac{a}{c}x + b$. Adding a constant to a function shifts that function up or down depending on the sign of the constant. (Think about how adding "$+C$" on the end of a definite integral changes the graph of the antiderivative.) This shift is shown in Figure 2.6.

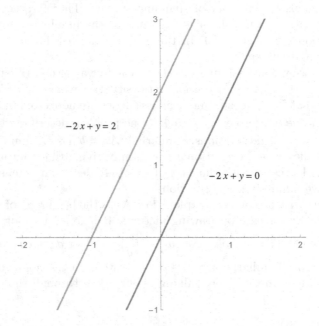

Figure 2.6: A corresponding pair of solution sets

The picture above suggests that for a fixed matrix A, the solution set of $A\vec{x} = \vec{b}$ is a shifted version of the solution set of $A\vec{x} = \vec{0}$. In our example above, the vector we added to create that shift is a point on the line which is the solution set of $A\vec{x} = \vec{b}$. For example, we could have used the point $\begin{bmatrix} 0 \\ 2 \end{bmatrix}$.

This relationship also holds more generally, i.e., the solution set of $A\vec{x} = \vec{b}$ is a shifted version of the solution set of $A\vec{x} = \vec{0}$ and that shift can be done by adding any particular solution to $A\vec{x} = \vec{b}$. In mathematical notation, if $\vec{v}$ is any solution to $A\vec{x} = \vec{b}$, i.e., $A\vec{v} = \vec{b}$, then the solution set to $A\vec{x} = \vec{b}$ is $\{\vec{x} + \vec{v} \mid A\vec{x} = \vec{0}\}$. This may sound a bit complicated, so let's do an example.

Example 6. Let $A = \begin{bmatrix} 3 & 2 & 0 & 4 & -1 \\ 0 & 2 & 6 & 0 & -8 \\ 1 & 1 & 1 & 4 & 1 \end{bmatrix}$. Use the fact that $\vec{v} = \begin{bmatrix} 0 \\ 3 \\ -1 \\ 1 \\ 2 \end{bmatrix}$ is a solution to $A\vec{x} = \vec{b}$ for $\vec{b} = \begin{bmatrix} 8 \\ -16 \\ 8 \end{bmatrix}$ to describe all solutions of $A\vec{x} = \vec{b}$.

Our matrix A is the same as in Example 5, and we know from there that the solution set of $A\vec{x} = \vec{0}$ is $\left\{ \begin{bmatrix} 2x_3 - x_5 \\ -3x_3 + 4x_5 \\ x_3 \\ -x_5 \\ x_5 \end{bmatrix} \right\}$. Since the solution set of $A\vec{x} = \vec{b}$ can be written as $\vec{v}$ plus the solution set to $A\vec{x} = \vec{0}$, we get that the solutions of $A\vec{x} = \vec{b}$ are

$$\left\{ \begin{bmatrix} 2x_3 - x_5 \\ -3x_3 + 4x_5 \\ x_3 \\ -x_5 \\ x_5 \end{bmatrix} + \begin{bmatrix} 0 \\ 3 \\ -1 \\ 1 \\ 2 \end{bmatrix} \right\} = \left\{ \begin{bmatrix} 2x_3 - x_5 \\ -3x_3 + 4x_5 + 3 \\ x_3 - 1 \\ -x_5 + 1 \\ x_5 + 2 \end{bmatrix} \right\}.$$

If we didn't already have a particular solution $\vec{v}$, we could also just solve directly for the solution set. It won't turn out to be a span as with $A\vec{x} = \vec{0}$, but will be almost a span.

Example 7. Find the solution set of $A\vec{x} = \vec{b}$ where $A = \begin{bmatrix} 1 & 0 & -1 & 2 \\ 0 & 1 & 3 & -3 \end{bmatrix}$ and $\vec{b} = \begin{bmatrix} -5 \\ 1 \end{bmatrix}$.

The augmented coefficient matrix of this equation is

$$\begin{bmatrix} 1 & 0 & -1 & 2 & | & -5 \\ 0 & 1 & 3 & -3 & | & 1 \end{bmatrix}$$

which is already in reduced echelon form. There is no leading 1 in the rightmost column, so $A\vec{x} = \vec{b}$ does have solutions. Our matrix gives us the equations $x_1 - x_3 + 2x_4 = -5$ and $x_2 + 3x_3 - 3x_4 = 1$, so our solution set can be written $x_1 = -5 + x_3 - 2x_4$ and $x_2 = 1 - 3x_3 + 3x_4$, with x_3 and x_4 free.

To try to make this look as much as possible like a span, we can follow the same general procedure as Example 5. First we rewrite our solution set as a vector to get

$$\begin{bmatrix} -5 + x_3 - 2x_4 \\ 1 - 3x_3 + 3x_4 \\ x_3 \\ x_4 \end{bmatrix}.$$

Each component of this vector is a sum of three terms: a constant, a multiple of x_3, and a multiple of x_4. Splitting it up as a sum of three vectors gives us

$$\begin{bmatrix} -5 \\ 1 \\ 0 \\ 0 \end{bmatrix} + \begin{bmatrix} x_3 \\ -3x_3 \\ x_3 \\ 0 \end{bmatrix} + \begin{bmatrix} -2x_4 \\ 3x_4 \\ 0 \\ x_4 \end{bmatrix}.$$

We can factor an x_3 out of the second vector and an x_4 out of the third to get

$$\begin{bmatrix} -5 \\ 1 \\ 0 \\ 0 \end{bmatrix} + x_3 \begin{bmatrix} 1 \\ -3 \\ 1 \\ 0 \end{bmatrix} + x_4 \begin{bmatrix} -2 \\ 3 \\ 0 \\ 1 \end{bmatrix}.$$

The second two terms are precisely the span of the second and third vectors, so this can be written as

$$\begin{bmatrix} -5 \\ 1 \\ 0 \\ 0 \end{bmatrix} + \text{Span} \left\{ \begin{bmatrix} 1 \\ -3 \\ 1 \\ 0 \end{bmatrix}, \begin{bmatrix} -2 \\ 3 \\ 0 \\ 1 \end{bmatrix} \right\}.$$

In this case we couldn't communicate our solution set simply by giving a list of spanning vectors, but we could give a list of spanning vectors along with the extra vector which needs to be added to that span to give the solution set. You can check that this meshes with our previous discussion, by checking that the constant vector is a solution to $A\vec{x} = \vec{b}$ and the span is the solution set to $A\vec{x} = \vec{0}$.

Another way to view the solution sets of matrix equations $A\vec{x} = \vec{b}$ is to connect them to the range of the function f with matrix A. (Since f is a matrix map, we could also refer to its range as the column space of A, $Col(A)$.) Recall that the range was all $\vec{b}$'s where $A\vec{x} = \vec{b}$ had a solution, i.e., the set of all vectors in the codomain that could be written as $A\vec{x}$ for some $\vec{x}$. One possible question we can ask is how large this range is, i.e., what is the dimension of the range of A's map. Since the range is also the span of the columns of A, we can find the dimension of the range by asking how many of A's columns are linearly independent, which prompts the following definition.

Definition. The *rank* of a matrix A, written $\text{rk}(A)$, is the number of linearly independent columns of A.

Example 8. Compute the rank of $A = \begin{bmatrix} 4 & 1 & 3 \\ 2 & 3 & -1 \\ -1 & 6 & -7 \end{bmatrix}$.

We saw in the last section that we could check the linear independence or linear dependence of a set of vectors by creating a matrix whose columns are our vectors and checking whether or not every column of its reduced echelon form had a leading 1. Here we're starting with a matrix and thinking of the vectors which are its columns, but otherwise this is the same procedure.

The reduced echelon form of A is $\begin{bmatrix} 1 & 0 & 1 \\ 0 & 1 & -1 \\ 0 & 0 & 0 \end{bmatrix}$. This doesn't have a leading 1 in every column, so the columns aren't linearly independent. This tells us that the rank of A isn't 3. Our free variable is x_3, which we saw in 1.2 can be interpreted as meaning that the third column is in the span of the other two columns. If we ignore A's third column, the other two columns do have leading 1s in their reduced echelon form. This means that they are linearly independent, so $rk(A) = 2$. This also tells us that A's column space has dimension 2.

The discussion before and during the last example showed us that the linearly independent columns of A were precisely the ones whose columns had a leading 1 in A's reduced echelon form and their number tells us the dimension of the range of A's map. This gives the following result.

Theorem 2. The *rank* of a matrix A equals both the number of columns of A's reduced echelon form which contain a leading 1 and the dimension of the columns space of A.

For example, this shows the column space of the matrix in Example 5 is a 2D subspace of $\mathbb{R}^5$.

If we combine this new idea of the rank of a matrix with our earlier discussion of the dimension of a matrix's null space, we get the following theorem, usually called the Rank-Nullity Theorem. (Some people call the dimension of a matrix's null space its *nullity*.)

Theorem 3. Let A be an $m \times n$ matrix. Then $rk(A) + dim(\mathrm{Nul}(A)) = n$.

This theorem is remarkable in that it operates on two levels: one where it is completely obvious that the theorem is true, and another where the theorem tells us something interesting. We'll start with the first layer to convince ourselves this equation always holds.

The easiest term to interpret in the Rank-Nullity Theorem's equation is n, which is just the number of columns of our matrix A. In our discussion of rank, we saw that the easiest way to compute the rank of a matrix is to count the number of columns with leading 1s in its reduced echelon form. This means we can view $rk(A)$ as the number of columns of A's reduced echelon form which contain a leading 1. In our discussion of solving $A\vec{x} = \vec{0}$, we saw that the dimension of the null space of A was the number of free variables. Since each free variable corresponds to a column of the reduced echelon form of A

which doesn't have a leading 1, $\dim(\text{Nul}(A))$ is the number of columns of A's reduced echelon form which don't contain a leading 1. From this perspective, our equation can be read as saying: the number of columns of A's reduced echelon form with a leading 1 plus the number of columns without a leading 1 equals the total number of columns. Since a column must either contain or not contain a leading 1, this is clearly true.

While the argument above tells us $\text{rk}(A) + \dim(\text{Nul}(A)) = n$ always holds, it doesn't provide any evidence that this result is interesting. To see why we should care about the Rank-Nullity Theorem, let's switch from thinking computationally about A's reduced echelon form to thinking geometrically about the linear function f with matrix A. Since A is an $m \times n$ matrix, we know $f : \mathbb{R}^n \to \mathbb{R}^m$, so the n on the right-hand side of our equation is the size of f's domain. The rank of A is the dimension of the column space of A, i.e., the dimension of the range of f. The null space of A is the kernel of f, so $\dim(\text{Nul}(A))$ is the dimension of f's kernel. Putting this all together, we get a sort of "conservation of dimension" law. The map f starts with n dimensions of inputs, and ends up with $\text{rk}(A)$ dimensions of outputs. What happened to the other dimensions of the domain? The Rank-Nullity Theorem says those dimensions must be collapsed down to $\vec{0}$ in $\mathbb{R}^m$, i.e., they are in the kernel. To me, conservation laws are very beautiful things, so I love having one here in linear algebra.

Example 9. Show the Rank-Nullity Theorem holds for $A = \begin{bmatrix} 4 & 1 & 3 \\ 2 & 3 & -1 \\ -1 & 6 & -7 \end{bmatrix}$.

This is the matrix from Example 8, so we know $\text{rk}(A) = 2$.

To find the dimension of the null space of A, we need to find a basis for $\text{Nul}(A)$ which we can do by writing the solution set of $A\vec{x} = \vec{0}$ as a span and remembering that the dimension of $\text{Nul}(A)$ is the number of spanning vectors for the solution set. We saw in Example 8 that the reduced echelon form of A is $\begin{bmatrix} 1 & 0 & 1 \\ 0 & 1 & -1 \\ 0 & 0 & 0 \end{bmatrix}$, which gives us the equations $x_1 + x_3 = 0$, and $x_2 - x_3 = 0$. Therefore the solution set of $A\vec{x} = \vec{0}$ is

$$\left\{ \begin{bmatrix} -x_3 \\ x_3 \\ x_3 \end{bmatrix} \right\} = \text{Span} \left\{ \begin{bmatrix} -1 \\ 1 \\ 1 \end{bmatrix} \right\}.$$

Since there is one spanning vector, we get that $\dim(\text{Nul}(A)) = 1$.

Putting this together, we see that

$$\text{rk}(A) + \dim(\text{Nul}(A)) = 2 + 1 = 3.$$

A has three columns, so $n = 3$ and the Rank-Nullity Theorem is satisfied.

Geometrically, this equation tells us that A's map f has domain $\mathbb{R}^3$, a two-dimensional range, and a one-dimensional kernel.

Exercises 2.8.

1. Which type of solution set do the following equations have?
   $x_1 + x_2 - 4x_3 = 0$, $-x_1 + 3x_3 = -1$, $-3x_1 + 4x_2 - 15x_3 = 5$.

2. Which type of solution set does the following vector equation have?
$$x_1 \begin{bmatrix} 2 \\ 0 \\ -4 \\ 1 \end{bmatrix} + x_2 \begin{bmatrix} -5 \\ 6 \\ 0 \\ 12 \end{bmatrix} + x_3 \begin{bmatrix} 3 \\ -1 \\ 2 \\ 7 \end{bmatrix} = \begin{bmatrix} 1 \\ 1 \\ 0 \\ 1 \end{bmatrix}$$

3. Find the solution set of the following equations or show they have no solution. $x_1 + 2x_2 + 3x_3 = 0$, $4x_1 + 5x_2 + 6x_3 = 3$, $7x_1 + 8x_2 + 9x_3 = 0$

4. Find the solution set of the following vector equation or show it has no solution. $x_1 \begin{bmatrix} -1 \\ 3 \\ 1 \end{bmatrix} + x_2 \begin{bmatrix} 1 \\ 0 \\ 2 \end{bmatrix} + x_3 \begin{bmatrix} 2 \\ 3 \\ 7 \end{bmatrix} = \begin{bmatrix} 0 \\ 0 \\ 0 \end{bmatrix}$

5. Find the solution set of the following matrix equation or show that it has no solution. $\begin{bmatrix} 2 & -2 & 0 \\ 0 & 5 & 10 \\ 0 & -4 & -7 \end{bmatrix} \vec{x} = \begin{bmatrix} 4 \\ 5 \\ -5 \end{bmatrix}$

6. Find the solution set of $f(\vec{x}) = \begin{bmatrix} 3 \\ 11 \end{bmatrix}$ or show it has no solution, where $f\left(\begin{bmatrix} x_1 \\ x_2 \end{bmatrix} \right) = \begin{bmatrix} x_1 + 2x_2 \\ 3x_1 + 4x_2 \end{bmatrix}$.

7. In 1.2's Exercise 7, we explored the chemical process of rusting iron, which starts when iron, Fe, oxygen gas, O_2, and water, H_2O, combine to form iron hydroxide, $Fe(OH)_3$. Let's match 1.2 and make the first entry in our vectors count iron atoms (Fe), the second count oxygen (O), and the third count hydrogen (H). If our reaction combines x_1 molecules of iron, x_2 molecules of oxygen, and x_3 molecules of water to create x_4 molecules of rust, how many of each kind of molecule must be present in this reaction? (Remember that chemists want their numbers of molecules to be positive integers!)

8. In 1.2's Exercise 8, we learned sodium chlorate, $NaClO_3$, is produced by the electrolysis of sodium chloride, $NaCl$, and water, H_2O, and that the reaction also produces hydrogen gas, H_2. Let's match 1.2 and make the first entry in our vectors count sodium atoms (Na), the second count chlorine (Cl), the third count oxygen (O), and the fourth count hydrogen (H). If our reaction combines x_1 molecules of sodium chloride with x_2 molecules of water to produce x_3 molecules of sodium chlorate and x_4 molecules of hydrogen gas, how many of each kind of molecule must be present in this reaction? (Remember that chemists want their numbers of molecules to be positive integers!)

9. (a) How can you tell from the reduced echelon form of an augmented coefficient matrix that the corresponding matrix equation has one unique solution?

   (b) Give an example of a 3×4 augmented coefficient matrix in reduced echelon form whose corresponding matrix equation has one unique solution.

10. For each option below, write down the reduced echelon form of an augmented coefficient matrix whose matrix equation has that type of solution set. What features of your reduced echelon form caused you to have that type of solution set?

   (a) no solution
   (b) 1 solution
   (c) infinitely many solutions

11. The solution set of the matrix equation $A\vec{x} = \vec{0}$ is: $x_1 = 3x_2 - 4x_4$, $x_3 = 5x_4$, and x_2 and x_4 free. Write this solution set as the span of a set of vectors.

12. The solution set of the matrix equation $A\vec{x} = \vec{0}$ is: $x_1 = 0$, $x_2 = 4x_3 - 5x_5$, $x_4 = 8x_5$ and x_3 and x_5 free. Write this solution set as the span of a set of vectors.

13. Let $A = \begin{bmatrix} 1 & -2 & 0 & 0 & 3 \\ 0 & 0 & 1 & 0 & -4 \\ 0 & 0 & 0 & 1 & 1 \end{bmatrix}$. Write the solution set to $A\vec{x} = \vec{0}$ as the span of a set of vectors.

14. Let $A = \begin{bmatrix} 1 & 1 & 0 & -1 \\ 0 & 0 & 1 & -2 \\ 0 & 0 & 0 & 0 \end{bmatrix}$. Write the solution set of $A\vec{x} = \vec{0}$ as the span of a set of vectors.

15. Let $A = \begin{bmatrix} 1 & 2 & 0 & -3 \\ 0 & 0 & 1 & -1 \\ 0 & 0 & 0 & 0 \end{bmatrix}$. Write the solution set of $A\vec{x} = \vec{0}$ as the span of a set of vectors.

16. Let $A = \begin{bmatrix} 1 & -1 & 0 & 2 \\ 0 & 0 & 1 & -3 \end{bmatrix}$. Write the solution set of $A\vec{x} = \vec{0}$ as the span of a set of vectors.

17. Let $A = \begin{bmatrix} 1 & 0 & 2 & -1 \\ 0 & 1 & 0 & 3 \end{bmatrix}$.

   (a) Write the solution set of $A\vec{x} = \vec{0}$ as the span of a set of vectors.
   (b) Fix some nonzero vector $\vec{b}$. If $A\vec{x} = \vec{b}$ has solutions, how is its solution set related to your answer to (a)?

18. Let $A = \begin{bmatrix} 1 & 0 & -1 & 4 \\ 0 & 1 & 3 & 0 \\ 0 & 0 & 0 & 0 \end{bmatrix}$.

(a) Write the solution set of $A\vec{x} = \vec{0}$ as the span of a set of vectors.

(b) Briefly describe the relationship between the solution set of $A\vec{x} = \vec{0}$ and the solution set of $A\vec{x} = \vec{b}$ where $\vec{b} = \begin{bmatrix} 2 \\ -3 \\ 9 \end{bmatrix}$.

19. Suppose $A\vec{x} = \vec{b}$ has at least one solution. Describe the geometric relationship between its solution set and the solution set of $A\vec{x} = \vec{0}$.

20. Can a linear system with five equations in seven variables have one unique solution? If you answer yes, give an example of such a system. If you answer no, briefly explain why this is impossible.

21. Is it possible to have a linear system of three equations in four variables which has one unique solution? If you answer yes, give an example of such a system. If you answer no, briefly explain why this is impossible.

22. Find the rank of $A = \begin{bmatrix} -3 & 4 & 0 \\ 1 & 2 & 1 \\ 2 & -6 & -1 \end{bmatrix}$.

23. Find the rank of $A = \begin{bmatrix} 2 & 5 & -1 \\ 3 & 8 & 4 \\ 1 & 9 & -2 \end{bmatrix}$.

24. Find the rank of $A = \begin{bmatrix} 1 & 2 & 4 \\ -1 & 3 & 5 \\ 0 & 1 & 0 \\ -2 & 6 & 9 \end{bmatrix}$.

25. Find the rank of $A = \begin{bmatrix} -5 & 0 & 2 & 1 \\ 2 & 0 & -4 & 6 \end{bmatrix}$.

26. Show the Rank-Nullity Theorem holds for $A = \begin{bmatrix} 1 & 2 & 3 \\ -1 & 3 & 2 \\ 0 & 1 & 1 \\ -2 & 6 & 4 \end{bmatrix}$.

27. Show the Rank-Nullity Theorem holds for $A = \begin{bmatrix} -3 & 1 & 2 \\ 4 & 2 & -6 \\ 0 & 1 & -1 \end{bmatrix}$.

28. Suppose $f : \mathbb{R}^5 \to \mathbb{R}^3$ is a linear map whose matrix has rank 2. What is the dimension of f's kernel?

29. Suppose $f : \mathbb{R}^6 \to \mathbb{R}^9$ is a linear map whose matrix has rank 4. What is the dimension of f's kernel?

2.9   Large Matrix Computations

In many modern applications, matrices are being used to store vast data sets worth of information. This may mean that our matrices are so large that even computers will be slowed down trying to do basic matrix operations or solve matrix equations. In this section we'll explore two possible tools to use in these types of situations that allow these vast problems to be reduced to easier or smaller cases.

A note before we begin our discussion: the major difficulty in trying to explore these ideas in a book is that we have the option to either do realistic examples which are too large to easily write down or to do examples which are so small that we should do them using the techniques developed in earlier sections. I've made the choice to use smaller examples here, so you'll have to use your imagination to extend them to the case where our matrices may have hundreds or even thousands of rows and columns.

The first idea we'll discuss helps us compute rA, $A + B$, or AB when A and B are very, very large. Our main trick here is to partition our matrices into smaller pieces and do our matrix operation piece by piece. This also has the advantage that we can choose to only compute part of rA, $A + B$, or AB rather than computing the whole matrix. Before we dive in, let's be very specific about what we mean by a partition of a matrix.

Definition.   A *partition* of a matrix A is a set of vertical and horizontal cuts which slice A into a collection of rectangular submatrices.

Example 1.   Partition the matrix $A = \begin{bmatrix} 2 & 1 & 0 & -5 \\ 7 & -3 & 9 & 0 \\ 1 & -1 & 6 & 8 \end{bmatrix}$ using a vertical cut between the third and fourth columns and a horizontal cut between the second and third rows.

The standard way to indicate where we are slicing our matrix to create our partition is to draw horizontal and vertical lines through the matrix to visually show each cut. Here we want to cut vertically between the third and fourth columns and horizontally between the second and third rows. This means we need to draw a vertical line between the third and fourth columns and a horizontal line between the second and third rows. This gives us the following visual illustration of our partitioned matrix

$$\left[\begin{array}{ccc|c} 2 & 1 & 0 & -5 \\ 7 & -3 & 9 & 0 \\ \hline 1 & -1 & 6 & 8 \end{array}\right].$$

This partition gives us a 2×3 top left submatrix, a 2×1 top right submatrix, a 3×1 bottom left submatrix, and a 1×1 bottom right submatrix.

Example 2. Explain why $\left[\begin{array}{c|cc} 3 & -1 & 4 \\ -2 & 1 & 5 \\ \hline 8 & 0 & 2 \end{array}\right]$ is not a partition of the matrix $A = \begin{bmatrix} 3 & -1 & 4 \\ -2 & 1 & 5 \\ 8 & 0 & 2 \end{bmatrix}$.

The vertical cut between the first and second columns is fine, but the horizontal cuts do not go all the way across the matrix. Therefore this is not a partition.

Now that we understand how to properly partition a matrix, let's look at how to do matrix operations on partitioned matrices. The easiest matrix operation here is multiplication by a scalar. Since we get rA by multiplying each entry of A by r, we can compute part of rA quite easily without computing the whole matrix. In the language of partitions, this means that if we've partitioned A then we can compute the submatrices of the same partition on rA one submatrix at a time by simply multiplying each submatrix by r. In other words, if we only want to know one piece of rA, we would choose a partition of A which isolates the part of rA we want to compute and then multiply it by r to get the piece of rA that we're interested in.

Example 3. Compute the submatrix of $-2A$ consisting of the top two rows and middle two columns where $A = \begin{bmatrix} 2 & 1 & 0 & -5 \\ 7 & -3 & 9 & 0 \\ 1 & -1 & 6 & 8 \end{bmatrix}$.

To do this, we first need to create a partition that isolates the requested submatrix of A. This means we need to do two vertical cuts, between the first and second columns and between the third and fourth columns, and one horizontal cut between the second and third rows. This looks like

$$\left[\begin{array}{c|cc|c} 2 & 1 & 0 & -5 \\ 7 & -3 & 9 & 0 \\ \hline 1 & -1 & 6 & 8 \end{array}\right].$$

The submatrix we're interested in is the top middle 2×2 piece. Since we're only interested in the corresponding part of $-2A$, we can ignore the rest of the matrix and just multiply that 2×2 submatrix by -2 to get

$$\begin{bmatrix} -2 & 0 \\ 6 & -18 \end{bmatrix}.$$

Matrix addition isn't much more complicated than scalar multiplication, since we compute each entry of $A + B$ by adding together the corresponding entries of A and B. This means that if we're interested in finding only part of $A + B$, we can compute only that part by using carefully selected partitions of A and B. However, we do have one additional restriction that we didn't have when computing rA: the sum $A + B$ only makes sense if A and B are the same size. This means that not only do we need to require A and B to be the same size, but also that we need to use the same partition on both of them and on $A + B$ so that we're adding submatrices of the same size. This may initially sound confusing, but will hopefully be cleared up in the example below.

Example 4. Compute the submatrix of $A + B$ consisting of the bottom two rows and left two columns where $A = \begin{bmatrix} 3 & -5 & 4 \\ 1 & 0 & -1 \\ 2 & 5 & -1 \end{bmatrix}$ and $B = \begin{bmatrix} 9 & -9 & 2 \\ 5 & 4 & -6 \\ 3 & -3 & 7 \end{bmatrix}$.

The submatrix of $A + B$ we're interested in is the 2×2 bottom left corner. We can create a partition which isolates this part of both A and B by cutting vertically between the second and third columns and horizontally between the first and second rows. This gives us

$$\left[\begin{array}{cc|c} 3 & -5 & 4 \\ 1 & 0 & -1 \\ 2 & 5 & -1 \end{array}\right] \text{ and } \left[\begin{array}{cc|c} 9 & -9 & 2 \\ 5 & 4 & -6 \\ 3 & -3 & 7 \end{array}\right].$$

Now we can compute the requested piece of $A + B$ by adding the 2×2 submatrices that form the bottom left corners of A and B. This gives us

$$\begin{bmatrix} 1 & 0 \\ 2 & 5 \end{bmatrix} + \begin{bmatrix} 5 & 4 \\ 3 & -3 \end{bmatrix} = \begin{bmatrix} 6 & 4 \\ 5 & 2 \end{bmatrix}.$$

The situation for matrix multiplication is a little more complicated, because both computing the entries of AB and choosing the right partitions for A and B are a little trickier. If we're interested in computing part of AB, we can't focus only on those entries of A and B as in matrix addition, because matrix multiplication uses the whole ith row of A and jth column of B to find the ijth entry of AB. Additionally, we can't simply use the same partition on A, B, and AB, because for matrix multiplication to make sense we need the number of columns of A to equal the number of rows of B and similarly for any submatrices of A and B that we want to multiply together. Let's tackle these problems in two stages, starting with the right choice of partitions.

To figure out the right partitions for A and B to compute a particular piece of AB, remember our rule that the product of an $m \times k$ and a $k \times n$ matrix is an $m \times n$ matrix. Reversing the order, we can see that to find an $m \times n$ submatrix of AB, we'll need to use an $m \times k$ partition of A and a $k \times n$ partition of B. In other words, the vertical cuts we use in our partition of A

need to match up with the horizontal cuts in our partition of B since these correspond to our choice of matching k. The placement of the horizontal cuts in our partition of A, which correspond to m, and the vertical cuts in our partition of B, which correspond to n, depend on the size of the submatrix of AB we want to compute.

Example 5. Find partitions of the matrices $A = \begin{bmatrix} -1 & 0 & 2 & 5 \\ 3 & 1 & -3 & 2 \\ 4 & -2 & 0 & 1 \end{bmatrix}$ and

$B = \begin{bmatrix} 0 & 7 & -2 & 4 & 1 \\ 6 & -4 & 2 & 0 & -1 \\ -9 & 5 & -2 & 4 & 3 \\ 3 & 5 & -1 & 1 & 2 \end{bmatrix}$ which could be used to compute the 2×3 submatrix in the top right corner of AB.

From our discussion above, we know that to create a 2×3 submatrix in the top right corner of AB we'll need to partition A with a horizontal cut between the second and third rows and B with a vertical cut between the second and third columns. This gives us

$$\left[\begin{array}{cccc} -1 & 0 & 2 & 5 \\ 3 & 1 & -3 & 2 \\ \hline 4 & -2 & 0 & 1 \end{array}\right] \text{ and } \left[\begin{array}{cc|ccc} 0 & 7 & -2 & 4 & 1 \\ 6 & -4 & 2 & 0 & -1 \\ -9 & 5 & -2 & 4 & 3 \\ 3 & 5 & -1 & 1 & 2 \end{array}\right].$$

Additionally, we'll need to partition A with a vertical cut between the kth and $k+1$st columns and B with a matching horizontal cut between the kth and $k+1$st rows. If there were any way to take advantage of blocks of zero entries in either A or B, the choice of k could matter more. However, since there aren't any zero blocks, let's divide the four rows of A and columns of B evenly, i.e., let $k = 2$. This means we're adding a vertical cut between the second and third columns of A and a horizontal cut between the second and third rows of B. This gives us

$$\left[\begin{array}{cc|cc} -1 & 0 & 2 & 5 \\ 3 & 1 & -3 & 2 \\ \hline 4 & -2 & 0 & 1 \end{array}\right] \text{ and } \left[\begin{array}{cc|ccc} 0 & 7 & -2 & 4 & 1 \\ 6 & -4 & 2 & 0 & -1 \\ \hline -9 & 5 & -2 & 4 & 3 \\ 3 & 5 & -1 & 1 & 2 \end{array}\right].$$

Notice that our partitions of A and B naturally give rise to a partition of AB as seen in the following example.

Example 6. Use the partitions on A and B in the previous example to create a partition on AB.

The partition on AB gets its horizontal cuts from the horizontal cuts in our partition of A and its vertical cuts from the vertical cuts in our partition of B. (I remember this by remembering that AB gets its number of rows from A and its number of columns from B so it seems natural to divide AB's rows as A's rows are divided and AB's columns as B's columns are divided.) Therefore we have one horizontal cut between the second and third rows of AB and one vertical cut between the second and third columns of AB. This gives us the partition below (with * used to denote an unknown matrix entry).

$$\left[\begin{array}{cc|ccc} * & * & * & * & * \\ * & * & * & * & * \\ \hline * & * & * & * & * \end{array}\right]$$

Now that we understand how to choose partitions for A and B and the partition this creates for AB, how can we compute the submatrices of AB's new partition? This ends up being easier than you might expect. In fact we can link it back to basic matrix multiplication by replacing each submatrix of A and B by a variable and then multiplying as if those variables were our matrix entries. This is illustrated below.

Example 7. Use the partitions of A and B from previous examples to compute the 2×3 submatrix in the top right corner of AB.

Recall that we had partitioned A and B as

$$\left[\begin{array}{cc|cc} -1 & 0 & 2 & 5 \\ 3 & 1 & -3 & 2 \\ \hline 4 & -2 & 0 & 1 \end{array}\right] \quad \text{and} \quad \left[\begin{array}{cc|ccc} 0 & 7 & -2 & 4 & 1 \\ 6 & -4 & 2 & 0 & -1 \\ \hline -9 & 5 & -2 & 4 & 3 \\ 3 & 5 & -1 & 1 & 2 \end{array}\right]$$

which gave us the following partition on AB

$$\left[\begin{array}{cc|ccc} * & * & * & * & * \\ * & * & * & * & * \\ \hline * & * & * & * & * \end{array}\right].$$

Each of our matrices has been divided into four submatrices, which we will call $A_1, \ldots, A_4$, $B_1, \ldots, B_4$, and $(AB)_1, \ldots, (AB)_4$ respectively. In this notation, our matrix multiplication becomes

$$\left[\begin{array}{c|c} A_1 & A_2 \\ \hline A_3 & A_4 \end{array}\right] \left[\begin{array}{c|c} B_1 & B_2 \\ \hline B_3 & B_4 \end{array}\right] = \left[\begin{array}{c|c} (AB)_1 & (AB)_2 \\ \hline (AB)_3 & (AB)_4 \end{array}\right].$$

Now our computation can be done as if we were multiplying two 2×2 matrices whose entries are our submatrices of A and B, i.e., $AB_1 = A_1 B_1 + A_2 B_3$, $AB_2 = A_1 B_2 + A_2 B_4$, $AB_3 = A_3 B_1 + A_4 B_3$, and $AB_4 = A_3 B_2 + A_4 B_4$.

We only want to compute $(AB)_2$, so we can ignore most of these submatrix products and just do $(AB)_2 = A_1 B_2 + A_2 B_4$. Going back to our matrix partitions, we get

$$A_1 = \begin{bmatrix} -1 & 0 \\ 3 & 1 \end{bmatrix}, A_2 = \begin{bmatrix} 2 & 5 \\ -3 & 2 \end{bmatrix}, B_2 = \begin{bmatrix} -2 & 4 & 1 \\ 2 & 0 & -1 \end{bmatrix}, B_4 = \begin{bmatrix} -2 & 4 & 3 \\ -1 & 1 & 2 \end{bmatrix}.$$

Plugging this into our formula above, we get that the 2×3 submatrix in the top right corner of AB is

$$\begin{bmatrix} -1 & 0 \\ 3 & 1 \end{bmatrix} \begin{bmatrix} -2 & 4 & 1 \\ 2 & 0 & -1 \end{bmatrix} + \begin{bmatrix} 2 & 5 \\ -3 & 2 \end{bmatrix} \begin{bmatrix} -2 & 4 & 3 \\ -1 & 1 & 2 \end{bmatrix}$$

$$= \begin{bmatrix} 2 & -4 & -1 \\ -4 & 12 & 2 \end{bmatrix} + \begin{bmatrix} -9 & 13 & 16 \\ 4 & -10 & -5 \end{bmatrix} = \begin{bmatrix} -7 & 9 & 15 \\ 0 & 2 & -3 \end{bmatrix}.$$

Clearly this represents significant computational savings over having to compute the entire matrix product AB. Even if you are interested in computing all of AB, you may still be able to get some work savings out of this method if either A or B has a block of entries which are all 0. If you can choose your partitions so that a submatrix of A or B is the zero matrix, then you've removed the need to compute any products with that submatrix.

Now that we've discussed how to break up matrix operations, let's turn our attention to solving the matrix equation $A\vec{x} = \vec{b}$ when A is terribly large. The main idea here will be to create a factorization of A as the product of two special matrices. For the discussion below we will assume that A is square, i.e., $n \times n$. It is possible to adapt this method to other cases, but we will opt for the simpler explanation and notation here. We start by noticing that there are two cases where solving $A\vec{x} = \vec{b}$ is especially easy.

Definition. A matrix A is *upper triangular* if it has $a_{ij} = 0$ for $i > j$. A matrix A is *lower triangular* if it has $a_{ij} = 0$ for $i < j$.

Note that the entries $a_{ij} = 0$ with $i > j$ are the entries below the diagonal of A while the entries $a_{ij} = 0$ with $i < j$ are the entries above the diagonal of A. I remember this by recognizing a triangular matrix as one which has a triangle of zeros either above or below its diagonal. The other entries can be freely chosen, and if these free entries are in the upper part of the matrix it is upper triangular. If the free entries are in the lower part of the matrix it is lower triangular. However you choose to remember this, it is important to note that the "upper" and "lower" in the definition above refer to the arbitrary entries rather than the entries which must be zero.

Example 8. Is $\begin{bmatrix} -4 & 7 & 2 \\ 0 & 0 & -5 \\ 0 & 0 & 1 \end{bmatrix}$ upper triangular or lower triangular?

This matrix has nonzero entries only on or above the diagonal, so it is upper triangular.

Example 9. Write down a 3×3 lower triangular matrix.

Here we want to create a 3×3 matrix whose nonzero entries are all on or below its diagonal. One possible way to do this is

$$\begin{bmatrix} 2 & 0 & 0 \\ -5 & 4 & 0 \\ 0 & 6 & 9 \end{bmatrix}.$$

Our factorization for A will be $A = LU$ where L is a lower triangular matrix and U is an upper triangular matrix. The reason for doing this is that solving a matrix equation is much easier when the matrix is either upper or lower triangular. Thus we'll be able to replace solving the general matrix equation $A\vec{x} = \vec{b}$ with a two-stage process where we first solve a matrix equation involving L and then solve a matrix equation involving U. Before we get into the details, let's explore how triangular matrices help us solve matrix equations more quickly.

If U is upper triangular, then we can solve $U\vec{x} = \vec{b}$ one variable at a time by starting at the bottom and working our way up. The bottom row of our augmented coefficient matrix corresponds to an equation where all variables have coefficient 0 except x_n, so we can easily solve for x_n. The next row up will have 0 coefficients on all by the last two variables. Since we already know the value of x_n, we can plug that in and easily solve for x_{n-1}. Repeating this process up the rows from the bottom, we soon find all of $\vec{x}$.

Example 10. Solve $\begin{bmatrix} -4 & 7 & 2 \\ 0 & -5 & 3 \\ 0 & 0 & -1 \end{bmatrix} \vec{x} = \begin{bmatrix} 4 \\ 2 \\ 6 \end{bmatrix}.$

The augmented coefficient matrix of this equation is

$$\begin{bmatrix} -4 & 7 & 2 & | & 4 \\ 0 & -5 & 3 & | & 2 \\ 0 & 0 & -1 & | & 6 \end{bmatrix}.$$

The bottom row gives us the equation $-x_3 = 6$, which we can quickly solve to get $x_3 = -6$. The next row up gives us the equation $-5x_2 + 3x_3 = 2$. Plugging in $x_3 = -6$, we get $-5x_2 - 18 = 2$ so $x_2 = -4$. The top row gives us the equation $-4x_1 + 7x_2 + 2x_3 = 4$. Plugging in $x_2 = -4$ and $x_3 = -6$, we get $-4x_1 - 28 - 12 = 4$ so $x_1 = -11$.

Solving $L\vec{x} = \vec{b}$ where L is lower triangular is very similar except that we start with the top row and work our way down. Here the top row of our augmented coefficient matrix corresponds to an equation where all but the first variable have 0 coefficients. Thus we can easily solve for x_1. The second row gives an equation with 0 coefficients except on the first two variables, so we can plug in the value of x_1 and solve for x_2. Repeating this process as we move down the rows of our matrix, we get $\vec{x}$.

Example 11. Solve $\begin{bmatrix} 2 & 0 & 0 \\ -5 & 4 & 0 \\ 1 & 6 & 9 \end{bmatrix} \vec{x} = \begin{bmatrix} 8 \\ -12 \\ 7 \end{bmatrix}$.

The augmented coefficient matrix of this equation is

$$\left[\begin{array}{ccc|c} 2 & 0 & 0 & 8 \\ -5 & 4 & 0 & -12 \\ 1 & 6 & 9 & 7 \end{array} \right].$$

The top row gives the equation $2x_1 = 8$, so $x_1 = 4$. The second row gives the equation $-5x_1 + 4x_2 = -12$. Plugging in $x_1 = 4$ gives $-20 + 4x_2 = -12$, so $x_2 = 2$. The bottom row gives the equation $x_1 + 6x_2 + 9x_3 = 7$. Plugging in $x_1 = 4$ and $x_2 = 2$ gives $4 + 12 + 9x_3 = 7$, so $x_3 = -1$.

In either case, it was much easier to solve a matrix equation where our matrix was triangular than it would have been for a general matrix.

Now suppose we have factored A into the product of a lower triangular matrix L and an upper triangular matrix U, so $A = LU$. Instead of solving $A\vec{x} = \vec{b}$ directly, we can first solve $L\vec{y} = \vec{b}$ and then solve $U\vec{x} = \vec{y}$. Then

$$A\vec{x} = (LU)\vec{x} = L(U\vec{x}) = L\vec{y} = \vec{b},$$

so $\vec{x}$ is our solution to $A\vec{x} = \vec{b}$.

Example 12. Use the fact that $A = \begin{bmatrix} 2 & 4 & 6 \\ -1 & -14 & 1 \\ -2 & -13 & 2 \end{bmatrix}$ has LU-factorization

$A = \begin{bmatrix} 2 & 0 & 0 \\ -1 & 4 & 0 \\ -2 & 3 & 1 \end{bmatrix} \begin{bmatrix} 1 & 2 & 3 \\ 0 & -3 & 1 \\ 0 & 0 & 5 \end{bmatrix}$ to solve $A\vec{x} = \begin{bmatrix} -2 \\ 13 \\ -4 \end{bmatrix}$.

We'll use the strategy above and start by solving the equation $L\vec{y} = \vec{b}$ which in our case is

$$\begin{bmatrix} 2 & 0 & 0 \\ -1 & 4 & 0 \\ -2 & 3 & 1 \end{bmatrix} \vec{y} = \begin{bmatrix} -2 \\ 13 \\ -4 \end{bmatrix}.$$

As in the previous example, we can start from the top and work our way down the rows. The top row says $2y_1 = -2$, so $y_1 = -1$. The second row tells us $-y_1 + 4y_2 = 13$. Plugging in $y_1 = -1$ gives $1 + 4y_2 = 13$, so $y_2 = 3$. The last row says $-2y_1 + 3y_2 + y_3 = -4$. Plugging in $y_1 = -1$ and $y_2 = 3$ gives $2 + 9 + y_3 = -4$, so $y_3 = -15$. This means $\vec{y} = \begin{bmatrix} -1 \\ 3 \\ -15 \end{bmatrix}$.

Next we need to solve the equation $U\vec{x} = \vec{y}$ which in our case is

$$\begin{bmatrix} 1 & 2 & 3 \\ 0 & -3 & 1 \\ 0 & 0 & 5 \end{bmatrix} \vec{x} = \begin{bmatrix} -1 \\ 3 \\ -15 \end{bmatrix}.$$

This is very similar to the process for $L\vec{y} = \vec{b}$ except that we'll start at the bottom row and work up. The bottom row says $5x_3 = -15$, so $x_3 = -3$. The second row says $-3x_2 + x_3 = 3$. Plugging in $x_3 = -3$ gives $-3x_2 - 3 = 3$, so $x_2 = -2$. The top row says $x_1 + 2x_2 + 3x_3 = -1$. Plugging in $x_2 = -2$ and $x_3 = -3$ gives us $x_1 - 4 - 9 = -1$, so $x_1 = 12$.

Thus the solution to our original matrix equation $A\vec{x} = \vec{b}$ is $\vec{x} = \begin{bmatrix} 12 \\ -2 \\ -3 \end{bmatrix}$.

Now that we understand why factoring A into the product of an upper and lower triangular matrix is so useful, let's discuss how to compute L and U. If you think about it, you'll realize that Part 1 of our row reduction algorithm actually produces an upper triangular matrix, because it starts from the upper left and creates zeros below each of our leading 1s. This means that we can find our upper triangular matrix U simply by running the first half of row reduction on A. The trickier part will therefore be finding L so that $A = LU$. To do this, we'll need a new way to view row operations as multiplication by a special type of matrix.

Definition. The *elementary matrix* of a row operation is the matrix we get by doing that row operation to I_n.

Example 13. Let $n = 3$. Find the elementary matrices of the row operations $r_3 - 7r_1$, $r_2 \leftrightarrow r_3$, and $\frac{1}{2}r_1$.

Since $n = 3$, we'll find the elementary matrices of these row operations by doing each one to I_3.

To find the elementary matrix of $r_3 - 7r_1$ we do

$$\begin{bmatrix} 1 & 0 & 0 \\ 0 & 1 & 0 \\ 0 & 0 & 1 \end{bmatrix} \to_{r_3 - 7r_1} \begin{bmatrix} 1 & 0 & 0 \\ 0 & 1 & 0 \\ -7 & 0 & 1 \end{bmatrix}$$

so the elementary matrix of $r_3 - 7r_1$ is $\begin{bmatrix} 1 & 0 & 0 \\ 0 & 1 & 0 \\ -7 & 0 & 1 \end{bmatrix}$.

To find the elementary matrix of $r_2 \leftrightarrow r_3$ we do

$$\begin{bmatrix} 1 & 0 & 0 \\ 0 & 1 & 0 \\ 0 & 0 & 1 \end{bmatrix} \rightarrow_{r_2 \leftrightarrow r_3} \begin{bmatrix} 1 & 0 & 0 \\ 0 & 0 & 1 \\ 0 & 1 & 0 \end{bmatrix}$$

so the elementary matrix of $r_2 \leftrightarrow r_3$ is $\begin{bmatrix} 1 & 0 & 0 \\ 0 & 0 & 1 \\ 0 & 1 & 0 \end{bmatrix}$.

To find the elementary matrix of $\frac{1}{2}r_1$ we do

$$\begin{bmatrix} 1 & 0 & 0 \\ 0 & 1 & 0 \\ 0 & 0 & 1 \end{bmatrix} \rightarrow_{\frac{1}{2}r_1} \begin{bmatrix} \frac{1}{2} & 0 & 0 \\ 0 & 1 & 0 \\ 0 & 0 & 1 \end{bmatrix}$$

so the elementary matrix of $\frac{1}{2}r_1$ is $\begin{bmatrix} \frac{1}{2} & 0 & 0 \\ 0 & 1 & 0 \\ 0 & 0 & 1 \end{bmatrix}$.

These elementary matrices link our row operations to matrix multiplication in the following way. Suppose E is the elementary matrix of some row operation and A is an $n \times n$ matrix. Then EA is the matrix we get by doing E's row operation to A. In other words, multiplication on the left by the elementary matrix of a row operation does that row operation to the matrix being multiplied.

Part 1 of our row reduction algorithm is just a series of row operations, so from this perspective we can write our upper triangular matrix U as

$$U = E_k E_{k-1} \cdots E_2 E_1 A$$

where E_i is the elementary matrix of the ith row operation we used. Note that since matrix multiplication is not commutative it is important to place E_1 closest to A since this corresponds to doing its row operation first, then E_2, etc. However, we want to write $A = LU$, so we need some way to move the E_i's to the other side of the equation and in particular to the left of U.

Luckily each row operation has another row operation which reverses it. If we added a multiple of one row to another row, we can subtract that same multiple. If we multiplied a row by a constant, we can divide by that constant. If we swapped two rows, we can swap those two rows back. Thus each E_i has another elementary matrix which undoes it. If we call this other matrix F_i, then what we're saying is that multiplication on the left by E_i and then F_i is the same as doing nothing, i.e., $F_i E_i = I_n$. If we find the $F_1, \ldots, F_k$ which undo $E_1, \ldots, E_k$, then we can start moving the E_i matrices to the other side of the equation $U = E_k E_{k-1} \cdots E_2 E_1 A$ to get something of the form $LU = A$. We start by multiplying both sides on the left by F_k to cancel off E_k. This gives us

$$F_k U = F_k E_k E_{k-1} \cdots E_2 E_1 A = I_n E_{k-1} \cdots E_2 E_1 A = E_{k-1} \cdots E_2 E_1 A.$$

Next we multiply on the left by F_{k-1} to cancel off E_{k-1}, which gives

$$F_{k-1} F_k U = E_{k-2} \cdots E_2 E_1 A.$$

Repeating this process with F_{k-2} through F_1 gives us

$$F_1 F_2 \cdots F_{k-1} F_k U = A.$$

This is very close to what we want, but to finish our factorization we will need to argue that $F_1 F_2 \cdots F_{k-1} F_k$ is a lower triangular matrix which we can use as our L.

The first step in seeing $F_1 F_2 \cdots F_{k-1} F_k$ is lower triangular is to see that each of the F_i matrices is lower triangular. Remember that each E_i was the elementary matrix of a row operation used in the first half of our row reduction algorithm to either create a leading 1 along the diagonal of A or use a leading 1 to create a zero below that leading 1. Thus the row operations we use only affect entries on or below the diagonal of A. To create the elementary matrix of a row operation, we perform our row operation on I_n. Since I_n is lower triangular and we're doing a row operation that changes only entries below the diagonal, the upper section of each of our E_i's will still be all 0s. Thus each E_i is lower triangular. Since each F_i is the reverse of the row operation of E_i, it also only affects entries below the diagonal. Therefore each F_i is also lower triangular. You will explore in the exercises that the product of lower triangular matrices is again lower triangular, so $F_1 F_2 \cdots F_{k-1} F_k$ is a lower triangular matrix, which we can call L. Therefore we can use careful row reduction and reversal of row operations to find L and U so that $A = LU$.

Example 14. Find an LU-factorization of $A = \begin{bmatrix} -4 & 12 & -20 \\ 0 & 2 & 4 \\ -1 & 6 & 0 \end{bmatrix}$.

The first step of this process is to use Part 1 of our row reduction algorithm to find our upper triangular matrix U. While we do this, we'll need to keep careful track of our row operations (in order) so we can use them to create our lower triangular matrix L. I'll use our usual notation to write down the row operation at each step.

$$A \to_{-\frac{1}{4} r_1} \begin{bmatrix} 1 & -3 & 5 \\ 0 & 2 & 4 \\ -1 & 6 & 0 \end{bmatrix} \to_{r_3 + r_1} \begin{bmatrix} 1 & -3 & 5 \\ 0 & 2 & 4 \\ 0 & 3 & 5 \end{bmatrix} \to_{\frac{1}{2} r_2} \begin{bmatrix} 1 & -3 & 5 \\ 0 & 1 & 2 \\ 0 & 3 & 5 \end{bmatrix}$$

$$\to_{r_3 - 3r_2} \begin{bmatrix} 1 & -3 & 5 \\ 0 & 1 & 2 \\ 0 & 0 & -1 \end{bmatrix} \to_{-r_3} \begin{bmatrix} 1 & -3 & 5 \\ 0 & 1 & 2 \\ 0 & 0 & 1 \end{bmatrix}.$$

This last matrix is our upper triangular matrix

$$U = \begin{bmatrix} 1 & -3 & 5 \\ 0 & 1 & 2 \\ 0 & 0 & 1 \end{bmatrix}.$$

Now we can use our row operations to create L. Remember that

$$L = F_1 F_2 \cdots F_{k-1} F_k$$

where F_i is the elementary matrix of the row operation that reverses the ith row operation used above.

Our first row operation was $-\frac{1}{4}r_1$, which is reversed by the row operation $-4r_1$. Doing $-4r_1$ to I_3 gives us the elementary matrix of $-4r_1$, so

$$F_1 = \begin{bmatrix} -4 & 0 & 0 \\ 0 & 1 & 0 \\ 0 & 0 & 1 \end{bmatrix}.$$

Our second row operation was $r_3 + r_1$, which is reversed by $r_3 - r_1$. This means

$$F_2 = \begin{bmatrix} 1 & 0 & 0 \\ 0 & 1 & 0 \\ -1 & 0 & 1 \end{bmatrix}.$$

Our third row operation was $\frac{1}{2}r_2$, which is reversed by $2r_2$. Thus

$$F_3 = \begin{bmatrix} 1 & 0 & 0 \\ 0 & 2 & 0 \\ 0 & 0 & 1 \end{bmatrix}.$$

Our fourth row operation was $r_3 - 3r_2$ which is reversed by $r_3 + 3r_2$, so

$$F_4 = \begin{bmatrix} 1 & 0 & 0 \\ 0 & 1 & 0 \\ 0 & 3 & 1 \end{bmatrix}.$$

Our fifth and last row operation was $-r_3$ which is reversed by $-r_3$, so

$$F_5 = \begin{bmatrix} 1 & 0 & 0 \\ 0 & 1 & 0 \\ 0 & 0 & -1 \end{bmatrix}.$$

This means

$$L = F_1 \cdots F_5 = \begin{bmatrix} -4 & 0 & 0 \\ 0 & 2 & 0 \\ -1 & 3 & -1 \end{bmatrix}.$$

Therefore our LU-factorization is

$$A = \begin{bmatrix} -4 & 0 & 0 \\ 0 & 2 & 0 \\ -1 & 3 & -1 \end{bmatrix} \begin{bmatrix} 1 & -3 & 5 \\ 0 & 1 & 2 \\ 0 & 0 & 1 \end{bmatrix}.$$

Notice that to use our LU-factorization method to solve $A\vec{x} = \vec{b}$, we need to find L and U which takes a bit of work. This means LU-factorization makes the most sense if we want to solve $A\vec{x} = \vec{b}$ for several different values of $\vec{b}$. In that case, we can find $A = LU$ once and then use our LU-factorization repeatedly to solve each matrix equation quickly. However, if we want to solve $A\vec{x} = \vec{b}$ for a single value of $\vec{b}$, it may be quicker to simply row reduce the augmented coefficient matrix.

This certainly isn't an exhaustive list of methods for dealing with very large matrices, and the matrices used in the examples and exercises of this book are small enough that we won't need these techniques again. However, given the increasing importance of large data sets stored in correspondingly large matrices, I hope you will keep these techniques in mind in case you need them in future applications.

Exercises 2.9.

1. Let A be a 3×3 matrix. Find the partition which isolates the entries a_{21} and a_{31} or say that no such partition exists.

2. Let A be a 3×4 matrix. Find the partition which isolates the entries a_{31}, a_{32}, and a_{34} or say that no such partition exists.

3. Let A be a 4×5 matrix. Find the partition which isolates the entries a_{22}, a_{23}, a_{32}, and a_{33} or say that no such partition exists.

4. Let A be a 5×6 matrix. Find the partition which isolates the entries a_{25}, a_{26}, a_{35}, a_{36}, a_{45}, and a_{46} or say that no such partition exists.

5. Let $A = \begin{bmatrix} -4 & 1 & 0 & 9 \\ 3 & -2 & 5 & 1 \\ 6 & 3 & -2 & 7 \end{bmatrix}$. Use a partition to compute the submatrix of $5A$ consisting of the bottom row and middle two columns.

6. Let $A = \begin{bmatrix} 5 & -4 & 2 \\ -1 & 8 & 0 \\ 2 & 3 & 4 \\ -1 & 0 & 6 \end{bmatrix}$. Use a partition to compute the submatrix of $-6A$ consisting of the middle two rows and left column.

7. Let $A = \begin{bmatrix} 12 & -7 & 4 \\ 3 & -1 & 4 \\ 9 & 2 & -3 \end{bmatrix}$. Use a partition to compute the submatrix of $4A$ consisting of the bottom two rows and right two columns.

8. Let $A = \begin{bmatrix} 7 & 1 & -5 & 3 \\ -1 & 0 & 3 & 6 \\ 5 & 2 & 3 & -1 \\ 9 & 3 & 0 & 1 \end{bmatrix}$. Use a partition to compute the submatrix of $-2A$ consisting of the middle two rows and left two columns.

9. Let $A = \begin{bmatrix} 1 & -1 & 3 \\ 5 & -2 & 9 \\ 4 & 1 & 2 \end{bmatrix}$ and $B = \begin{bmatrix} 12 & -7 & 4 \\ 3 & -1 & 4 \\ 9 & 2 & -3 \end{bmatrix}$. Use a partition to compute the submatrix of $A + B$ consisting of the middle row and right two columns.

10. Let $A = \begin{bmatrix} 1 & 6 & -7 \\ -3 & 4 & -10 \\ 2 & -8 & 1 \end{bmatrix}$ and $B = \begin{bmatrix} 6 & -2 & 8 \\ 0 & 1 & -1 \\ 3 & 5 & 2 \end{bmatrix}$. Use a partition to compute the submatrix of $A + B$ consisting of the top two rows and middle column.

11. Let $A = \begin{bmatrix} 4 & -2 & 4 \\ 7 & 1 & -1 \\ 1 & 1 & 5 \end{bmatrix}$ and $B = \begin{bmatrix} 3 & -9 & 2 \\ 0 & -2 & 4 \\ -5 & 1 & 3 \end{bmatrix}$. Use a partition to compute the submatrix of $A + B$ consisting of the bottom two rows and left two columns.

12. Let $A = \begin{bmatrix} 1 & 0 & -1 \\ 0 & 3 & -1 \\ -3 & 1 & 0 \\ 1 & 1 & -1 \end{bmatrix}$ and $B = \begin{bmatrix} 6 & 1 & 2 \\ 1 & 3 & 4 \\ 0 & 2 & 1 \\ 5 & 2 & 3 \end{bmatrix}$. Use a partition to compute the submatrix of $A + B$ consisting of the middle two rows and right two columns.

13. Let A be a 5×3 matrix and B be a 3×4 matrix. Find partitions of A and B that you could use to compute the submatrix of AB consisting of the top two rows and right column.

14. Let A be a 3×2 matrix and B be a 2×4 matrix. Find partitions of A and B that you could use to compute the submatrix of AB consisting of the bottom row and left two columns.

15. Let A be a 3×3 matrix and B be a 3×4 matrix. Find partitions of A and B that you could use to compute the submatrix of AB consisting of the middle row and middle two columns.

16. Let A be a 5×2 matrix and B be a 2×3 matrix. Find partitions of A and B that you could use to compute the submatrix of AB consisting of the top three rows and left two columns.

17. Let $A = \begin{bmatrix} 1 & -1 & 3 \\ 2 & 0 & -2 \\ 4 & 1 & 1 \\ -1 & 2 & 1 \end{bmatrix}$ and $B = \begin{bmatrix} 3 & 1 \\ -1 & 2 \\ 4 & -2 \end{bmatrix}$. Use a partition to compute the submatrix of AB consisting of the middle two rows and right column.

18. Let $A = \begin{bmatrix} 6 & 1 & 2 \\ -1 & 3 & 4 \\ 0 & 2 & 1 \\ 5 & 2 & 3 \end{bmatrix}$ and $B = \begin{bmatrix} -4 & 1 & 0 & 9 \\ 3 & -2 & 5 & 1 \\ 6 & 3 & -2 & 7 \end{bmatrix}$. Use a partition

to compute the submatrix of AB consisting of the bottom two rows and right two columns.

19. Let $A = \begin{bmatrix} -4 & 1 & 0 & 9 \\ 3 & -2 & 5 & 1 \\ 6 & 3 & -2 & 7 \end{bmatrix}$ and $B = \begin{bmatrix} 6 & 1 & 2 \\ -1 & 3 & 4 \\ 0 & 2 & 1 \\ 5 & 2 & 3 \end{bmatrix}$. Use a partition

to compute the submatrix of AB consisting of the top two rows and left two columns.

20. Let $A = \begin{bmatrix} 2 & -1 & 3 & 0 \\ 1 & 1 & -1 & 3 \\ 4 & 2 & 0 & -1 \\ 5 & 2 & 1 & -2 \end{bmatrix}$ and $B = \begin{bmatrix} 4 & 2 & -1 & 9 \\ 3 & -1 & 2 & 1 \\ 0 & 1 & 1 & 5 \\ -2 & 1 & -1 & 6 \end{bmatrix}$. Use a

partition to compute the submatrix of AB consisting of the middle two rows and middle two columns.

21. Use the factorization $\begin{bmatrix} -2 & 0 & 10 \\ 1 & 8 & -1 \\ -3 & 2 & 14 \end{bmatrix} = \begin{bmatrix} 2 & 0 & 0 \\ -1 & 4 & 0 \\ 3 & 1 & 1 \end{bmatrix} \begin{bmatrix} -1 & 0 & 5 \\ 0 & 2 & 1 \\ 0 & 0 & -2 \end{bmatrix}$

to solve $\begin{bmatrix} -2 & 0 & 10 \\ 1 & 8 & -1 \\ -3 & 2 & 14 \end{bmatrix} \vec{x} = \begin{bmatrix} 12 \\ -6 \\ 14 \end{bmatrix}$.

22. Use the factorization $\begin{bmatrix} -6 & -3 & 12 \\ 10 & 4 & -23 \\ 0 & 1 & 9 \end{bmatrix} = \begin{bmatrix} -3 & 0 & 0 \\ 5 & -1 & 0 \\ 0 & 1 & 3 \end{bmatrix} \begin{bmatrix} 2 & 1 & -4 \\ 0 & 1 & 3 \\ 0 & 0 & 2 \end{bmatrix}$

to solve $\begin{bmatrix} -6 & -3 & 12 \\ 10 & 4 & -23 \\ 0 & 1 & 9 \end{bmatrix} \vec{x} = \begin{bmatrix} 33 \\ -66 \\ 35 \end{bmatrix}$.

23. Use the factorization $\begin{bmatrix} 2 & -4 & -1 \\ -4 & 8 & 5 \\ 6 & -12 & 2 \end{bmatrix} = \begin{bmatrix} -1 & 0 & 0 \\ 2 & 1 & 0 \\ -3 & 2 & -1 \end{bmatrix} \begin{bmatrix} -2 & 4 & 1 \\ 0 & 0 & 3 \\ 0 & 0 & 1 \end{bmatrix}$

to solve $\begin{bmatrix} 2 & -4 & -1 \\ -4 & 8 & 5 \\ 6 & -12 & 2 \end{bmatrix} \vec{x} = \begin{bmatrix} 11 \\ -25 \\ 28 \end{bmatrix}$.

24. Use the factorization $\begin{bmatrix} 12 & 4 & -8 \\ 0 & 3 & -1 \\ -6 & -5 & 3 \end{bmatrix} = \begin{bmatrix} 4 & 0 & 0 \\ 0 & -1 & 0 \\ -2 & 1 & 2 \end{bmatrix} \begin{bmatrix} 3 & 1 & -2 \\ 0 & -3 & 1 \\ 0 & 0 & -1 \end{bmatrix}$

to solve $\begin{bmatrix} 12 & 4 & -8 \\ 0 & 3 & -1 \\ -6 & -5 & 3 \end{bmatrix} \vec{x} = \begin{bmatrix} -8 \\ 13 \\ -13 \end{bmatrix}$.

25. Show that multiplying on the left by the elementary matrix $\begin{bmatrix} 1 & 0 \\ 0 & -2 \end{bmatrix}$ multiplies the second row by -2.

26. Show that multiplying on the left by the elementary matrix $\begin{bmatrix} 0 & 1 \\ 1 & 0 \end{bmatrix}$ swaps the first and second rows.

27. Find the 2×2 elementary matrix which adds 6 times the first row to the second row.

28. Find the 2×2 elementary matrix which multiplies the first row by 7.

29. Find the 3×3 elementary matrix which swaps the first and third rows.

30. Find the 3×3 elementary matrix which adds -3 times the third row to the second row.

31. Find the elementary matrix which undoes $\begin{bmatrix} 4 & 0 \\ 0 & 1 \end{bmatrix}$.

32. Find the elementary matrix which undoes $\begin{bmatrix} 1 & -5 \\ 0 & 1 \end{bmatrix}$.

33. Find the elementary matrix which undoes $\begin{bmatrix} 1 & 0 & 0 \\ 0 & 0 & 1 \\ 0 & 1 & 0 \end{bmatrix}$.

34. Find the elementary matrix which undoes $\begin{bmatrix} 1 & 0 & 0 \\ 0 & 1 & 9 \\ 0 & 0 & 1 \end{bmatrix}$.

35. Find an LU-factorization of $A = \begin{bmatrix} -1 & 1 & 3 \\ -5 & 8 & 39 \\ 1 & 1 & 3 \end{bmatrix}$.

36. Find an LU-factorization of $A = \begin{bmatrix} 2 & -8 & 6 \\ 3 & -15 & 6 \\ 1 & -5 & 3 \end{bmatrix}$.

37. Find an LU-factorization of $A = \begin{bmatrix} 1 & -2 & 4 \\ 0 & 2 & -2 \\ 1 & 1 & 5 \end{bmatrix}$.

38. Find an LU-factorization of $A = \begin{bmatrix} 1 & 6 & -7 \\ 1 & 4 & -13 \\ -2 & -8 & 21 \end{bmatrix}$.

39. (a) Show that the product of two 3×3 lower triangular matrices is again lower triangular.
    (b) Use your result from (a) to explain why any finite product of 3×3 lower triangular matrices is lower triangular.

40. Repeat the previous exercise with $n \times n$ lower triangular matrices.

2.10   Invertibility

So far, we've solved $f(\vec{x}) = \vec{b}$ for $\vec{x}$ using row reduction. However, from your earlier study of functions, you know that some functions can be quickly undone using another function called the inverse function. For example, $f(x) = 2x$ has $f^{-1}(x) = \frac{1}{2}x$. On the other hand, some functions don't have an inverse, like $f(x) = x^2$. In this section we'll develop a technique to simultaneously determine whether or not a given linear function from $\mathbb{R}^n$ to $\mathbb{R}^m$ has an inverse and to find its inverse function if it has one.

Intuitively, the reason $f(x) = 2x$ and $f^{-1}(x) = \frac{1}{2}x$ are inverses is that they undo each other. What we mean by that is that if we compose these two functions in either order we get the identity function. In other words, $f(f^{-1}(x)) = f\left(\frac{1}{2}x\right) = 2\left(\frac{1}{2}x\right) = x$ and $f^{-1}(f(x)) = f^{-1}(2x) = \frac{1}{2}(2x) = x$ The general versions of these two equations give us our formal definition of invertibility.

Definition. A function $f : \mathbb{R}^n \to \mathbb{R}^m$ is *invertible* if it has an inverse function $f^{-1} : \mathbb{R}^m \to \mathbb{R}^n$ so that $f^{-1}(f(\vec{x})) = \vec{x}$ for every $\vec{x}$ in $\mathbb{R}^n$ and $f(f^{-1}(\vec{b})) = \vec{b}$ for every $\vec{b}$ in $\mathbb{R}^m$.

This definition looks a little complicated, but the first condition is saying that f^{-1} undoes f while the second says that f undoes f^{-1}. This means we get a bonus fact: f^{-1} is also invertible and has inverse function f.

Theorem 1. A function f is invertible if and only if it is both 1-1 and onto.

This makes sense, because to be able to reverse a function we need everything in the domain and codomain to match up in unique pairs. If our function isn't onto then there are elements of the codomain which aren't matched up with anything from the domain, so f can't be invertible. If our function isn't 1-1, then multiple elements from the domain map to some single element of the codomain, so f also won't be invertible.

From our work in 2.5, we know that a linear function $f : \mathbb{R}^n \to \mathbb{R}^m$ can be 1-1 only if $n \leq m$ and can be onto only if $m \leq n$. Therefore it is only possible to have both 1-1 and onto when $m = n$.

For the rest of this section, we will assume that all functions have the same domain and codomain, i.e., $f : \mathbb{R}^n \to \mathbb{R}^n$ and all matrices are $n \times n$.

Since we are only working with linear functions from $\mathbb{R}^n$ to $\mathbb{R}^n$, we know f's matrix will be $n \times n$. Our inverse function f^{-1} will also have an $n \times n$ matrix, which we'll call A^{-1}. This means we can reformulate our definition of invertibility in terms of matrices as follows. (This reformulation uses the fact that the matrix of the identity map $f : \mathbb{R}^n \to \mathbb{R}^n$ by $f(\vec{x}) = \vec{x}$ is the identity matrix I_n.)

Definition. A matrix A is *invertible* if it has an inverse matrix A^{-1} so that $A^{-1}A = I_n$ and $AA^{-1} = I_n$.

Note that we need to check $A^{-1}A = I_n$ and $AA^{-1} = I_n$ separately, since we've seen that matrix multiplication doesn't always commute.

This definition allows us to rephrase our twin problems of deciding whether or not a linear function is invertible and finding its inverse function (if possible) as deciding whether or not a matrix is invertible and finding its inverse matrix. This translation may seem semantic, but it allows us to bring row reduction to bear on the problem which will turn out to be a big help.

Let's tackle these two problems one at a time, starting with how to tell whether or not a given $n \times n$ matrix A is invertible. We know A is invertible if its corresponding linear function is both 1-1 and onto. We can check both of these conditions simultaneously by row reducing A and looking at where the leading 1s lie in its reduced echelon form. As we saw in 2.5, if we have a leading 1 in every column, then A's function is 1-1. If we have a leading 1 in every row, then A's function is onto. Therefore to get both 1-1 and onto, which together imply invertibility, we need to see a leading 1 in every row and column of A's reduced echelon form. Since A is a square matrix, this means its reduced echelon form must have a stripe of leading 1s along the diagonal, i.e., A's reduced echelon form must be I_n.

Example 1. $A = \begin{bmatrix} 2 & 1 & -2 \\ 18 & 2 & 3 \\ 2 & -1 & 4 \end{bmatrix}$ is not invertible.

To see this we compute the reduced echelon form of A, which is

$$\begin{bmatrix} 1 & 0 & \frac{1}{2} \\ 0 & 1 & -3 \\ 0 & 0 & 0 \end{bmatrix}.$$

Since the reduced echelon form is not I_3, A is not invertible.

Example 2. $f : \mathbb{R}^3 \to \mathbb{R}^3$ by $f\left(\begin{bmatrix} x_1 \\ x_2 \\ x_3 \end{bmatrix}\right) = \begin{bmatrix} 4x_1 + x_2 - x_3 \\ x_2 + 3x_3 \\ -4x_1 - x_2 \end{bmatrix}$ is invertible.

To see this we first find the matrix of f, which is

$$A = \begin{bmatrix} 4 & 1 & -1 \\ 0 & 1 & 3 \\ -4 & -1 & 0 \end{bmatrix}.$$

(See 2.2 if you need a reminder of how to find the matrix of a linear function.) Next we compute the reduced echelon form of A, which is

$$I_3 = \begin{bmatrix} 1 & 0 & 0 \\ 0 & 1 & 0 \\ 0 & 0 & 1 \end{bmatrix}.$$

Therefore f is invertible.

If it turns out that f and A are not invertible then we are done, since we can't find f^{-1} and A^{-1}. However, if f and A are invertible, we'd like a method for finding f^{-1} and A^{-1}. To do this, we'll start by considering the equation $AA^{-1} = I_n$. We don't know how to solve an equation like this, so we'll have to reduce it to a set of simpler equations of the form $A\vec{x} = \vec{b}$. In 2.3, we saw that the jth column of a matrix product AB can be found by multiplying A by the jth column of B. In our equation $B = A^{-1}$, so we can find the n columns of A^{-1} by solving the n matrix equations

$$A\vec{x} = \vec{e}_1, A\vec{x} = \vec{e}_2, \dots, A\vec{x} = \vec{e}_n$$

where $\vec{e}_j$ is the jth column of I_n and so is all 0s except for a 1 as its jth entry. (We know that each equation will have one unique solution because A's reduced echelon form is I_n.)

We could solve these equations by row reducing the n augmented coefficient matrices

$$[A \,|\, \vec{e}_1], [A \,|\, \vec{e}_2], \dots, [A \,|\, \vec{e}_n]$$

however that's a lot of redundant work since we'd have to row reduce A repeatedly. Additionally, since A's reduced echelon form is I_n, we know that whatever $\vec{e}_j$ becomes in the reduced echelon form of $[A \,|\, \vec{e}_j]$ is the answer to $A\vec{x} = \vec{e}_j$. A more efficient way to find A^{-1} is to solve for all its columns at once by row reducing $[A \,|\, I_n]$. The left-hand side will become I_n, while each column of the right hand side becomes a column of A^{-1}. Thus the reduced echelon form of $[A \,|\, I_n]$ is $[I_n \,|\, A^{-1}]$.

The best part of this is that the process for finding A^{-1} and the process for deciding if A is invertible can be done at once! Simply row reduce $[A \,|\, I_n]$. If the left half turns into I_n, then A is invertible and A^{-1} is the right half of the reduced echelon form. If the left half doesn't turn into A^{-1}, then A isn't invertible.

Example 3. Determine whether or not $A = \begin{bmatrix} 3 & 7 & 0 \\ 2 & 5 & 0 \\ 0 & 1 & 1 \end{bmatrix}$ is invertible. If it is invertible, find A^{-1}.

To simultaneously decide whether or not A is invertible and compute A^{-1} if it exists, we row reduce

$$[A \mid I_3] = \begin{bmatrix} 3 & 7 & 0 & 1 & 0 & 0 \\ 2 & 5 & 0 & 0 & 1 & 0 \\ 0 & 1 & 1 & 0 & 0 & 1 \end{bmatrix}.$$

This gives us

$$\begin{bmatrix} 1 & 0 & 0 & 5 & -7 & 0 \\ 0 & 1 & 0 & -2 & 3 & 0 \\ 0 & 0 & 1 & 2 & -3 & 1 \end{bmatrix}.$$

Since the left half of our row reduced matrix is I_3, A is invertible. Its inverse is the right half of our row reduced matrix, so

$$A^{-1} = \begin{bmatrix} 5 & -7 & 0 \\ -2 & 3 & 0 \\ 2 & -3 & 1 \end{bmatrix}.$$

Example 4. Find the inverse function of $f \left(\begin{bmatrix} x_1 \\ x_2 \\ x_3 \end{bmatrix} \right) = \begin{bmatrix} 4x_1 + x_2 - x_3 \\ x_2 + 3x_3 \\ -4x_1 - x_2 \end{bmatrix}$.

This is our function from Example 2, so we already know it is invertible and that f corresponds to the matrix

$$A = \begin{bmatrix} 4 & 1 & -1 \\ 0 & 1 & 3 \\ -4 & -1 & 0 \end{bmatrix}.$$

To find f's inverse function, we'll use the fact that f^{-1} has matrix A^{-1}. We'll do that as in the last example by row reducing $[A \mid I_3]$. In our case this is

$$\begin{bmatrix} 4 & 1 & -1 & 1 & 0 & 0 \\ 0 & 1 & 3 & 0 & 1 & 0 \\ -4 & -1 & 0 & 0 & 0 & 1 \end{bmatrix}$$

which row reduces to

$$\begin{bmatrix} 1 & 0 & 0 & -\frac{3}{4} & -\frac{1}{4} & -1 \\ 0 & 1 & 0 & 3 & 1 & 3 \\ 0 & 0 & 1 & -1 & 0 & -1 \end{bmatrix}.$$

The right half of this matrix gives us

$$A^{-1} = \begin{bmatrix} -\frac{3}{4} & -\frac{1}{4} & -1 \\ 3 & 1 & 3 \\ -1 & 0 & -1 \end{bmatrix}.$$

This means f^{-1} is the function with $f^{-1}(\vec{x}) = A^{-1}\vec{x}$. If we want to write f^{-1} in the same way we originally wrote f, we can compute

$$A^{-1}\vec{x} = \begin{bmatrix} -\frac{3}{4} & -\frac{1}{4} & -1 \\ 3 & 1 & 3 \\ -1 & 0 & -1 \end{bmatrix} \begin{bmatrix} x_1 \\ x_2 \\ x_3 \end{bmatrix} = \begin{bmatrix} -\frac{3}{4}x_1 - \frac{1}{4}x_2 - x_3 \\ 3x_1 + x_2 + 3x_3 \\ -x_1 - x_3 \end{bmatrix}$$

to get

$$f^{-1}\left(\begin{bmatrix} x_1 \\ x_2 \\ x_3 \end{bmatrix}\right) = \begin{bmatrix} -\frac{3}{4}x_1 - \frac{1}{4}x_2 - x_3 \\ 3x_1 + x_2 + 3x_3 \\ -x_1 - x_3 \end{bmatrix}.$$

Note that there is still one gap here: we know that the matrix we're calling A^{-1} satisfies $AA^{-1} = I_n$, but we haven't shown that it satisfies $A^{-1}A = I_n$. Suppose that you have a matrix A which you know is invertible, so there is some A^{-1} with $AA^{-1} = I_n$ and $A^{-1}A = I_n$. We've found a matrix which for the moment I'll call B so that $AB = I_n$. Multiplying both sides of this equation on the left by A^{-1} gives us $A^{-1}(AB) = A^{-1}(I_n)$ which can be simplified to $B = A^{-1}$. Therefore the matrix we've been solving for by row reducing $[A \,|\, I_n]$ is actually A^{-1}.

Recall that in Chapter 0's Example 4 we talked about multiple linear regression as a search for an equation $y = \beta_0 + \beta_1 x_1 + \cdots \beta_p x_p$ which can predict the value of a variable y in terms of a set of variables $x_1, \ldots, x_p$. Now that we understand how to compute the inverse of a matrix, we can actually solve for the βs to find these equations.

These regression models are based on data sets where we know the values of all the variables, including y. This allows us to plug in the values of y and $x_1, \ldots, x_p$ from each data point one at a time to create a set of linear equations in $\beta_0, \beta_1, \ldots, \beta_p$. (If the constant term bothers you, pretend there is a variable x_0 whose value for every data point is 1.) If we have n data points, this gives us the equations

$$y_1 = \beta_0 + \beta_1 x_{11} + \cdots \beta_p x_{1p}$$
$$y_2 = \beta_0 + \beta_1 x_{21} + \cdots \beta_p x_{2p}$$
$$\vdots$$
$$y_n = \beta_0 + \beta_1 x_{n1} + \cdots \beta_p x_{np}$$

where x_{ij} is the value of the jth variable and y_i is the value of y for the ith data point. (Don't let the unusual notation here fool you, we know the values of the xs and are going to solve for the values of the βs.)

Usually there isn't a single value for each β that will solve this system of equations, so we introduce an error term ϵ_i onto the end of each equation. We can rewrite this as the matrix equation $\vec{y} = X\vec{\beta} + \vec{\epsilon}$ for the $n \times (p+1)$ matrix

$$X = \begin{bmatrix} 1 & x_{11} & \cdots & x_{1p} \\ 1 & x_{21} & \cdots & x_{2p} \\ \vdots & \vdots & & \vdots \\ 1 & x_{n1} & \cdots & x_{np} \end{bmatrix}.$$

Our goal is to find the vector $\vec{\beta}$ that minimizes the errors. I'll skip over the derivation of the solution formula which involves tools from statistics and calculus. However, we will need one new definition: the transpose of a matrix.

Definition. The *transpose* of an $m \times n$ matrix A is the $n \times m$ matrix A^T where $a_{ij}^T = a_{ji}$.

In other words, the transpose is the matrix formed by reflecting A's entries across the diagonal.

Example 5. Find the transpose of $A = \begin{bmatrix} 0 & 4 \\ -1 & 0 \\ 2 & 3 \end{bmatrix}$.

Since A is 3×2, its transpose, A^T, will be 2×3. Using A's columns as A^T's rows (or reflecting A's entries across the diagonal) gives us $A = \begin{bmatrix} 0 & -1 & -2 \\ 4 & 0 & -3 \end{bmatrix}$.

Using our new notation, the coefficients that minimize the errors in the regression equation $y = \beta_0 + \beta_1 x_1 + \cdots + \beta_p x_p$ are given by the formula

$$\vec{\beta} = (X^T X)^{-1} X^T \vec{y}.$$

You may worry about taking the inverse of $X^T X$ since X wasn't a square matrix, however, X is $n \times (p+1)$ so X^T is $(p+1) \times n$ which makes $X^T X$ as square $((p+1) \times n) \cdot (n \times (p+1)) = (p+1) \times (p+1)$ matrix.

Example 6. A chain of toy stores is trying to predict quarterly sales by city (in thousands of dollars) based on each city's population under 12 years old (in thousands) and the average amount of money people have each month to spend on nonessential items. They collect data for four of the cities where they have stores and find the data given in the table below. Use this data set to find a regression equation to predict sales based on population under 12 and average nonessential spending money.

City Code	Sales	Population Under 12	Nonessential Spending
1	$174.4	68.5	$167
2	$164.4	45.2	$168
3	$244.2	91.3	$182
4	$154.6	47.8	$163

We want to predict quarterly sales, so that is our variable y. We will base our prediction on population under 12, x_1, and average monthly nonessential spending money, x_2. This means our regression equation will have the form

$$y = \beta_0 + \beta_1 x_1 + \beta_2 x_2.$$

To solve for our βs, we need to set up our matrix solution formula. First, our solution vector will be $\vec{\beta} = \begin{bmatrix} \beta_0 \\ \beta_1 \\ \beta_2 \end{bmatrix}$. We have four data points, so $n = 4$.

Since y_i is the quarterly sales in the ith city, we have $\vec{y} = \begin{bmatrix} y_1 \\ y_2 \\ y_3 \\ y_4 \end{bmatrix} = \begin{bmatrix} 174.4 \\ 164.4 \\ 244.2 \\ 154.6 \end{bmatrix}$.

The x_{i1} entry of our matrix X is the population under 12 and the x_{i2} entry is the nonessential spending for the ith city. This means

$$X = \begin{bmatrix} 1 & 68.5 & 167 \\ 1 & 45.2 & 168 \\ 1 & 91.3 & 182 \\ 1 & 47.8 & 163 \end{bmatrix}.$$

We want to find $\vec{\beta} = \left(X^T X\right)^{-1} X^T \vec{y}$, and we can start by finding X's transpose.

$$X^T = \begin{bmatrix} 1 & 1 & 1 & 1 \\ 68.5 & 45.2 & 91.3 & 47.8 \\ 167 & 168 & 182 & 163 \end{bmatrix}.$$

This means

$$X^T X = \begin{bmatrix} 1 & 1 & 1 & 1 \\ 68.5 & 45.2 & 91.3 & 47.8 \\ 167 & 168 & 182 & 163 \end{bmatrix} \begin{bmatrix} 1 & 68.5 & 167 \\ 1 & 45.2 & 168 \\ 1 & 91.3 & 182 \\ 1 & 47.8 & 163 \end{bmatrix}$$

$$= \begin{bmatrix} 4 & 252.8 & 680 \\ 252.8 & 17,355.8 & 43,441.1 \\ 680 & 43,441.1 & 115,806 \end{bmatrix}.$$

To find $\left(X^T X\right)^{-1}$, we row reduce $[X^T X \mid I_3]$ which gives us (with some rounding)

$$\begin{bmatrix} 1 & 0 & 0 & 453.212 & 0.975 & -12.322 \\ 0 & 1 & 0 & 0.975 & 0.003 & -0.007 \\ 0 & 0 & 1 & -3.027 & -0.007 & 0.020 \end{bmatrix}.$$

This means $X^T X$ is invertible, and we have

$$\left(X^T X\right)^{-1} = \begin{bmatrix} 453.212 & 0.975 & -3.027 \\ 0.975 & 0.003 & -0.007 \\ -3.027 & -0.007 & 0.020 \end{bmatrix}.$$

Plugging these elements into our formula $\vec{\beta} = \left(X^T X\right)^{-1} X^T \vec{y}$ gives us

$$\begin{bmatrix} \beta_0 \\ \beta_1 \\ \beta_2 \end{bmatrix} = \begin{bmatrix} 453.212 & 0.975 & -3.027 \\ 0.975 & 0.003 & -0.007 \\ -3.027 & -0.007 & 0.020 \end{bmatrix} \begin{bmatrix} 1 & 1 & 1 & 1 \\ 68.5 & 45.2 & 91.3 & 47.8 \\ 167 & 168 & 182 & 163 \end{bmatrix} \begin{bmatrix} 174.4 \\ 164.4 \\ 244.2 \\ 154.6 \end{bmatrix}$$

$$= \begin{bmatrix} -445.48 \\ 0.60 \\ 3.48 \end{bmatrix}$$

so our regression equation is $y = -445.48 + 0.60x_1 + 3.48x_2$.

Exercises 2.10.

1. Let $A = \begin{bmatrix} 7 & 0 \\ -14 & 1 \end{bmatrix}$. Verify that $A^{-1} = \begin{bmatrix} \frac{1}{7} & 0 \\ 2 & 1 \end{bmatrix}$.

2. Let $A = \begin{bmatrix} 2 & 1 & -1 \\ 3 & 1 & 4 \\ -1 & 0 & -2 \end{bmatrix}$. Verify that $A^{-1} = \begin{bmatrix} \frac{2}{3} & -\frac{2}{3} & -\frac{5}{3} \\ \frac{2}{3} & \frac{5}{3} & \frac{11}{3} \\ -\frac{1}{3} & \frac{1}{3} & \frac{1}{3} \end{bmatrix}$.

3. Find the inverse of $A = \begin{bmatrix} 1 & -2 & 1 \\ -3 & 7 & -6 \\ 2 & -3 & 0 \end{bmatrix}$.

4. Find the inverse of $A = \begin{bmatrix} -3 & 4 & 0 \\ 1 & 2 & 1 \\ 2 & 0 & 1 \end{bmatrix}$.

5. Let $A = \begin{bmatrix} 12 & -4 \\ -6 & 2 \end{bmatrix}$. Find A^{-1} or show that A is not invertible.

6. Let $A = \begin{bmatrix} 5 & 1 \\ 4 & 1 \end{bmatrix}$. Find A^{-1} or show that A is not invertible.

7. Let $A = \begin{bmatrix} 2 & 0 & 4 \\ 0 & -1 & 0 \\ -2 & 6 & -3 \end{bmatrix}$. Find A^{-1} or show that A is not invertible.

8. Let $A = \begin{bmatrix} 1 & -1 & 2 \\ 0 & 1 & -2 \\ 2 & 0 & -1 \end{bmatrix}$. Find A^{-1} or show that A is not invertible.

9. Let $A = \begin{bmatrix} 0 & -3 & -6 \\ -1 & -2 & -1 \\ -2 & -3 & 0 \end{bmatrix}$. Find A^{-1} or show that A is not invertible.

10. Let $A = \begin{bmatrix} 1 & -4 & 2 \\ 0 & -1 & 3 \\ 0 & 1 & -2 \end{bmatrix}$. Find A^{-1} or show that A is not invertible.

11. Let $A = \begin{bmatrix} 1 & 2 & 1 \\ 1 & 5 & 2 \\ 2 & 0 & 1 \end{bmatrix}$. Find A^{-1} or show that A is not invertible.

12. Let $A = \begin{bmatrix} 1 & 0 & 6 \\ 2 & 2 & -9 \\ 1 & 1 & -3 \end{bmatrix}$. Find A^{-1} or show that A is not invertible.

13. Let $f\left(\begin{bmatrix} x_1 \\ x_2 \\ x_3 \end{bmatrix} \right) = \begin{bmatrix} -x_1 + 2x_2 + x_3 \\ 2x_2 + x_3 \\ x_1 + x_2 \end{bmatrix}$. Find f^{-1} or show that f is not invertible.

14. Let $f\left(\begin{bmatrix} x_1 \\ x_2 \\ x_3 \end{bmatrix} \right) = \begin{bmatrix} 5x_1 + x_2 + 3x_3 \\ 2x_1 - 3x_2 + 6x_3 \\ 3x_1 + 4x_2 - 3x_3 \end{bmatrix}$. Find f^{-1} or show that f is not invertible.

15. Let $f\left(\begin{bmatrix} x_1 \\ x_2 \\ x_3 \end{bmatrix} \right) = \begin{bmatrix} -x_1 + 2x_2 + x_3 \\ -2x_2 + 2x_3 \\ x_2 - x_3 \end{bmatrix}$. Find f^{-1} or show that f is not invertible.

16. Let $f\left(\begin{bmatrix} x_1 \\ x_2 \\ x_3 \end{bmatrix} \right) = \begin{bmatrix} x_1 + x_2 + 3x_3 \\ x_1 - 4x_3 \\ x_2 - x_3 \end{bmatrix}$. Find f^{-1} or show that f is not invertible.

17. Suppose A is an invertible $n \times n$ matrix. Explain why A^{-1} is invertible.

18. Why can't the function with matrix $A = \begin{bmatrix} 7 & -3 & 0 \\ 1 & -5 & 2 \end{bmatrix}$ be invertible?

19. In Example 2 from 2.7, we solved the matrix equation $(I_3 - A)\vec{x} = \vec{b}$ to figure out our production levels. Use matrix inverses to solve this equation. (You should still get the same answer!)

20. What does your answer to the previous problem say about which net production levels, $\vec{b}$, are possible to achieve for the three industries from Example 2 in 2.7?

21. Use the method from Example 6 and the data in the table below to find the regression equation which predicts the percentage of people in a given city who have heart disease based on the percentage of people in that city who bike to work and the percentage who smoke.

City Code	Heart Disease	Bikers	Smokers
1	17	25	37
2	12	40	30
3	19	15	40
4	18	20	35

22. Use the method from Example 6 and the data in the table below to find the regression equation which predicts a student's percentage on the final exam based on their percentages on the two midterms.

Student Code	Final	Midterm 1	Midterm 2
1	68	78	73
2	75	74	76
3	85	82	79
4	94	90	96
5	86	87	90
6	90	90	92
7	86	83	95
8	68	72	69
9	55	68	67
10	69	69	70

2.11   The Invertible Matrix Theorem

In this section, we'll do two things at once: link invertibility to a huge list of other conditions, and sneakily review the majority of the concepts we've learned about matrix equations and linear functions.

Note: All matrices discussed in this section are $n \times n$ matrices!

Theorem 1. Suppose A is an $n \times n$ matrix. Then the following conditions are equivalent:

1. A is invertible.
2. The reduced echelon form of A is I_n.
3. The reduced echelon form of A contains n leading 1s.
4. $rk(A) = n$.
5. The columns of A are linearly independent.
6. The equation $A\vec{x} = \vec{0}$ has one unique solution.
7. $Nul(A) = \{\vec{0}\}$.
8. The linear function $f(\vec{x}) = A\vec{x}$ is 1-1.
9. The columns of A span $\mathbb{R}^n$.
10. The equation $A\vec{x} = \vec{b}$ always has a solution.
11. The linear function $f(\vec{x}) = A\vec{x}$ is onto.
12. $Col(A) = \mathbb{R}^n$.
13. There is a matrix C with $AC = I_n$.
14. There is a matrix D with $DA = I_n$.

Two notes before we dive in:

One good way to think about "the following are equivalent" theorems is as mathematical "k for the price of 1" deals. If you know one condition on this list, then you get all the rest as well. Similarly, if you know one condition on this list fails, then all the others fail as well.

It might seem like we need to explain why each condition implies every other condition, but that would require $14 * 13 = 182$ different explanations. Instead, think of each explanation as an arrow, where $P \to Q$ means we've explained that if condition P holds then condition Q must also hold. (Note that the arrow and explanation modeled above are one-way.) Our goal is to provide just enough connections between our conditions so that we could start at any one of these 14 statements and travel along arrows to any of the other statements. This more efficient approach will allow us to get away with only 20 arrows, two of which are left as exercises.

The bird's-eye view of the 18 arrows we'll establish together is given in Figure 2.7.

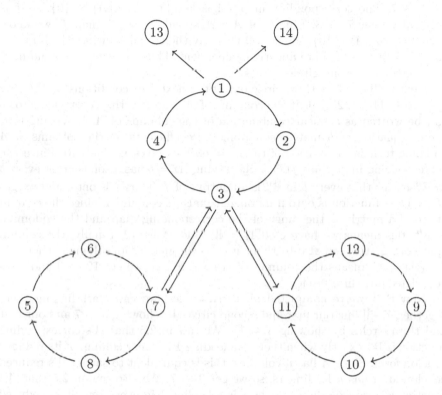

Figure 2.7: A road map of our proof

We'll start by establishing that $1 \to 2 \to 3 \to 4 \to 1$, so the first four conditions are linked in a circle. We just established in 2.10 that the reduced echelon form of an invertible matrix is I_n, so $1 \to 2$. Since it is clear that I_n has n leading 1s, we get $2 \to 3$. The rank of a matrix equals the number of leading 1s in its reduced echelon form, so it is clear that a matrix with n leading 1s has rank n which gives us $3 \to 4$. Finally, if a matrix has rank n, then its reduced echelon form has n leading 1s. Since leading 1s can't share a row or column, this means we must have our leading 1s down the diagonal forming I_n. We saw in 2.10 that if the reduced echelon form of a matrix is I_n, then A is invertible. Hence $4 \to 1$ is established, and we've completed our circle.

Our next stage is to create a second circle out of the next four conditions by showing $5 \to 6 \to 7 \to 8 \to 5$. If the columns of A are linearly independent, then we know from our second definition of linear independence that $A\vec{x} = \vec{0}$ has only the trivial solution. This gives us $5 \to 6$. The equation $A\vec{x} = \vec{0}$ always has the solution $\vec{x} = \vec{0}$, so if $A\vec{x} = \vec{0}$ has one unique solution, that solution is $\vec{x} = \vec{0}$. Since the null space of A is the solution set of $A\vec{x} = \vec{0}$, this gives us $6 \to 7$. The corresponding map f has $\ker(f) = Nul(A) = \{\vec{0}\}$, so f is 1-1 and we get $7 \to 8$. Finally, if A corresponds to a 1-1 map f, we know $\ker(f) = Nul(A) = \{\vec{0}\}$, so $A\vec{x} = \vec{0}$ has only the trivial solution which means that the columns of A are linearly independent. Thus we have $8 \to 5$ and have completed our second circle.

Next we'll create a third circle out of the next four conditions by showing $9 \to 10 \to 11 \to 12 \to 9$. If the columns of A span $\mathbb{R}^n$, then every n-vector $\vec{b}$ can be written as a linear combination of the columns of A. Rewriting that vector equation as a matrix equation, the coefficients on the columns of A combine to form a solution $\vec{x}$ to $A\vec{x} = \vec{b}$. This gives us $9 \to 10$. Since the corresponding map f has $f(\vec{x}) = A\vec{x}$, saying $A\vec{x} = \vec{b}$ has a solution for every $\vec{b}$ in $\mathbb{R}^n$ means that every $\vec{b}$ in $\mathbb{R}^n$ is an output of f. Thus f is onto, and we get $10 \to 11$. A function is onto if its range equals its codomain. Since the column space of a matrix is the range of the corresponding map and the codomain is $\mathbb{R}^n$, this means we have $Col(A) = \mathbb{R}^n$ and $11 \to 12$. Finally, the column space of a matrix can also be thought of as the span of its columns. Therefore $Col(A) = \mathbb{R}^n$ means the columns of A span $\mathbb{R}^n$, so we get $12 \to 9$ and have completed our third circle.

Now that we've completed all three circles, we can start linking them together. We'll link our first and second circles by showing $3 \leftrightarrows 7$ and our first and third circles by showing $3 \leftrightarrows 11$. We saw in 2.5 that A's corresponding function is 1-1 exactly when there is a leading 1 in every column of its reduced echelon form. Since A has n columns, this is equivalent to saying A's reduced echelon form has n leading 1s, so we get $3 \leftrightarrows 7$. We also saw in 2.5 that A's function is onto exactly when there is a leading 1 in every row of its reduced echelon form. Since A has n rows, this is equivalent to saying A's reduced echelon form has n leading 1s, so we get $3 \leftrightarrows 11$.

At this point, we've established the equivalence of the conditions $1 - 12$. To finish up our explanation of this theorem, we'll need to link in the last two statements. I'll provide two of those arrows here, and leave the other two as exercises. If our matrix is invertible, it has an inverse matrix A^{-1} with $AA^{-1} = I_n$ and $A^{-1}A = I_n$. This means we can let $A^{-1} = C$ to get $1 \to 13$ and $A^{-1} = D$ to get $1 \to 14$. Going back from 13 or 14 to 1 is hard. We did something similar sounding in 2.10, but it was under the assumption that A was invertible, which we can't assume if we're starting with 13 or 14. Instead I'd suggest showing $13 \to 10$, and $14 \to 6$. These are Exercises 8 and 9 respectively. With these last two conditions connected to the rest, we are done.

Now that we've established our theorem, let's see some examples of how it can be used.

Example 1. $A = \begin{bmatrix} 0 & 1 \\ 1 & 0 \end{bmatrix}$ is invertible.

The linear function f given by $f(\vec{x}) = A\vec{x}$ can be thought of as reflection about the line $y = x$. Geometrically, it isn't too hard to convince yourself that this function is onto. In fact, since doing a reflection twice brings you back where you started, we get $f(f(\vec{x})) = \vec{x}$. In other words, for every vector $\vec{x}$, we know $f(\vec{x})$ maps to $\vec{x}$, so f is onto.

This means A satisfies condition 11 of the Invertible Matrix Theorem, so A is invertible.

Example 2. $A = \begin{bmatrix} -3 & 1 & 0 \\ 4 & 2 & -7 \\ 1 & 0 & 9 \end{bmatrix}$ is invertible.

This matrix has reduced echelon form I_3, so it satisfies condition 2 of the Invertible Matrix Theorem and is therefore invertible.

Example 3. $A = \begin{bmatrix} 1 & -5 & 4 \\ 2 & -7 & 3 \\ -2 & 1 & 7 \end{bmatrix}$ is not invertible.

In Example 5 from 2.7, we saw that $A\vec{x} = \vec{b}$ didn't have a solution for $\vec{b} = \begin{bmatrix} -3 \\ -2 \\ -1 \end{bmatrix}$. This means $A\vec{x} = \vec{b}$ doesn't have a solution for every $\vec{b}$.

Thus A fails to satisfy condition 10 of the Invertible Matrix Theorem, and therefore isn't invertible.

Exercises 2.11.

1. Give a direct explanation of the arrow $1 \to 6$.

2. Give a direct explanation of the arrow $1 \to 10$.

3. Give a direct explanation of the arrow $11 \to 2$.

4. Give a direct explanation of the arrow $3 \to 9$.

5. Our explanation of the Invertible Matrix Theorem didn't include a direct explanation of why a matrix whose function is 1-1 has rank n, i.e., $8 \to 4$. Use the explanations/arrows in our proof of this theorem to explain how you can get from condition 8 to condition 4.

6. Use the Rank-Nullity Theorem from 2.8 to give a direct explanation of the arrows $7 \leftrightarrows 12$.

7. Let $A = \begin{bmatrix} -3 & 0 & 1 & 4 \\ 1 & 0 & -5 & 8 \\ 7 & 0 & -2 & 6 \\ 2 & 0 & 4 & -1 \end{bmatrix}$. Which of the Invertible Matrix

Theorem's thirteen conditions do you think most easily shows A is not invertible? Give an explanation of how your choice shows A^{-1} does not exist.

8. Let A be an $n \times n$ matrix. Show that if there is a matrix C with $AC = I_n$, then the equation $A\vec{x} = \vec{b}$ always has a solution. (This is the missing arrow $13 \to 10$.)

9. Let A be an $n \times n$ matrix. Show that if there is a matrix D with $DA = I_n$, then the equation $A\vec{x} = \vec{0}$ has one unique solution. (This is the missing arrow $14 \to 6$.)

10. Show that if A is not invertible then AB is not invertible

11. Show that if B is not invertible then AB is not invertible.

12. Suppose A is an invertible $n \times n$ matrix. Explain why A satisfies the Rank-Nullity Theorem from 2.8.

3

Vector Spaces

3.1 Basis and Coordinates

In 2.6, we developed row reduction to help us work with $\mathbb{R}^n$ and its linear functions. However, we showed in 2.3 that M_{mn} is not just the space of linear functions from $\mathbb{R}^n$ to $\mathbb{R}^m$, but also a vector space in its own right. This means we can ask the same types of questions in M_{mn} as in $\mathbb{R}^n$, but we don't have the same computational toolbox at our disposal. In this section, we'll develop a system for creating a map from any vector space V to $\mathbb{R}^n$ in order to transfer our questions about V over to $\mathbb{R}^n$. This will allow us to answer those questions in $\mathbb{R}^n$ using row reduction if that seems easier. We'll start in the only concrete case we have where $V = M_{mn}$.

Our first step in linking M_{mn} and $\mathbb{R}^k$ is to figure out a way to match each matrix from M_{mn} up with a vector in $\mathbb{R}^k$. (I'm using $\mathbb{R}^k$ instead of $\mathbb{R}^n$ since we've already used n in specifying M_{mn}.) We'll do this by finding a set of matrices which span M_{mn} and identifying each $m \times n$ matrix with the coefficients used to write it as a linear combination of our spanning set.

Example 1. Show $B_1 = \begin{bmatrix} 1 & 1 \\ 0 & 0 \end{bmatrix}$, $B_2 = \begin{bmatrix} 0 & 0 \\ 1 & 1 \end{bmatrix}$, $B_3 = \begin{bmatrix} 1 & 0 \\ 1 & 0 \end{bmatrix}$, $B_4 = \begin{bmatrix} 0 & 1 \\ 0 & 1 \end{bmatrix}$, and $B_5 = \begin{bmatrix} 1 & 0 \\ 0 & 1 \end{bmatrix}$ span M_{22}. Use this spanning set to identify the matrix $A = \begin{bmatrix} 2 & 1 \\ 1 & 2 \end{bmatrix}$ with a vector $\vec{x}$ in $\mathbb{R}^5$.

Before we get started, notice that our vector $\vec{x}$ is in $\mathbb{R}^5$ because we have five matrices in our spanning set, which means A's linear combination will have five coefficients.

To see that B_1, B_2, B_3, B_4, and B_5 span M_{22}, let's pick a generic 2×2 matrix $\begin{bmatrix} a & b \\ c & d \end{bmatrix}$ and try to write it as a linear combination of the Bs. To do this, we need to solve

$$x_1 \begin{bmatrix} 1 & 1 \\ 0 & 0 \end{bmatrix} + x_2 \begin{bmatrix} 0 & 0 \\ 1 & 1 \end{bmatrix} + x_3 \begin{bmatrix} 1 & 0 \\ 1 & 0 \end{bmatrix} + x_4 \begin{bmatrix} 0 & 1 \\ 0 & 1 \end{bmatrix} + x_5 \begin{bmatrix} 1 & 0 \\ 0 & 1 \end{bmatrix} = \begin{bmatrix} a & b \\ c & d \end{bmatrix}.$$

(Here we're assuming we know a, b, c, d and are solving for $x_1, \ldots, x_5$.) We can simplify the left side of this equation to get

$$\begin{bmatrix} x_1 + x_3 + x_5 & x_1 + x_4 \\ x_2 + x_3 & x_2 + x_4 + x_5 \end{bmatrix} = \begin{bmatrix} a & b \\ c & d \end{bmatrix}.$$

Identifying corresponding matrix entries gives us

$$x_1 + x_3 + x_5 = a$$
$$x_1 + x_4 = b$$
$$x_2 + x_3 = c$$
$$x_2 + x_4 + x_5 = d.$$

This is a much more familiar problem! In fact it is equivalent to asking when the matrix equation

$$\begin{bmatrix} 1 & 0 & 1 & 0 & 1 \\ 1 & 0 & 0 & 1 & 0 \\ 0 & 1 & 1 & 0 & 0 \\ 0 & 1 & 0 & 1 & 1 \end{bmatrix} \vec{x} = \vec{b}$$

has solutions for every 4-vector $\vec{b}$. When we discussed finding the range of a linear function in 2.5, we realized this is equivalent to the matrix having a leading 1 in every row of its reduced echelon form. The reduced echelon form of

$$\begin{bmatrix} 1 & 0 & 1 & 0 & 1 \\ 1 & 0 & 0 & 1 & 0 \\ 0 & 1 & 1 & 0 & 0 \\ 0 & 1 & 0 & 1 & 1 \end{bmatrix}$$

is

$$\begin{bmatrix} 1 & 0 & 0 & 1 & 0 \\ 0 & 1 & 0 & 1 & 0 \\ 0 & 0 & 1 & -1 & 0 \\ 0 & 0 & 0 & 0 & 1 \end{bmatrix}.$$

We do have a leading 1 in every row, so our matrix equation always has a solution, and $B_1, \ldots, B_5$ span M_{22}.

Since the Bs span M_{22}, our matrix A must be a linear combination of these five matrices. Notice that $A = B_1 + B_2 + B_5$, so our linear combination has coefficients 1 on B_1, B_2, and B_5 and coefficients 0 on B_3 and B_4. This means we identify A with the 5-vector $\vec{x}$ which has 1s in the 1st, 2nd, and 5th spots and 0s in the 3rd and 4th spots, so

$$\vec{x} = \begin{bmatrix} 1 \\ 1 \\ 0 \\ 0 \\ 1 \end{bmatrix}.$$

There are two possible issues with this process of identifying a matrix with its vector of coefficients. The first is that our vector of coefficients depends heavily on the order in which we listed our spanning vectors. For instance, in the example above we could swap the order of B_2 and B_3. This would mean $A = B_1 + B_3 + B_5$, so we'd identify A with the 5-vector $\vec{x} = \begin{bmatrix} 1 \\ 0 \\ 1 \\ 0 \\ 1 \end{bmatrix}$. We can fix this problem by being careful to list our spanning set in the particular order we want and preserving this order throughout the process. I'll always specify an order in this book, and I recommend that you always specify an order for your spanning sets when doing problems on your own.

The second possible problem occurs when we can write our matrix as a linear combination of the spanning matrices in more than one way, i.e., using two different sets of coefficients. This can certainly happen, and means we don't have a unique vector to match up with our matrix. For instance, in the example above we can also write $A = B_3 + B_4 + B_5$ which would mean identifying A with the 5-vector $\vec{y} = \begin{bmatrix} 0 \\ 0 \\ 1 \\ 1 \\ 1 \end{bmatrix}$. Clearly this isn't good, because we want to create a function which maps from M_{22} to $\mathbb{R}^n$ and this example's function appears to be multi-valued. Let's explore this situation further to see if we can figure out how to avoid it.

Suppose A is in the span of $B_1, \ldots, B_k$ with two different sets of coefficients. This means we have two sets of scalars $x_1, \ldots, x_k$ and $y_1, \ldots, y_k$ with $A = x_1 B_1 + \cdots x_k B_k$ and $A = y_1 B_1 + \cdots y_k B_k$. If we subtract one of these linear combinations from the other we get

$$(x_1 B_1 + \cdots x_k B_k) - (y_1 B_1 + \cdots y_k B_k) = A - A = \vec{0}_{M_{mn}}.$$

We can combine like terms of the left-hand side to get

$$(x_1 - y_1)B_1 + \cdots (x_k - y_k)B_k = \vec{0}_{M_{mn}}.$$

Since our two sets of coefficients are different, we must have at least one place where $x_i - y_i \neq 0$. I'll assume for ease of notation that this happens at $i = 1$, i.e., $x_1 - y_1 \neq 0$. In that case, we can subtract $(x_1 - y_1)B_1$ from both sides of our equation to get

$$(x_2 - y_2)B_2 + \cdots (x_k - y_k)B_k = -(x_1 - y_1)B_1$$

Since $-(x_1 - y_1) \neq 0$, we can divide all our coefficients by it to see that B_1 is in the span of the other Bs. Aha! That's a condition we've discussed before.

What we've just seen is that the coefficients used to write a matrix as a linear combination of our spanning set aren't unique exactly when the spanning matrices are linearly dependent. This means that we can avoid this situation by requiring that our spanning set be linearly independent, which leads to the following definition. Note that there is nothing in our exploration above that doesn't generalize to a spanning set in any vector space, so I'll state the definition in those general terms.

Definition. A set of vectors $\mathcal{B} = \{\vec{b}_1, \ldots, \vec{b}_n\}$ is a *basis* for a vector space V if $V = \text{Span}\{\vec{b}_1, \ldots, \vec{b}_n\}$ and $\vec{b}_1, \ldots, \vec{b}_n$ are linearly independent.

We've actually already seen this before when we constructed a spanning set for the solution set of a homogeneous matrix equation.

Example 2. Find a basis for the null space of $A = \begin{bmatrix} 3 & 2 & 0 & 4 & -1 \\ 0 & 2 & 6 & 0 & -8 \\ 1 & 1 & 1 & 4 & 1 \end{bmatrix}$.

The null space of A is the same as the solution set of $A\vec{x} = \vec{0}$, and we saw in Example 5 from 2.8, that

$$Nul(A) = \text{Span} \left\{ \begin{bmatrix} 2 \\ -3 \\ 1 \\ 0 \\ 0 \end{bmatrix}, \begin{bmatrix} -1 \\ 4 \\ 0 \\ -1 \\ 1 \end{bmatrix} \right\}.$$

If you look at the 3rd and 5th entries of these vectors corresponding to the free variables x_3 and x_5, you'll see that all vectors have a 0 in that spot except the vector containing the coefficients of that free variable which has a 1. This means we cannot form a nontrivial linear combination of these vectors with all coefficients equal to zero. Thus our spanning vectors are linearly independent and so form a basis for $Nul(A)$.

Note that there was nothing special about the matrix A and its null space in the example above. Although we didn't know it at the time, our method for finding a spanning set for $Nul(A)$ actually finds a basis.

Example 3. The matrices $\begin{bmatrix} 1 & 0 \\ 0 & 0 \end{bmatrix}$, $\begin{bmatrix} 0 & 1 \\ 0 & 0 \end{bmatrix}$, $\begin{bmatrix} 0 & 0 \\ 1 & 0 \end{bmatrix}$, and $\begin{bmatrix} 0 & 0 \\ 0 & 1 \end{bmatrix}$ are a basis for M_{22}.

To see that these four matrices form a basis for M_{22}, we need to check that they span M_{22} and are linearly independent. We can check that they span in the same way we did in Example 1, by setting a linear combination of them equal to a generic 2×2 matrix. This means we want to solve

$$x_1 \begin{bmatrix} 1 & 0 \\ 0 & 0 \end{bmatrix} + x_2 \begin{bmatrix} 0 & 1 \\ 0 & 0 \end{bmatrix} + x_3 \begin{bmatrix} 0 & 0 \\ 1 & 0 \end{bmatrix} + x_4 \begin{bmatrix} 0 & 0 \\ 0 & 1 \end{bmatrix} = \begin{bmatrix} a & b \\ c & d \end{bmatrix}.$$

Simplifying the left side of this equation, we get

$$\begin{bmatrix} x_1 & x_2 \\ x_3 & x_4 \end{bmatrix} = \begin{bmatrix} a & b \\ c & d \end{bmatrix}.$$

This can be always be solved by letting $x_1 = a, x_2 = b, x_3 = c, x_4 = d$, so our four matrices span M_{22}.

To see that these four matrices are linearly independent, we need to show that

$$x_1 \begin{bmatrix} 1 & 0 \\ 0 & 0 \end{bmatrix} + x_2 \begin{bmatrix} 0 & 1 \\ 0 & 0 \end{bmatrix} + x_3 \begin{bmatrix} 0 & 0 \\ 1 & 0 \end{bmatrix} + x_4 \begin{bmatrix} 0 & 0 \\ 0 & 1 \end{bmatrix} = \begin{bmatrix} 0 & 0 \\ 0 & 0 \end{bmatrix}$$

has only the solution $x_1 = x_2 = x_3 = x_4 = 0$. As above, the left-hand side simplifies to give us

$$\begin{bmatrix} x_1 & x_2 \\ x_3 & x_4 \end{bmatrix} = \begin{bmatrix} 0 & 0 \\ 0 & 0 \end{bmatrix}.$$

Identifying entries gives us that all the x_is are zero, so our matrices are linearly independent. Since they also span, they are a basis for M_{22}.

We can form a similar basis for M_{mn} for any m and n by using mn matrices each of which has a 1 in one entry and zeros everywhere else. This is what is usually called the standard basis for M_{mn}. The only occasional point of confusion is how to order these matrices. Most mathematicians order them by the position of the 1 as in the example above, i.e., from left to right along each row from the top to the bottom.

Since spanning and linear independence don't depend on the order of the set of matrices, this basis also gives us several more bases for M_{22} by simply reordering these four matrices. However, there are also a vast array of other bases for M_{22}.

Example 4. The matrices $\begin{bmatrix} 1 & 0 \\ 0 & -1 \end{bmatrix}$, $\begin{bmatrix} 0 & 2 \\ -2 & 0 \end{bmatrix}$, $\begin{bmatrix} 0 & 0 \\ 0 & 2 \end{bmatrix}$, $\begin{bmatrix} 0 & 3 \\ 0 & 0 \end{bmatrix}$ are also a basis for M_{22}.

We'll start by checking linear independence using the equation

$$x_1 \begin{bmatrix} 1 & 0 \\ 0 & -1 \end{bmatrix} + x_2 \begin{bmatrix} 0 & 2 \\ -2 & 0 \end{bmatrix} + x_3 \begin{bmatrix} 0 & 0 \\ 0 & 2 \end{bmatrix} + x_4 \begin{bmatrix} 0 & 3 \\ 0 & 0 \end{bmatrix} = \begin{bmatrix} 0 & 0 \\ 0 & 0 \end{bmatrix}.$$

This can be simplified to

$$\begin{bmatrix} x_1 & 2x_2 + 3x_4 \\ -2x_2 & -x_1 + 2x_3 \end{bmatrix} = \begin{bmatrix} 0 & 0 \\ 0 & 0 \end{bmatrix}.$$

Identifying matrix entries tells us that $x_1 = 0$, $2x_2 + 3x_4 = 0$, $-2x_2 = 0$, and $-x_1 + 2x_3 = 0$. Since $x_1 = 0$, the fourth equation says $2x_3 = 0$ which means $x_3 = 0$. The third equation tells us $x_2 = 0$, and plugging that into the second equation gives $3x_4 = 0$ or $x_4 = 0$. Since all coefficients in our linear combination must be zero, our four matrices are linearly independent.

To see that our four matrices span M_{22}, we'll proceed as in the first example and set a linear combination of them equal to a generic 2×2 matrix. This gives us

$$x_1 \begin{bmatrix} 1 & 0 \\ 0 & -1 \end{bmatrix} + x_2 \begin{bmatrix} 0 & 2 \\ -2 & 0 \end{bmatrix} + x_3 \begin{bmatrix} 0 & 0 \\ 0 & 2 \end{bmatrix} + x_4 \begin{bmatrix} 0 & 3 \\ 0 & 0 \end{bmatrix} = \begin{bmatrix} a & b \\ c & d \end{bmatrix}$$

(As in Example 1, remember that we're assuming a, b, c, d are known, and we're solving for the xs.) Simplifying gives us

$$\begin{bmatrix} x_1 & 2x_2 + 3x_4 \\ -2x_2 & -x_1 + 2x_3 \end{bmatrix} = \begin{bmatrix} a & b \\ c & d \end{bmatrix}$$

which can be rewritten as the list of equations below.

$$x_1 = a$$
$$2x_2 + 3x_4 = b$$
$$-2x_2 = c$$
$$-x_1 + 2x_3 = d$$

The first equation clearly tells us that $x_1 = a$, and we can solve the third equation for x_2 to get $x_2 = -\frac{1}{2}c$. Plugging $x_1 = a$ into the fourth equation gives us $-a + 2x_3 = d$. Solving for x_3 now gives us $x_3 = \frac{1}{2}d - \frac{1}{2}a$. Plugging $x_2 = -\frac{1}{2}c$ into the second equation gives us $2(-\frac{1}{2}c) + 3x_4 = b$ which simplifies to $-c + 3x_4 = b$. Solving for x_4, we get $x_4 = \frac{1}{3}b + \frac{1}{3}c$. Since these solutions make sense no matter what values of a, b, c, d we used, our four matrices span M_{22} and are therefore a basis.

Even though we already have our usual computational tools available in the case where $V = \mathbb{R}^n$, we will sometimes still want to use a basis for $\mathbb{R}^n$. There is certainly nothing in our definition that prevents us from applying this concept there.

Example 5. The (standard) basis for $\mathbb{R}^n$ is $\begin{bmatrix} 1 \\ 0 \\ 0 \\ \vdots \\ 0 \end{bmatrix}, \begin{bmatrix} 0 \\ 1 \\ 0 \\ \vdots \\ 0 \end{bmatrix}, \ldots, \begin{bmatrix} 0 \\ 0 \\ \vdots \\ 0 \\ 1 \end{bmatrix}$.

In $\mathbb{R}^n$, we have more tools for checking linear independence and spanning.

In particular, remember that we can apply the Invertible Matrix Theorem from 2.11. Since we have n n-vectors, we can view them as the columns of an $n \times n$ matrix. In the case of these particular vectors, they are the columns of the $n \times n$ identity matrix I_n. This means the Invertible Matrix Theorem automatically tells us that they are linearly independent and span all of $\mathbb{R}^n$. Therefore they are a basis.

As with M_{mn}, our standard basis is by no means the only basis for $\mathbb{R}^n$.

Example 6. Another basis for $\mathbb{R}^2$ is $\begin{bmatrix} 1 \\ 1 \end{bmatrix}$ and $\begin{bmatrix} 1 \\ -1 \end{bmatrix}$.

While we could use the Invertible Matrix Theorem to check that these two vectors are a basis for $\mathbb{R}^2$, let's practice a more geometric approach.

If we sketch a picture of these two vectors and the lines they span in the plane, it looks like

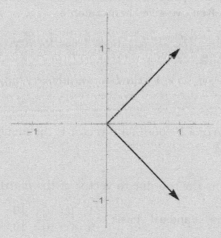

From this picture, it is clear that neither of these vectors lies along the line which is the span of the other. This means they are linearly independent. To see that they span all of $\mathbb{R}^2$, let's think about the dimension of

$$\text{Span}\left\{ \begin{bmatrix} 1 \\ 1 \end{bmatrix}, \begin{bmatrix} 1 \\ -1 \end{bmatrix} \right\}.$$

The two spanning vectors are linearly independent, so their span is 2D, i.e., a plane. Since $\mathbb{R}^2$ is a plane, these two vectors must span $\mathbb{R}^2$. Therefore $\begin{bmatrix} 1 \\ 1 \end{bmatrix}$ and $\begin{bmatrix} 1 \\ -1 \end{bmatrix}$ are a basis for $\mathbb{R}^2$.

The picture above also suggests a reason we might choose to work with this basis instead of the standard basis for $\mathbb{R}^2$: working with reflection about line $y = x$. Since this line is the span of $\begin{bmatrix} 1 \\ 1 \end{bmatrix}$, our first basis vector is fixed

by that reflection. The other basis vector $\begin{bmatrix} 1 \\ -1 \end{bmatrix}$ is sent to its negative by that

reflection. This means that if we put a general vector in terms of this basis it becomes very easy to see what happens to it under reflection about $y = x$. In particular, if $\vec{v} = a \begin{bmatrix} 1 \\ 1 \end{bmatrix} + b \begin{bmatrix} 1 \\ -1 \end{bmatrix}$, then $\vec{v}$'s reflection is $a \begin{bmatrix} 1 \\ 1 \end{bmatrix} - b \begin{bmatrix} 1 \\ -1 \end{bmatrix}$.

Now that we have a solid understanding of how a basis works, we can use it to help us match elements of any vector space V with vectors in $\mathbb{R}^n$ so we can transfer questions about V as we discussed at the start of this section. We do this by matching a vector $\vec{v}$ in V with a vector made up of the coefficients used to write it as a linear combination of the basis vectors which we've guaranteed is both possible and unique by having a basis both span and be linearly independent. Since we'll be discussing these vectors of basis coefficients so often, we give them a name.

Definition. Let $\mathcal{B} = \{\vec{b}_1, \ldots, \vec{b}_n\}$ be a basis for a vector space V. For a vector $\vec{v} = a_1 \vec{b}_1 + \cdots + a_n \vec{b}_n$ in V, the $\mathcal{B}$-*coordinate vector of* $\vec{v}$ is $[\vec{v}]_{\mathcal{B}} = \begin{bmatrix} a_1 \\ \vdots \\ a_n \end{bmatrix}$.

Note that the size of a $\mathcal{B}$-coordinate vector is the number of vectors in the basis $\mathcal{B}$.

Example 7. What is the coordinate vector of the matrix $A = \begin{bmatrix} 1 & -1 & 5 \\ 2 & 0 & 8 \end{bmatrix}$ with respect to the standard basis $\begin{bmatrix} 1 & 0 & 0 \\ 0 & 0 & 0 \end{bmatrix}, \begin{bmatrix} 0 & 1 & 0 \\ 0 & 0 & 0 \end{bmatrix}, \begin{bmatrix} 0 & 0 & 1 \\ 0 & 0 & 0 \end{bmatrix},$ $\begin{bmatrix} 0 & 0 & 0 \\ 1 & 0 & 0 \end{bmatrix}, \begin{bmatrix} 0 & 0 & 0 \\ 0 & 1 & 0 \end{bmatrix}, \begin{bmatrix} 0 & 0 & 0 \\ 0 & 0 & 1 \end{bmatrix}$ for M_{23}?

Since our coordinate vector is made up of the coefficients used to write A as a linear combination of the basis vectors, we'll start by solving for those coefficients. This means solving the equation

$$x_1 \begin{bmatrix} 1 & 0 & 0 \\ 0 & 0 & 0 \end{bmatrix} + x_2 \begin{bmatrix} 0 & 1 & 0 \\ 0 & 0 & 0 \end{bmatrix} + x_3 \begin{bmatrix} 0 & 0 & 1 \\ 0 & 0 & 0 \end{bmatrix} + x_4 \begin{bmatrix} 0 & 0 & 0 \\ 1 & 0 & 0 \end{bmatrix}$$

$$+ x_5 \begin{bmatrix} 0 & 0 & 0 \\ 0 & 1 & 0 \end{bmatrix} + x_6 \begin{bmatrix} 0 & 0 & 0 \\ 0 & 0 & 1 \end{bmatrix} = \begin{bmatrix} 1 & -1 & 5 \\ 2 & 0 & 8 \end{bmatrix}.$$

The left side simplifies to give us

$$\begin{bmatrix} x_1 & x_2 & x_3 \\ x_4 & x_5 & x_6 \end{bmatrix} = \begin{bmatrix} 1 & -1 & 5 \\ 2 & 0 & 8 \end{bmatrix}$$

which has solution $x_1 = 1$, $x_2 = -1$, $x_3 = 5$, $x_4 = 2$, $x_5 = 0$, $x_6 = 8$. This

means A has $\mathcal{B}$-coordinate vector $[A]_\mathcal{B} = \begin{bmatrix} 1 \\ -1 \\ 5 \\ 2 \\ 0 \\ 8 \end{bmatrix}$.

If you noticed that the entries of this coordinate vector were the same as the entries of the matrix A read off from left to right across each row starting with the first row and working downward, you have hit on the easiest way to compute coordinate vectors with respect to the standard basis for M_{23}. In fact, this method works for any matrix in any M_{mn} as long as you are using the standard basis. The ease of finding coordinate vectors with respect to the standard basis for M_{mn} is why this is the basis most commonly used, and hence called standard.

Example 8. Find the coordinate vector of matrix $A = \begin{bmatrix} 2 & 6 \\ -3 & 14 \end{bmatrix}$ with respect to the basis $\mathcal{B} = \left\{ \begin{bmatrix} 1 & 0 \\ 0 & -1 \end{bmatrix}, \begin{bmatrix} 0 & 2 \\ -2 & 0 \end{bmatrix}, \begin{bmatrix} 0 & 0 \\ 0 & 2 \end{bmatrix}, \begin{bmatrix} 0 & 3 \\ 0 & 0 \end{bmatrix} \right\}$ for M_{22}.

Here we are not using the standard basis for M_{22}, so we can't just read off matrix entries and will have to use the general method of finding a linear combination of the basis vectors which equals our matrix A. We could go back to Example 4 and plug in our values for a, b, c, d, but I'll do this problem from scratch to more accurately reflect the solution process we'd use if we hadn't done that previous work.

Let's start by writing our matrix A as a linear combination of our basis via the equation

$$x_1 \begin{bmatrix} 1 & 0 \\ 0 & -1 \end{bmatrix} + x_2 \begin{bmatrix} 0 & 2 \\ -2 & 0 \end{bmatrix} + x_3 \begin{bmatrix} 0 & 0 \\ 0 & 2 \end{bmatrix} + x_4 \begin{bmatrix} 0 & 3 \\ 0 & 0 \end{bmatrix} = \begin{bmatrix} 2 & 6 \\ -3 & 14 \end{bmatrix}$$

which can be simplified to

$$\begin{bmatrix} x_1 & 2x_2 + 3x_4 \\ -2x_2 & -x_1 + 2x_3 \end{bmatrix} = \begin{bmatrix} 2 & 6 \\ -3 & 14 \end{bmatrix}.$$

This gives us the equations below.

$$x_1 = 2$$
$$2x_2 + 3x_4 = 6$$
$$-2x_2 = -3$$
$$-x_1 + 2x_3 = 14$$

These equations have augmented coefficient matrix

$$\left[\begin{array}{cccc|c} 1 & 0 & 0 & 0 & 2 \\ 0 & 2 & 0 & 3 & 6 \\ 0 & -2 & 0 & 0 & -3 \\ -1 & 0 & 2 & 0 & 14 \end{array}\right]$$

which has reduced echelon form

$$\left[\begin{array}{cccc|c} 1 & 0 & 0 & 0 & 2 \\ 0 & 1 & 0 & 0 & \frac{3}{2} \\ 0 & 0 & 1 & 0 & 8 \\ 0 & 0 & 0 & 1 & 1 \end{array}\right].$$

This means $x_1 = 2$, $x_2 = \frac{3}{2}$, $x_3 = 8$, and $x_4 = 1$, so our $\mathcal{B}$-coordinate vector is

$$[A]_{\mathcal{B}} = \begin{bmatrix} 2 \\ \frac{3}{2} \\ 8 \\ 1 \end{bmatrix}.$$

We can also find coordinate vectors in $\mathbb{R}^n$. You can convince yourself that if we use the standard basis for $\mathbb{R}^n$ discussed above then the coordinate vector of any vector $\vec{v}$ is just that vector itself. However, we will sometimes want to work in terms of another basis for $\mathbb{R}^n$, especially in Chapter 4.

Example 9. If we use the basis $\mathcal{B} = \left\{ \begin{bmatrix} 1 \\ 1 \end{bmatrix}, \begin{bmatrix} 1 \\ -1 \end{bmatrix} \right\}$ for $\mathbb{R}^2$, what is the $\mathcal{B}$-coordinate vector of $\vec{v} = \begin{bmatrix} -3 \\ 7 \end{bmatrix}$?

To find this coordinate vector, we need to solve

$$x_1 \begin{bmatrix} 1 \\ 1 \end{bmatrix} + x_2 \begin{bmatrix} 1 \\ -1 \end{bmatrix} = \begin{bmatrix} -3 \\ 7 \end{bmatrix}.$$

The left side simplifies to give us

$$\begin{bmatrix} x_1 + x_2 \\ x_1 - x_2 \end{bmatrix} = \begin{bmatrix} -3 \\ 7 \end{bmatrix}$$

so we need to solve $x_1 + x_2 = -3$, $x_1 - x_2 = 7$ which has augmented coefficient matrix

$$\left[\begin{array}{cc|c} 1 & 1 & -3 \\ 1 & -1 & 7 \end{array}\right].$$

This row reduces to

$$\left[\begin{array}{cc|c} 1 & 0 & 2 \\ 0 & 1 & -5 \end{array}\right]$$

so our solution is $x_1 = 2$ and $x_2 = -5$. Thus $[\vec{v}]_\mathcal{B} = \begin{bmatrix} 2 \\ -5 \end{bmatrix}$.

To check our answer is correct, we can compute

$$2\begin{bmatrix} 1 \\ 1 \end{bmatrix} - 5\begin{bmatrix} 1 \\ -1 \end{bmatrix} = \begin{bmatrix} 2-5 \\ 2+5 \end{bmatrix} = \begin{bmatrix} -3 \\ 7 \end{bmatrix}.$$

Since we got $\vec{v}$, our coordinate vector is right.

We came up with the idea of coordinate vectors in order to link any vector space V to $\mathbb{R}^n$ by matching a vector from $\vec{v}$ with its coordinate vector with respect to some basis $\mathcal{B}$ for V. We can formalize this by creating a map from V to $\mathbb{R}^n$ which sends each vector from V to its $\mathcal{B}$-coordinate vector. Since a coordinate vector has as many entries as there are vectors in the basis, this map goes to $\mathbb{R}^n$ where n is the number of vectors in V's basis $\mathcal{B}$.

Definition. Let $\mathcal{B} = \{\vec{b}_1, \ldots, \vec{b}_n\}$ be a basis for a vector space V, then the $\mathcal{B}$-*coordinate map* is the function $f_\mathcal{B} : V \to \mathbb{R}^n$ by $f(\vec{v}) = [\vec{v}]_\mathcal{B}$.

Example 10. Take the basis $\mathcal{B} = \left\{ \begin{bmatrix} 1 & 0 \\ 0 & -1 \end{bmatrix}, \begin{bmatrix} 0 & 2 \\ -2 & 0 \end{bmatrix}, \begin{bmatrix} 0 & 0 \\ 0 & 2 \end{bmatrix}, \begin{bmatrix} 0 & 3 \\ 0 & 0 \end{bmatrix} \right\}$ for M_{22}. Find n so that $f_\mathcal{B} : M_{22} \to \mathbb{R}^n$, and compute $f_\mathcal{B}\left(\begin{bmatrix} 2 & 6 \\ -3 & 14 \end{bmatrix} \right)$.

The first part of this question is a fairly straightforward counting problem. The $\mathcal{B}$-coordinate map connects V, which in our case is M_{22} with $\mathbb{R}^n$ where n is the number of elements in our basis $\mathcal{B}$. Since $\mathcal{B}$ contains four matrices, we have $n = 4$. Thus $f_\mathcal{B} : M_{22} \to \mathbb{R}^4$. This also provides a sanity check for our computation of $f_\mathcal{B}(A)$, because we now know our answer must be a 4-vector.

The $\mathcal{B}$-coordinate map sends any matrix to its $\mathcal{B}$-coordinate vector, i.e., $f_\mathcal{B}(A) = [A]_\mathcal{B}$. This is the same basis $\mathcal{B}$ and matrix A as in Example 8, so

$$f_\mathcal{B}(A) = [A]_\mathcal{B} = \begin{bmatrix} 2 \\ \frac{3}{2} \\ 8 \\ 1 \end{bmatrix}.$$

No matter which basis $\mathcal{B}$ we chose to create $f_\mathcal{B}$, this map always has several nice properties. The first is that every $f_\mathcal{B}$ is 1-1. To see this, remember that 1-1 means we cannot have $\vec{v} \neq \vec{w}$ with $f_\mathcal{B}(\vec{v}) = f_\mathcal{B}(\vec{w})$. In this case, that would mean $\vec{v} \neq \vec{w}$ with $[\vec{v}]_\mathcal{B} = [\vec{w}]_\mathcal{B}$. If two vectors have the same $\mathcal{B}$-coordinate vector, that means they are both equal to the same linear combination of the basis vectors and are therefore equal. Thus having $f_\mathcal{B}(\vec{v}) = f_\mathcal{B}(\vec{w})$ means $\vec{v} = \vec{w}$, so $f_\mathcal{B}$ is 1-1.

The coordinate map $f_\mathcal{B}$ is also onto for every $\mathcal{B}$. For any vector $\vec{x}$ in $\mathbb{R}^n$, we can find $\vec{v}$ in V which has $\vec{x} = [\vec{v}]_\mathcal{B}$ by letting $\vec{v} = x_1\vec{b}_1 + \cdots + x_n\vec{b}_n$. Thus every element of the codomain is also in the range, i.e., $f_\mathcal{B}$ is onto.

Since $f_\mathcal{B}$ is both 1-1 and onto, it is invertible. This means our coordinate map has created an exact correspondence between the set of vectors in V and the set of vectors in $\mathbb{R}^n$, so it can be used to go either direction between the two vector spaces. The only thing we might worry about is that $f_\mathcal{B}$ doesn't create a similar correspondence between the operations of V and the operations of $\mathbb{R}^n$. However, life is as good as possible, i.e., $f_\mathcal{B}$ is a linear function.

To see that $f_\mathcal{B} : V \to \mathbb{R}^n$ is linear, we need to check that for every $\vec{v}$ and $\vec{w}$ in V and every scalar r we have

$$f_\mathcal{B}(\vec{v} + \vec{w}) = f_\mathcal{B}(\vec{v}) + f_\mathcal{B}(\vec{w}) \text{ and } f_\mathcal{B}(r \cdot \vec{v}) = r \cdot f_\mathcal{B}(\vec{v}).$$

This is equivalent to checking that

$$[\vec{v} + \vec{w}]_\mathcal{B} = [\vec{v}]_\mathcal{B} + [\vec{w}]_\mathcal{B} \text{ and } [r \cdot \vec{v}]_\mathcal{B} = r \cdot [\vec{v}]_\mathcal{B}.$$

(Notice that the "+" and "$\cdot$" on the left-hand sides of these equations are the operations from V, while the $+$ and $\cdot$ on the right-hand sides are our usual vector operations from $\mathbb{R}^n$.)

Suppose

$$\vec{v} = x_1\vec{b}_1 + \cdots + x_n\vec{b}_n \text{ and } \vec{w} = y_1\vec{b}_1 + \cdots + y_n\vec{b}_n$$

so

$$[\vec{v}]_\mathcal{B} = \begin{bmatrix} x_1 \\ \vdots \\ x_n \end{bmatrix} \text{ and } [\vec{w}]_\mathcal{B} = \begin{bmatrix} y_1 \\ \vdots \\ y_n \end{bmatrix}.$$

Then

$$\vec{v} + \vec{w} = x_1\vec{b}_1 + \cdots + x_n\vec{b}_n + y_1\vec{b}_1 + \cdots + y_n\vec{b}_n$$
$$= (x_1 + y_1)\vec{b}_1 + \cdots + (x_n + y_n)\vec{b}_n$$

so

$$[\vec{v} + \vec{w}]_\mathcal{B} = \begin{bmatrix} x_1 + y_1 \\ \vdots \\ x_n + y_n \end{bmatrix}.$$

Now it is easy to see that

$$[\vec{v} + \vec{w}]_\mathcal{B} = \begin{bmatrix} x_1 + y_1 \\ \vdots \\ x_n + y_n \end{bmatrix} = \begin{bmatrix} x_1 \\ \vdots \\ x_n \end{bmatrix} + \begin{bmatrix} y_1 \\ \vdots \\ y_n \end{bmatrix} = [\vec{v}]_\mathcal{B} + [\vec{w}]_\mathcal{B}.$$

Similarly,

$$r \cdot \vec{v} = r(x_1\vec{b}_1 + \cdots + x_n\vec{b}_n) = rx_1\vec{b}_1 + \cdots + rx_n\vec{b}_n$$

so

$$[r \cdot \vec{v}]_\mathcal{B} = \begin{bmatrix} rx_1 \\ \vdots \\ rx_n \end{bmatrix}.$$

Therefore

$$[r \cdot \vec{v}]_\mathcal{B} = \begin{bmatrix} rx_1 \\ \vdots \\ rx_n \end{bmatrix} = r \begin{bmatrix} x_1 \\ \vdots \\ x_n \end{bmatrix} = r \cdot [\vec{v}]_\mathcal{B}$$

so $f_\mathcal{B}$ is a linear function.

We can summarize this discussion as follows.

Theorem 1. If $\mathcal{B}$ is a basis for a vector space V, then $f_\mathcal{B}$ is a 1-1 and onto (and hence invertible) linear function.

Note that if we are in the special case where $V = \mathbb{R}^n$, then $f_\mathcal{B}$ is an invertible linear function from $\mathbb{R}^n$ to $\mathbb{R}^n$. This means $f_\mathcal{B}$ corresponds to an invertible $n \times n$ matrix, so it also has all the other equivalent conditions from the Invertible Matrix Theorem in 2.11.

These good properties of $f_\mathcal{B}$ allow us to finally do what we set out to at the beginning of this section: answer a question about a vector space V by translating our question to $\mathbb{R}^n$ and using row reduction. We do this by choosing a basis $\mathcal{B}$ for V and mapping all the vectors from V in our question over to $\mathbb{R}^n$ with $f_\mathcal{B}$. Then we ask the same question about the resulting vectors in $\mathbb{R}^n$. Because $f_\mathcal{B}$ is a 1-1, onto, linear map, the answer we get in $\mathbb{R}^n$ will be the same as the answer we would have gotten if we'd just answered the original question in V.

Example 11. Are $A_1 = \begin{bmatrix} 2 & 4 \\ -1 & 0 \end{bmatrix}$, $A_2 = \begin{bmatrix} 1 & -3 \\ -3 & 1 \end{bmatrix}$, and $A_3 = \begin{bmatrix} 1 & 1 \\ 1 & 2 \end{bmatrix}$ linearly independent in M_{22}?

We could answer this question directly in M_{22} by seeing if

$$x_1 \begin{bmatrix} 2 & 4 \\ -1 & 0 \end{bmatrix} + x_2 \begin{bmatrix} 1 & -3 \\ -3 & 1 \end{bmatrix} + x_3 \begin{bmatrix} 1 & 1 \\ 1 & 2 \end{bmatrix} = \begin{bmatrix} 0 & 0 \\ 0 & 0 \end{bmatrix}$$

has only the trivial solution. However, now that we have the idea of coordinate maps, we can also answer it by translating the problem into $\mathbb{R}^n$ and solving it there. I'll show this newer method here, but feel free to go back to 2.4 to see the older method being used.

To use a coordinate map, I first need to choose a basis for M_{22}. To make things as easy as possible, I'll use the standard basis

$$\mathcal{B} = \left\{ \begin{bmatrix} 1 & 0 \\ 0 & 0 \end{bmatrix}, \begin{bmatrix} 0 & 1 \\ 0 & 0 \end{bmatrix}, \begin{bmatrix} 0 & 0 \\ 1 & 0 \end{bmatrix}, \begin{bmatrix} 0 & 0 \\ 0 & 1 \end{bmatrix} \right\}.$$

Since we're using the standard basis, we can find the coordinate vectors of our three original matrices by reading left to right across each row starting at the top. This gives us

$$[A_1]_\mathcal{B} = \begin{bmatrix} 2 \\ 4 \\ -1 \\ 0 \end{bmatrix}, \; [A_2]_\mathcal{B} = \begin{bmatrix} 1 \\ -3 \\ -3 \\ 1 \end{bmatrix}, \text{ and } [A_3]_\mathcal{B} = \begin{bmatrix} 1 \\ 1 \\ 1 \\ 2 \end{bmatrix}.$$

These three coordinate vectors are linearly independent exactly when A_1, A_2, and A_3 are linearly independent, but it is much easier to check linear independence of vectors in $\mathbb{R}^4$ using row reduction. The matrix whose columns are the three coordinate vectors is

$$\begin{bmatrix} 2 & 1 & 1 \\ 4 & -3 & 1 \\ -1 & -3 & 1 \\ 0 & 1 & 2 \end{bmatrix}$$

which has reduced echelon form

$$\begin{bmatrix} 1 & 0 & 0 \\ 0 & 1 & 0 \\ 0 & 0 & 1 \\ 0 & 0 & 0 \end{bmatrix}.$$

Since each column of the reduced echelon form contains a leading 1, the three coordinate vectors and hence the three matrices A_1, A_2, and A_3 are linearly independent.

Now that we have $f_\mathcal{B} : V \to \mathbb{R}^n$ for any basis $\mathcal{B} = \{\vec{b}_1, \ldots, \vec{b}_n\}$ for V, there is still one important question left to be answered: Is it possible to have two different bases for V which have different numbers of vectors in them? If so, this would link V to $\mathbb{R}^n$ and $\mathbb{R}^m$ for $n \neq m$. This sounds fairly strange since the coordinate map is supposed to preserve everything about V, while we intuitively think of $\mathbb{R}^n$ and $\mathbb{R}^m$ as different sized spaces for $n \neq m$. Therefore our next goal is to show that this cannot happen.

Suppose we have a vector space V with two different bases $\mathcal{B} = \{\vec{b}_1, \ldots, \vec{b}_n\}$ and $\mathcal{C} = \{\vec{c}_1, \ldots, \vec{c}_m\}$. We know that $\vec{b}_1, \ldots, \vec{b}_n$ are linearly independent vectors, so their $\mathcal{C}$-coordinate vectors in $\mathbb{R}^m$ must be linearly independent. This means the $m \times n$ matrix with columns $\left[\vec{b}_1\right]_\mathcal{C}, \ldots, \left[\vec{b}_n\right]_\mathcal{C}$ must have a reduced echelon form with leading 1s in every column which is only possible if $n \leq m$. Similarly, $\vec{c}_1, \ldots, \vec{c}_m$ are linearly independent vectors, so their $\mathcal{B}$-coordinate vectors in $\mathbb{R}^n$ must also be linearly independent. This means the $n \times m$ matrix whose columns are $[\vec{c}_1]_\mathcal{B}, \ldots, [\vec{c}_m]_\mathcal{B}$ must have a reduced echelon form with leading 1s in every column. For this to be possible, we must have

$m \leq n$. Therefore we must have $m = n$, i.e., our two bases must contain the same number of vectors.

One way to think about this number n of basis vectors in any basis for V is that to write down any vector in V, we can choose n different numbers to use as coefficients to use in the linear combination of our basis vectors. This means that V allows us n independent choices. This reminds me of our discussion of the geometric idea of dimension at the start of 1.3 where we said a n-dimensional object allowed us n different directions of motion. In either case, we're specifying a particular vector/point in a space which allows us n independent choices/directions. We'll use this connection to create our linear algebra definition of dimension.

Definition. The *dimension* of a vector space V, written $\dim(V)$, is the number of vectors in any basis for V.

Since we already have an established idea that the dimension of $\mathbb{R}^n$ is n, our first job is to make sure our two notions of $\dim(\mathbb{R}^n)$ agree with each other.

Example 12. $\dim(\mathbb{R}^n) = n$

Since every basis contains the same number of vectors, it doesn't matter which basis we use to compute the dimension. The standard basis

$$\begin{bmatrix} 1 \\ 0 \\ 0 \\ \vdots \\ 0 \end{bmatrix}, \begin{bmatrix} 0 \\ 1 \\ 0 \\ \vdots \\ 0 \end{bmatrix}, \ldots, \begin{bmatrix} 0 \\ 0 \\ \vdots \\ 0 \\ 1 \end{bmatrix}$$

contains n vectors – one for each entry in an n-vector. Thus the dimension of $\mathbb{R}^n$ is n.

Example 13. $\dim(M_{mn}) = mn$

Again, we might as well compute the dimension of M_{mn} by counting the number of matrices in the standard basis. Since the standard basis for M_{mn} has one matrix for each entry in an $m \times n$ matrix, it contains mn matrices. Therefore $\dim(M_{mn}) = mn$.

Example 14. Find the dimension of the null space of $A = \begin{bmatrix} 3 & 2 & 0 & 4 & -1 \\ 0 & 2 & 6 & 0 & -8 \\ 1 & 1 & 1 & 4 & 1 \end{bmatrix}$.

This is the matrix from Example 2, where we found that $Nul(A)$ had basis $\begin{bmatrix} 2 \\ -3 \\ 1 \\ 0 \\ 0 \end{bmatrix}, \begin{bmatrix} -1 \\ 4 \\ 0 \\ -1 \\ 1 \end{bmatrix}$. Since there are two basis vectors, the null space of A is two-dimensional.

The most straightforward way to find a basis is to start with a spanning set and whittle it down to a basis. If your spanning set is already linearly independent, as in the case of our spanning sets for $Nul(A)$, it is already a basis and you are done. If not, there is some vector in the spanning set which is in the span of the others. Remove that vector, which doesn't change the overall span. Now check linear independence again. Keep repeating this process until you are left with a linearly independent spanning set, i.e., a basis for your original span. As long as you start with a finite set of spanning vectors, this process will always terminate. (This is very similar to our process from 1.3 for finding the dimension of a span, which makes sense because basis is so closely tied to dimension.)

Example 15. Find a basis for Span $\{B_1, B_2, B_3, B_4, B_5\}$ where $B_1 = \begin{bmatrix} 1 & 1 \\ 0 & 0 \end{bmatrix}$, $B_2 = \begin{bmatrix} 0 & 0 \\ 1 & 1 \end{bmatrix}$, $B_3 = \begin{bmatrix} 1 & 0 \\ 1 & 0 \end{bmatrix}$, $B_4 = \begin{bmatrix} 0 & 1 \\ 0 & 1 \end{bmatrix}$, and $B_5 = \begin{bmatrix} 1 & 0 \\ 0 & 1 \end{bmatrix}$.

These are the five 2×2 matrices from Example 1 of this section, which we saw spanned all of M_{22}. This means we're really looking for a new basis for M_{22} which is a subset of B_1, B_2, B_3, B_4, B_5. Our first step here is to see if these matrices are linearly independent and already form a basis. Here it is clear that they are not linearly independent, since $B_1 + B_2 = B_3 + B_4$ which means $B_1 = -B_2 + B_3 + B_4$, i.e., B_1 is in the span of the other matrices. According to our algorithm's instructions, we'll remove B_1 from our set. The span is still the same, so Span$\{B_2, B_3, B_4, B_5\} = M_{22}$, and our new question is whether these four matrices are linearly independent. We can check that by solving

$$x_2 \begin{bmatrix} 0 & 0 \\ 1 & 1 \end{bmatrix} + x_3 \begin{bmatrix} 1 & 0 \\ 1 & 0 \end{bmatrix} + x_4 \begin{bmatrix} 0 & 1 \\ 0 & 1 \end{bmatrix} + x_5 \begin{bmatrix} 1 & 0 \\ 0 & 1 \end{bmatrix} = \begin{bmatrix} 0 & 0 \\ 0 & 0 \end{bmatrix}$$

to see if we have nontrivial solutions. This can be simplified to

$$\begin{bmatrix} x_3 + x_5 & x_4 \\ x_2 + x_3 & x_2 + x_4 + x_5 \end{bmatrix} = \begin{bmatrix} 0 & 0 \\ 0 & 0 \end{bmatrix}$$

which can be rewritten as $x_3 + x_5 = 0$, $x_4 = 0$, $x_2 + x_3 = 0$, $x_2 + x_4 + x_5 = 0$. Since $x_4 = 0$, this reduces to $x_3 + x_5 = 0$, $x_2 + x_3 = 0$, $x_2 + x_5 = 0$. The first and third equations now tell us that $x_3 = -x_5 = x_2$, which means the middle equation can be rewritten as $-2x_5 = 0$. Thus $x_5 = 0$ and hence $x_2 = x_3 = 0$ as well. This means B_2, B_3, B_4, B_5 are linearly independent and therefore a basis for their span, M_{22}.

We've seen something similar to this algorithm when we discussed the rank of a matrix in 2.8 as the number of linearly independent vectors in the spanning set for the column space of a matrix. Let's take those ideas to their logical conclusion by finding a basis for $Col(A)$ below.

Example 16. Find a basis for the column space of $A = \begin{bmatrix} 1 & -4 & -2 & -6 \\ -4 & -2 & 2 & 6 \\ 3 & 3 & -1 & 7 \end{bmatrix}$.

Since the column space is the span of A's columns, our spanning set is the four 3-vectors which are the columns of A. We can use our algorithm above to reduce this spanning set down to a basis for $Col(A)$.

The first step is to check if they are linearly independent, which we can do by row reducing and seeing if we have a leading 1 in every column of the reduced echelon form. Our matrix A row reduces to

$$\begin{bmatrix} 1 & 0 & -\frac{2}{3} & \frac{2}{3} \\ 0 & 1 & \frac{1}{3} & \frac{5}{3} \\ 0 & 0 & 0 & 0 \end{bmatrix}$$

which doesn't have a leading 1 in every column. This means our spanning set isn't a basis.

In the case of a general spanning set, we'd find one of our vectors to remove and start the process over. However, here in $\mathbb{R}^n$ with row reduction at our disposal we can be more efficient. The first two columns of the reduced echelon form are the only ones which have leading 1s in them. This means the first two columns of A are linearly independent and the other two are not. This allows us to skip ahead and simply pick the columns of A which produced leading 1s in the reduced echelon form as our basis for $Col(A)$. Thus a basis for $Col(A)$ is

$$\begin{bmatrix} 1 \\ -4 \\ 3 \end{bmatrix}, \begin{bmatrix} -4 \\ -2 \\ 3 \end{bmatrix}.$$

This procedure works in general, so we can find a basis for the column space of a matrix by simply finding the reduced echelon form and choosing as our basis for $Col(A)$ the columns of A which correspond to the columns of the reduced echelon form containing leading 1s. Note that it is very important to use the columns of A as your basis rather than the columns of the reduced echelon form!

Here's another fact about dimension that should seem reasonable.

Theorem 2. If W is a subspace of V, then we have $\dim(W) < \dim(V)$.

You can see this illustrated in the example above for $W = Col(A)$ and $V = \mathbb{R}^3$, since $\dim(W) = 2$ and $\dim(\mathbb{R}^3) = 3$. This theorem is true because a basis for W is linearly independent not only as a set of vectors in W, but also as a set of vectors in V. By an argument very similar to the one where we showed that every basis of a vector space has the same size, you can now see that our basis for W can contain at most $\dim(V)$ vectors.

To finish up this section, we can leverage the ability to translate problems from V to $\mathbb{R}^n$ to create a shortcut for checking that a set of vectors is a basis for V. The Invertible Matrix Theorem tells us that the two conditions necessary for a basis of $\mathbb{R}^n$ are equivalent as long as we have n n-vectors to form the columns of a square $n \times n$ matrix. (See our discussion of the standard basis for $\mathbb{R}^n$.) In the theorem below, we'll generalize that idea to general vector spaces.

Theorem 3. Suppose $\dim(V) = n$, and $\mathcal{B} = \{\vec{b}_1, \ldots, \vec{b}_n\}$. Then the following are equivalent:

     1. $\vec{b}_1, \ldots, \vec{b}_n$ are linearly independent

     2. $\text{Span}\{\vec{b}_1, \ldots, \vec{b}_n\} = V$

In other words, if we know that $\mathcal{B}$ contains the correct number of vectors to be a basis for V, then we only need to check one of the two conditions for being a basis and we automatically get the other one for free!

To see that this is true, take coordinate vectors of $\vec{b}_1, \ldots, \vec{b}_n$ with respect to some other basis to translate the whole situation to $\mathbb{R}^n$. This means we have $\vec{v}_1, \ldots, \vec{v}_n$ in $\mathbb{R}^n$. If we use these n vectors as the columns in an $n \times n$ matrix, we immediately get that these two conditions are equivalent, since they are both conditions from the invertible matrix theorem. Thus having $\vec{b}_1, \ldots, \vec{b}_n$ linearly independent and having $\text{Span}\{\vec{b}_1, \ldots, \vec{b}_n\} = V$ are equivalent.

Example 17. Is $\mathcal{B} = \left\{ \begin{bmatrix} 1 & 0 \\ 0 & 0 \end{bmatrix}, \begin{bmatrix} 1 & 1 \\ 0 & 0 \end{bmatrix}, \begin{bmatrix} 1 & 1 \\ 1 & 0 \end{bmatrix}, \begin{bmatrix} 1 & 1 \\ 1 & 1 \end{bmatrix} \right\}$ a basis for M_{22}?

The dimension of M_{22} is $2 \cdot 2 = 4$, so this set contains the correct number of matrices to be a basis. (If it didn't contain four matrices, we'd automatically know it couldn't be a basis for M_{22}.) Using the theorem above, we can decide whether or not our four matrices from $\mathcal{B}$ are a basis either by checking if they span M_{22} or if they are linearly independent. Whichever condition we choose to check, we can do it in M_{22} or use a coordinate map to check it in $\mathbb{R}^n$. In this case, I'll use coordinate vectors with respect to the standard basis for M_{22} to show that $\mathcal{B}$ spans M_{22}.

With respect to the standard basis, our matrices from $\mathcal{B}$ have coordinate vectors

$$\begin{bmatrix} 1 \\ 0 \\ 0 \\ 0 \end{bmatrix}, \begin{bmatrix} 1 \\ 1 \\ 0 \\ 0 \end{bmatrix}, \begin{bmatrix} 1 \\ 1 \\ 1 \\ 0 \end{bmatrix}, \begin{bmatrix} 1 \\ 1 \\ 1 \\ 1 \end{bmatrix}.$$

These coordinate vectors span $\mathbb{R}^4$ exactly when the matrices from $\mathcal{B}$ span M_{22}. Four 4-vectors span $\mathbb{R}^4$ if and only if the matrix which has them as columns is invertible. In our case, this would be the matrix

$$\begin{bmatrix} 1 & 1 & 1 & 1 \\ 0 & 1 & 1 & 1 \\ 0 & 0 & 1 & 1 \\ 0 & 0 & 0 & 1 \end{bmatrix}.$$

This matrix has reduced echelon form I_4, so by the invertible matrix theorem the coordinate vectors span $\mathbb{R}^n$. This means our four matrices from $\mathcal{B}$ span M_{22} and hence are a basis.

Exercises 3.1.

1. Do $\begin{bmatrix} 2 \\ 0 \\ 1 \end{bmatrix}$, $\begin{bmatrix} -1 \\ 1 \\ 4 \end{bmatrix}$, and $\begin{bmatrix} 3 \\ -1 \\ -3 \end{bmatrix}$ form a basis for $\mathbb{R}^3$?

2. Do $\begin{bmatrix} 3 \\ 0 \\ 4 \end{bmatrix}$, $\begin{bmatrix} 4 \\ -1 \\ 5 \end{bmatrix}$, and $\begin{bmatrix} 1 \\ 2 \\ -3 \end{bmatrix}$ form a basis for $\mathbb{R}^3$?

3. Do $\begin{bmatrix} 1 \\ 2 \\ 3 \\ 4 \end{bmatrix}$, $\begin{bmatrix} 3 \\ -1 \\ 5 \\ 0 \end{bmatrix}$, $\begin{bmatrix} 0 \\ 4 \\ -6 \\ 1 \end{bmatrix}$, $\begin{bmatrix} -2 \\ 0 \\ 1 \\ 2 \end{bmatrix}$ form a basis for $\mathbb{R}^4$?

4. Do $\begin{bmatrix} -6 \\ 4 \\ 0 \\ 2 \end{bmatrix}$, $\begin{bmatrix} 1 \\ 0 \\ -5 \\ 3 \end{bmatrix}$, $\begin{bmatrix} -2 \\ 3 \\ 1 \\ 1 \end{bmatrix}$, and $\begin{bmatrix} 0 \\ 8 \\ 4 \\ 0 \end{bmatrix}$ form a basis for $\mathbb{R}^4$?

5. Use the Invertible Matrix Theorem to redo Example 6.

6. Could there be a basis for $\mathbb{R}^3$ consisting of 5 vectors? Briefly explain why or why not.

7. Show that $\begin{bmatrix} 1 & 1 \\ 0 & 0 \end{bmatrix}$, $\begin{bmatrix} 0 & 0 \\ 1 & 1 \end{bmatrix}$, $\begin{bmatrix} 1 & 0 \\ 0 & 1 \end{bmatrix}$, and $\begin{bmatrix} 0 & 1 \\ 0 & 0 \end{bmatrix}$ form a basis for M_{22}.

8. Show that $\begin{bmatrix} 2 & -1 \\ 0 & 0 \end{bmatrix}$, $\begin{bmatrix} -1 & 0 \\ 0 & 2 \end{bmatrix}$, $\begin{bmatrix} 0 & 0 \\ 2 & -1 \end{bmatrix}$, and $\begin{bmatrix} 0 & 2 \\ -1 & 0 \end{bmatrix}$ form a basis for M_{22}.

9. Do $\begin{bmatrix} 1 & 0 \\ 0 & 1 \end{bmatrix}$, $\begin{bmatrix} 0 & 1 \\ 1 & 0 \end{bmatrix}$, $\begin{bmatrix} 1 & 0 \\ 0 & 0 \end{bmatrix}$ and $\begin{bmatrix} 0 & 0 \\ 1 & 0 \end{bmatrix}$ form a basis for M_{22}?

10. Do $\begin{bmatrix} 1 & 0 & 0 \\ 1 & 0 & 0 \end{bmatrix}$, $\begin{bmatrix} 0 & 1 & 0 \\ 0 & 1 & 0 \end{bmatrix}$, $\begin{bmatrix} 0 & 0 & 1 \\ 0 & 0 & 1 \end{bmatrix}$, $\begin{bmatrix} 1 & 1 & 1 \\ 0 & 0 & 0 \end{bmatrix}$, $\begin{bmatrix} 0 & 0 & 0 \\ 1 & 1 & 1 \end{bmatrix}$, and

$\begin{bmatrix} 1 & 0 & 0 \\ 0 & 0 & 1 \end{bmatrix}$ form a basis for M_{23}?

11. Find a basis for the column space of $A = \begin{bmatrix} 1 & -2 & 0 & 4 & 5 \\ 0 & 0 & 2 & 1 & 3 \\ 0 & 0 & 0 & -1 & -1 \\ 0 & 0 & 1 & 2 & 3 \end{bmatrix}$.

12. Find a basis for the column space of $A = \begin{bmatrix} 1 & 3 & 0 \\ -2 & 2 & -8 \\ 0 & -1 & 1 \end{bmatrix}$.

13. Find a basis for the column space of $A = \begin{bmatrix} 2 & 0 & 4 & -2 \\ 0 & 1 & -3 & 1 \end{bmatrix}$.

14. Find a basis for the column space of $A = \begin{bmatrix} 2 & -4 & 0 & 2 \\ 0 & 1 & -1 & 2 \\ 0 & -1 & 2 & -3 \end{bmatrix}$.

15. Find a basis for the null space of $A = \begin{bmatrix} 1 & -4 & 0 & 1 & 0 & 0 \\ 0 & 0 & 1 & -2 & 0 & -1 \\ 0 & 0 & 0 & 0 & 1 & 5 \end{bmatrix}$.

16. Find a basis for the null space of $A = \begin{bmatrix} 0 & 2 & 1 & 0 \\ 1 & 0 & 1 & -1 \\ 0 & 0 & 0 & 9 \end{bmatrix}$.

17. Find a basis for the null space of $A = \begin{bmatrix} 1 & 0 & 2 & -1 & 0 \\ 0 & -1 & 3 & 2 & -1 \\ 0 & 2 & -6 & -3 & 4 \end{bmatrix}$.

18. Find a basis for the null space of $A = \begin{bmatrix} 1 & -1 & 0 & 2 \\ -2 & 2 & 1 & 8 \\ -1 & 1 & 1 & 10 \end{bmatrix}$.

19. Do $\begin{bmatrix} 2 & 0 & 0 \\ 0 & 0 & 0 \\ 0 & 0 & 1 \end{bmatrix}$, $\begin{bmatrix} 1 & 0 & 0 \\ 0 & 0 & 0 \\ 0 & 0 & 0 \end{bmatrix}$, and $\begin{bmatrix} -1 & 0 & 0 \\ 0 & 1 & 0 \\ 0 & 0 & -1 \end{bmatrix}$ form a basis for the

set $W = \left\{ \begin{bmatrix} a & 0 & 0 \\ 0 & b & 0 \\ 0 & 0 & c \end{bmatrix} \right\}$?

20. Do $\begin{bmatrix} 1 & 0 & 0 \\ 0 & 1 & 0 \\ 0 & 0 & 1 \end{bmatrix}$, $\begin{bmatrix} 1 & 0 & 0 \\ 0 & -1 & 0 \\ 0 & 0 & 0 \end{bmatrix}$, and $\begin{bmatrix} 0 & 0 & 0 \\ 0 & 1 & 0 \\ 0 & 0 & -1 \end{bmatrix}$ form a basis for the

set $W = \left\{ \begin{bmatrix} a & 0 & 0 \\ 0 & b & 0 \\ 0 & 0 & c \end{bmatrix} \right\}$?

21. What is the dimension of M_{49}?

22. What is the dimension of M_{25}?

23. What is the dimension of the null space of $\begin{bmatrix} -6 & -5 & 1 & 4 \\ 1 & -4 & -5 & 9 \\ 4 & 1 & -3 & 2 \\ -1 & 1 & 2 & -3 \end{bmatrix}$?

24. What is the dimension of the column space of $\begin{bmatrix} 2 & 6 & -1 & -7 \\ 1 & 3 & -2 & -2 \\ -4 & -12 & 0 & 16 \end{bmatrix}$?

25. What is the codomain of the coordinate map $f_\mathcal{B}$ for any basis $\mathcal{B}$ of M_{34}?

26. Why was it so important to us that every basis of a vector space contain the same number of vectors?

27. Let $\mathcal{B} = \left\{ \begin{bmatrix} 5 \\ 10 \\ 0 \end{bmatrix}, \begin{bmatrix} 2 \\ 1 \\ 4 \end{bmatrix}, \begin{bmatrix} -1 \\ 0 \\ 1 \end{bmatrix} \right\}$ be a basis for $\mathbb{R}^3$.

    (a) Find the vector $\vec{v}$ where $[\vec{v}]_\mathcal{B} = \begin{bmatrix} 3 \\ -2 \\ 6 \end{bmatrix}$.

    (b) Find the coordinate vector of $\vec{w} = \begin{bmatrix} -4 \\ 0 \\ -7 \end{bmatrix}$ with respect to $\mathcal{B}$.

28. Let $\mathcal{B} = \left\{ \begin{bmatrix} 0 \\ 3 \\ -4 \end{bmatrix}, \begin{bmatrix} 1 \\ -1 \\ 2 \end{bmatrix}, \begin{bmatrix} -2 \\ 0 \\ 2 \end{bmatrix} \right\}$ be a basis for $\mathbb{R}^3$.

    (a) Find the vector $\vec{v}$ where $[\vec{v}]_\mathcal{B} = \begin{bmatrix} 4 \\ -1 \\ 2 \end{bmatrix}$.

    (b) Find the coordinate vector of $\vec{w} = \begin{bmatrix} 6 \\ -7 \\ 10 \end{bmatrix}$ with respect to $\mathcal{B}$.

29. Let $\mathcal{B} = \left\{ \begin{bmatrix} 3 \\ 2 \\ 0 \\ 1 \end{bmatrix}, \begin{bmatrix} -1 \\ 2 \\ 5 \\ 0 \end{bmatrix}, \begin{bmatrix} 0 \\ 1 \\ 7 \\ -2 \end{bmatrix}, \begin{bmatrix} 4 \\ -6 \\ 0 \\ 3 \end{bmatrix} \right\}$ be a basis for $\mathbb{R}^4$.

(a) Find the vector $\vec{v}$ where $[\vec{v}]_B = \begin{bmatrix} 3 \\ 0 \\ -2 \\ -1 \end{bmatrix}$.

(b) Find the coordinate vector of $\vec{w} = \begin{bmatrix} -12 \\ 13 \\ 12 \\ -9 \end{bmatrix}$ with respect to B.

30. Let $B = \left\{ \begin{bmatrix} 1 \\ -1 \\ 0 \\ -1 \end{bmatrix}, \begin{bmatrix} -1 \\ 0 \\ 1 \\ -1 \end{bmatrix}, \begin{bmatrix} 1 \\ 1 \\ 0 \\ -1 \end{bmatrix}, \begin{bmatrix} -1 \\ 0 \\ -1 \\ -1 \end{bmatrix} \right\}$ be a basis for $\mathbb{R}^4$.

(a) Find the vector $\vec{v}$ where $[\vec{v}]_B = \begin{bmatrix} -2 \\ 3 \\ 1 \\ -4 \end{bmatrix}$.

(b) Find the coordinate vector of $\vec{w} = \begin{bmatrix} -3 \\ 2 \\ 7 \\ 13 \end{bmatrix}$ with respect to B.

31. Let $B = \left\{ \begin{bmatrix} 2 & 0 \\ 0 & 2 \end{bmatrix}, \begin{bmatrix} 0 & 0 \\ 0 & 1 \end{bmatrix}, \begin{bmatrix} 0 & -1 \\ 0 & 0 \end{bmatrix}, \begin{bmatrix} 0 & 1 \\ 1 & 0 \end{bmatrix} \right\}$ be a basis for M_{22}.

(a) Find the vector $\vec{v}$ where $[\vec{v}]_B = \begin{bmatrix} 3 \\ 2 \\ -1 \\ 4 \end{bmatrix}$.

(b) Find the coordinate vector of $\vec{w} = \begin{bmatrix} 1 & 2 \\ 3 & 4 \end{bmatrix}$ with respect to B.

32. Let $B = \left\{ \begin{bmatrix} 1 & 0 \\ 0 & 0 \end{bmatrix}, \begin{bmatrix} 1 & 0 \\ 0 & 1 \end{bmatrix}, \begin{bmatrix} 0 & 1 \\ 0 & 0 \end{bmatrix}, \begin{bmatrix} 0 & 1 \\ 1 & 0 \end{bmatrix} \right\}$ be a basis for M_{22}.

(a) The vector $\vec{v}$ whose B-coordinate vector is $\begin{bmatrix} 1 \\ -2 \\ 3 \\ -1 \end{bmatrix}$.

(b) Find the B-coordinate vector of $\vec{v} = \begin{bmatrix} 2 & 4 \\ 6 & 8 \end{bmatrix}$.

33. Let $B = \left\{ \begin{bmatrix} 2 & -1 \\ 0 & 0 \end{bmatrix}, \begin{bmatrix} -1 & 0 \\ 0 & 2 \end{bmatrix}, \begin{bmatrix} 0 & 0 \\ 2 & -1 \end{bmatrix}, \begin{bmatrix} 0 & 2 \\ -1 & 0 \end{bmatrix} \right\}$ be a basis for M_{22}.

(a) Find the vector $\vec{v}$ where $[\vec{v}]_\mathcal{B} = \begin{bmatrix} 2 \\ -1 \\ 3 \\ 1 \end{bmatrix}$.

(b) Find the coordinate vector of $\vec{w} = \begin{bmatrix} 5 & -2 \\ -3 & 7 \end{bmatrix}$ with respect to $\mathcal{B}$.

34. Let $\mathcal{B} = \left\{ \begin{bmatrix} 1 & 1 & 1 \\ 1 & 1 & 0 \end{bmatrix}, \begin{bmatrix} 1 & 1 & 1 \\ 1 & 0 & 1 \end{bmatrix}, \begin{bmatrix} 1 & 1 & 1 \\ 0 & 1 & 1 \end{bmatrix}, \begin{bmatrix} 1 & 1 & 0 \\ 1 & 1 & 1 \end{bmatrix}, \begin{bmatrix} 1 & 0 & 1 \\ 1 & 1 & 1 \end{bmatrix}, \begin{bmatrix} 0 & 1 & 1 \\ 1 & 1 & 1 \end{bmatrix} \right\}$ be a basis for M_{23}.

(a) Find the vector $\vec{v}$ where $[\vec{v}]_\mathcal{B} = \begin{bmatrix} -1 \\ 2 \\ 0 \\ 1 \\ -3 \\ 4 \end{bmatrix}$.

(b) Find the coordinate vector of $\vec{w} = \begin{bmatrix} -3 & 3 & -4 \\ 0 & -6 & 5 \end{bmatrix}$ with respect to $\mathcal{B}$.

35. Use a coordinate map to check whether $\vec{v}_1 = \begin{bmatrix} 1 & 2 \\ 3 & 4 \end{bmatrix}, \vec{v}_2 = \begin{bmatrix} 4 & 3 \\ 2 & -1 \end{bmatrix}$ and $\vec{v}_3 = \begin{bmatrix} -2 & 1 \\ 4 & 9 \end{bmatrix}$ are linearly independent or linearly dependent.

36. Use a coordinate map to check whether $\vec{v}_1 = \begin{bmatrix} 2 & 0 \\ 0 & 2 \end{bmatrix}, \vec{v}_2 = \begin{bmatrix} 1 & 1 \\ 0 & 1 \end{bmatrix}$, and $\vec{v}_3 = \begin{bmatrix} 0 & 1 \\ 1 & 0 \end{bmatrix}$ in M_{22} are linearly independent or linearly dependent.

37. Use a coordinate map to check whether $\vec{v}_1 = \begin{bmatrix} 1 & 2 \\ 3 & 4 \end{bmatrix}, \vec{v}_2 = \begin{bmatrix} 4 & 3 \\ 2 & -1 \end{bmatrix}$, $\vec{v}_3 = \begin{bmatrix} -1 & 0 \\ 4 & 1 \end{bmatrix}$, and $\vec{v}_4 = \begin{bmatrix} -2 & 1 \\ 4 & 9 \end{bmatrix}$ span M_{22}.

38. Use a coordinate map to check whether $\vec{v}_1 = \begin{bmatrix} 2 & 0 \\ 0 & 2 \end{bmatrix}, \vec{v}_2 = \begin{bmatrix} 1 & 1 \\ 0 & 1 \end{bmatrix}$, $\vec{v}_3 = \begin{bmatrix} 0 & 2 \\ 2 & 0 \end{bmatrix}$, and $\vec{v}_4 = \begin{bmatrix} 0 & 1 \\ 1 & 0 \end{bmatrix}$ span M_{22}.

3.2   Polynomial Vector Spaces

In this section, we'll discuss a new family of commonly used vector spaces whose vectors are polynomials. While it is possible to have polynomials with many variables, here we'll restrict to the case of one variable which I'll call x which leaves us with familiar objects like $3x^2 - 7x$ and $x + 2$.

Note that I used the word objects here very intentionally. You may be more used to treating polynomials as functions where you're plugging in numbers or solving for x. Resist that temptation here! Remember that our original description of a vector space was a set of mathematical objects with two operations which followed certain rules. If you find yourself treating a polynomial as a function in this section there is a very high probability that you're making a mistake.

Definition.   The set of all polynomials in x is P.

From earlier algebra classes, we already know how to add two polynomials together by adding the coefficients on corresponding powers of x and how to multiply a polynomial by a scalar by multiplying each coefficient by that scalar. Although polynomials don't look like matrices or vectors, we can use these operations to view P as a vector space.

Theorem 1. P is a vector space using polynomial addition as its $+$ and multiplication of a polynomial by a scalar as its $\cdot$.

We can check this by looking at each of the conditions in the definition of a vector space introduced in 2.4. I'll number the checks to correspond with the numbering of the conditions in our definition.

1. The first condition is closure of addition. If we add together two polynomials $p(x)$ and $q(x)$, it is clear that $p(x) + q(x)$ is still a polynomial, and therefore is in P.

2. Next up is commutativity of addition. If we are adding two polynomials $p(x)$ and $q(x)$, it doesn't matter what order we add them so
$$p(x) + q(x) = q(x) + p(x)$$
which shows polynomial addition is commutative.

3. Similarly, if we're adding three polynomials, it doesn't matter which pair we add first. In other words
$$(p(x) + q(x)) + t(x) = p(x) + (q(x) + t(x))$$
so polynomial addition is associative.

4. To show P has an additive identity we need to find a polynomial to act as our zero vector $\vec{0}_P$. This polynomial needs to have the property $p(x) + \vec{0} = p(x)$ for any polynomial $p(x)$ in P. Since we add polynomials by adding corresponding coefficients (and assuming that any missing terms have a coefficient of zero), we can construct our $\vec{0}_P$ polynomial by setting all its coefficients equal to zero. This means our additive identity is simply $\vec{0}_P = 0$. If it is hard to see this as a polynomial, you can think of it as $0 + 0x + 0x^2 + \cdots$.

5. Now that we know that $\vec{0}_P = 0$, we can look for additive inverses in P. For each $p(x)$, we want to find a $-p(x)$ so that $-p(x) + p(x) = \vec{0}_P$ or $-p(x) + p(x) = 0$. To cancel out a polynomial, we need to cancel out each of its coefficients, so $-p(x)$ is $p(x)$ with the sign of each coefficient switched.

6. If we multiply a polynomial by a scalar, we get another polynomial, so P is closed under scalar multiplication.

7. If we are planning to multiply a polynomial $p(x)$ by two scalars r and s, we have
$$r(sp(x)) = (rs)p(x)$$
so scalar multiplication is associative.

8 Multiplying any polynomial by 1 leaves the polynomial unchanged.

9. If we have two scalars r and s and a polynomial $p(x)$, we know
$$(r + s)p(x) = rp(x) + sp(x).$$

10. If we have two polynomials $p(x)$ and $q(x)$ and a scalar r, we know
$$r \cdot (p(x) + q(x)) = rp(x) + rq(x)$$
so scalar multiplication distributes over addition.

Since all ten properties check out, P with the usual polynomial operations is a vector space.

Often we don't want to consider polynomials of arbitrarily large degree, but instead want to work with a smaller set like all cubics or all quadratics. We'll tackle this whole family of bounded degree polynomial vector spaces at once by leaving the bound on the degree generic and calling it n.

Definition. For any positive integer n, P_n is the set of all polynomials in x of degree at most n.

Note that P_n also includes polynomials whose degrees are smaller than n. If that bothers you, think of those polynomials as having coefficient 0 on the missing powers of the x.

Theorem 2. P_n is a vector space using polynomial addition as its $+$ and multiplication of a polynomial by a scalar as its $\cdot$.

We could check this using our original vector space definition, but a much easier way to approach this is to show instead that P_n is a subspace of P. After all, the operations are the same and the polynomials in a given P_n are certainly a subset of P. This means we need to show that $\vec{0}_P$ is in P_n and that P_n is closed under polynomial addition and scalar multiplication.

The zero vector of P is $\vec{0}_P = 0$. Since 0 is a constant, its degree is 0 which is less than any positive n. Therefore $\vec{0}_P$ is in P_n for any n.

To show closure of addition, suppose we have two polynomials $p(x)$ and $q(x)$ in P_n. We need to show $p(x) + q(x)$ is also in P_n. The sum of two polynomials in x is another polynomial in x. Since $p(x)$ and $q(x)$ are both in P_n, their degrees are both less than or equal to n. Adding polynomials never raises their degree, so $p(x) + q(x)$ will have a degree at most the larger of $\deg(p)$ and $\deg(q)$. Therefore $p(x) + q(x)$'s degree is at most n which means $p(x) + q(x)$ is in P_n.

To show closure of scalar multiplication, suppose we have a polynomial $p(x)$ in P_n and a scalar r. We need to show $rp(x)$ is also in P_n. A scalar multiple of a polynomial in x is another polynomial in x. As above, we have $\deg(p) \leq n$. Multiplying a polynomial by a constant never raises its degree, so $\deg(rp) \leq \deg(p) \leq n$ which means $rp(x)$ is in P_n.

Thus P_n is a subspace of P, and therefore a vector space in its own right.

Now that we know P and P_n are vector spaces, we can develop the same ideas and tools for working with them that we have for $\mathbb{R}^n$ and M_{mn}. We'll start with the idea of linear independence.

Our general definition of linear independence involved the vector equation $x_1\vec{v}_1 + \cdots + x_k\vec{v}_k = \vec{0}$. Here we've already got an x as part of our polynomials, so I'll call our coefficients $a_1, \ldots, a_k$ to avoid confusion. This will mean that we'll be solving the equation $a_1\vec{v}_1 + \cdots + a_k\vec{v}_k = \vec{0}$ for the a_i's not for the x that's part of our polynomials!

Example 1. Are $x^2 - 2x + 1$, $x^3 - 2x + 1$, and $6x - 3$ linearly independent in P_3?

To check whether or not these polynomials are linearly independent, we need to determine if there is a solution to the equation

$$a_1(x^2 - 2x + 1) + a_2(x^3 - 2x + 1) + a_3(6x - 3) = 0$$

other than $a_1 = a_2 = a_3 = 0$. Notice how I carefully put each polynomial in parentheses to ensure that each term is multiplied by its coefficient a_i.

We can expand our equation by multiplying each polynomial through by a_i to get

$$a_1x^2 - 2a_1x + a_1 + a_2x^3 - 2a_2x + a_2 + 6a_3x - 3a_3 = 0.$$

Collecting like terms gives us

$$a_2x^3 + a_1x^2 + (-2a_1 - 2a_2 + 6a_3)x + (a_1 + a_2 - 3a_3) = 0.$$

For a polynomial to equal the zero polynomial, we need each of its coefficients to be zero. This gives us four equations: $a_2 = 0$ from the x^3 term, $a_1 = 0$ from the x^2 term, $-2a_1 - 2a_2 + 6a_3 = 0$ from the x term, and $a_1 + a_2 - 3a_3 = 0$ from the constant term. Since the first two equations tell us $a_1 = a_2 = 0$, we can plug that into either the third or fourth equation to see $a_3 = 0$.

Since all coefficients in our equation must be 0, our polynomials are linearly independent.

We already saw that P_n is a subspace of P, but we can also find many other subspaces of P and P_n. In the following example we see one such subspace and also explore the span of a set of polynomials.

Example 2. Show $W = \{ax^2 + bx + (a+b) \mid a, b \in \mathbb{R}\}$ is a subspace of P_4.

We could check this using the subspace test along the lines of our check that P_n is a subspace of P, but I'll instead write W as the span of a set of polynomials. As we saw in 2.3, spans are always automatically subspaces.

To write W as a span, we need to find a set of polynomials whose linear combinations give us all of W. In other words, we need to rewrite our generic element of W as a linear combination of our spanning polynomials. Notice that there are two important pieces to any polynomial in W, the part controlled by a and the part controlled by b. Separating our generic element of W into the sum of those two parts gives us

$$ax^2 + bx + (a+b) = (ax^2 + a) + (bx + b).$$

We can pull an a out of the first part and a b out of the second part to get

$$ax^2 + bx + (a+b) = a(x^2 + 1) + b(x + 1).$$

Aha! Now our generic element of W is written as a linear combination of $x^2 + 1$ and $x + 1$. This means

$$W = \text{Span}\{x^2 + 1, x + 1\}$$

so W is automatically a subspace of P_4.

Now that we understand how to work within P_n, let's explore the option of mapping problems to $\mathbb{R}^n$ using a basis by finding standard bases for P and P_n along with their coordinate maps.

The standard basis vectors for P_n are $x^n, x^{n-1}, \ldots, x^2, x, 1$. However, there is no widespread consensus about whether this list should start with 1 and go up to x^n or start at x^n and go down to 1, so you should always be careful to specify which order you're using.

Example 3. $\mathcal{B} = \{x^n, x^{n-1}, \ldots, x^2, x, 1\}$ is our standard basis for P_n.

To see that $\mathcal{B}$ is a basis for P_n, we need to check that it spans P_n and is linearly independent.

We can check linear independence by arguing that none of the polynomials from $\mathcal{B}$ are in the span of the rest. This is true because each polynomial in $\mathcal{B}$ has a unique degree. Multiplying by a nonzero scalar doesn't change the degree of a polynomial, and multiplication by zero just gives us 0 which isn't helpful. Adding polynomials of different degrees can't cancel out the term of the largest degree, so will never have a sum of 0. Therefore $\mathcal{B}$ is linearly independent.

To check that $\mathcal{B}$ spans P_n, consider a generic element $a_n x^n + \cdots + a_1 x + a_0$ of P_n. On closer inspection, it is already written as a linear combination of the polynomials from $\mathcal{B}$. Therefore $\mathcal{B}$ spans P_n, so it is a basis for P_n.

Since the dimension of a vector space is defined to be the number of vectors in any basis, we get the following.

Theorem 3. $\dim(P_n) = n + 1$

If you are tempted to think the dimension of P_n is n, remember that a polynomial of degree n has $n + 1$ terms: n powers of x plus the constant term. This means when we write down a polynomial with degree n we are choosing $n+1$ different coefficients, and we've previously described dimension intuitively as the number of choices needed to specify a particular object/point.

Now that we have our standard basis $\mathcal{B}$ for P_n we can take advantage of the corresponding coordinate map $f_{\mathcal{B}} : P_n \to \mathbb{R}^{n+1}$. For example, $f_{\mathcal{B}}$ gives us a new way to answer the following question.

Example 4. Are $x^2 - 2x + 1$, $x^3 - 2x + 1$, and $6x - 3$ linearly independent in P_3?

We could solve this problem in P_3 as we did in Example 1, but instead let's use our coordinate map with respect to the standard basis $\mathcal{B} = \{x^3, x^2, x, 1\}$. Since the entries of a $\mathcal{B}$-coordinate vector are just the coefficients of the polynomial (in order of degree), we get

$$[x^2 - 2x + 1]_{\mathcal{B}} = \begin{bmatrix} 0 \\ 1 \\ -2 \\ 1 \end{bmatrix}, \; [x^3 - 2x + 1]_{\mathcal{B}} = \begin{bmatrix} 1 \\ 0 \\ -2 \\ 1 \end{bmatrix}, \; [6x - 3]_{\mathcal{B}} = \begin{bmatrix} 0 \\ 0 \\ 6 \\ -3 \end{bmatrix}.$$

The matrix with these three columns is

$$\begin{bmatrix} 0 & 1 & 0 \\ 1 & 0 & 0 \\ -2 & -2 & 6 \\ 1 & 1 & 3 \end{bmatrix}$$

which has reduced echelon form

$$\begin{bmatrix} 1 & 0 & 0 \\ 0 & 1 & 0 \\ 0 & 0 & 1 \\ 0 & 0 & 0 \end{bmatrix}.$$

Since we have a leading 1 in each column of the reduced echelon form, our three $\mathcal{B}$-coordinate vectors and hence $x^2 - 2x + 1$, $x^3 - 2x + 1$, and $6x - 3$ are linearly independent.

As with $\mathbb{R}^n$ and M_{mn}, the standard basis is by no means the only basis for P_n.

Example 5. Show that $x^3 - 3, x^2 - 2, x - 1$, and $x^3 + x^2 + x + 1$ form a basis for P_3.

Since we have four polynomials and $\dim(P_3) = 4$, we can invoke Theorem 3 from 3.1 to show this is a basis by showing that our four polynomials span P_3. (We could also have shown that they are linearly independent.) We could do this directly in P_3, but as in the previous example I'll use the coordinate map with respect to the standard basis $\mathcal{B} = \{x^3, x^2, x, 1\}$ to do this in $\mathbb{R}^4$.

Our four polynomials have the following $\mathcal{B}$-coordinate vectors:

$$[x^3 - 3]_\mathcal{B} = \begin{bmatrix} 1 \\ 0 \\ 0 \\ -3 \end{bmatrix}, [x^2 - 2]_\mathcal{B} = \begin{bmatrix} 0 \\ 1 \\ 0 \\ -2 \end{bmatrix}, [x - 1]_\mathcal{B} = \begin{bmatrix} 0 \\ 0 \\ 1 \\ -1 \end{bmatrix}, [x^3 + x^2 + x + 1]_\mathcal{B} = \begin{bmatrix} 1 \\ 1 \\ 1 \\ 1 \end{bmatrix}.$$

The matrix whose columns are our four vectors is

$$\begin{bmatrix} 1 & 0 & 0 & 1 \\ 0 & 1 & 0 & 1 \\ 0 & 0 & 1 & 1 \\ -3 & -2 & -1 & 1 \end{bmatrix}$$

which has reduced echelon form I_4. Since each column of the reduced echelon form has a leading 1, our four coordinate vectors span $\mathbb{R}^4$ and hence our four polynomials span P_3. Therefore $x^3 - 3$, $x^2 - 2$, $x - 1$, and $x^3 + x^2 + x + 1$ form a basis for P_3.

The situation for P is a little stranger. Since there is no upper limit on the degree of polynomials in P we need to include x^k for every positive integer k.

Example 6. The standard basis of P is $1, x, x^2, x^3, \ldots$.

You can convince yourself that this is a basis using a very similar argument to that for P_n.

Since our basis for P has infinitely many basis vectors, we are forced to conclude the following.

Theorem 4. $\dim(P) = \infty$

While we could conceivably still use a coordinate map for P, we'd have to map to $\mathbb{R}^\infty$ (which we haven't defined) so we'll just stick with working in P.

We can also explore linear functions with P or P_n as their domain or codomain. While these functions can't be written as matrix multiplication, we can still check linearity and compute their kernels and ranges.

Example 7. Let $f : P_2 \to M_{22}$ by $f(ax^2 + bx + c) = \begin{bmatrix} a+c & b-2c \\ 0 & a+c \end{bmatrix}$. Show that f is linear and compute its kernel and range.

To check that f is a linear function, we need to show that

$$f(p(x) + q(x)) = f(p(x)) + f(q(x)) \text{ and } f(r \cdot p(x)) = r \cdot f(p(x))$$

for all polynomials $p(x)$ and $q(x)$ and all scalars r. Let's fix the notation $p(x) = ax^2 + bx + c$ and $q(x) = \alpha x^2 + \beta x + \gamma$.

Our first check is that f splits up over addition.

$$
\begin{aligned}
f(p(x) + q(x)) &= f(ax^2 + bx + c + \alpha x^2 + \beta x + \gamma) \\
&= f((a+\alpha)x^2 + (b+\beta)x + (c+\gamma)) \\
&= \begin{bmatrix} (a+\alpha) + (c+\gamma) & (b+\beta) - 2(c+\gamma) \\ 0 & (a+\alpha) + (c+\gamma) \end{bmatrix} \\
&= \begin{bmatrix} (a+c) + (\alpha+\gamma) & (b-2c) + (\beta - 2\gamma) \\ 0 & (a+c) + (\alpha+\gamma) \end{bmatrix} \\
&= \begin{bmatrix} a+c & b-2c \\ 0 & a+c \end{bmatrix} + \begin{bmatrix} \alpha+\gamma & \beta-2\gamma \\ 0 & \alpha+\gamma \end{bmatrix} \\
&= f(ax^2 + bx + c) + f(\alpha x^2 + \beta x + \gamma) = f(p(x)) + f(q(x))
\end{aligned}
$$

Next we'll check that f splits up over scalar multiplication.

$$f(r \cdot p(x)) = f(r(ax^2 + bx + c)) = f((ra)x^2 + (rb)x + (rc))$$

$$= \begin{bmatrix} (ra) + (rc) & (rb) - 2(rc) \\ 0 & (ra) + (rc) \end{bmatrix} = \begin{bmatrix} r(a+c) & r(b-2c) \\ 0 & r(a+c) \end{bmatrix}$$

$$= r \begin{bmatrix} a+c & b-2c \\ 0 & a+c \end{bmatrix} = r \cdot f(ax^2 + bx + c) = r \cdot f(p(x))$$

Since f satisfies both conditions, it is a linear function.

The kernel of f is all polynomials which are mapped to the zero vector of M_{22}. Since $\vec{0}_{M_{22}}$ is the 2×2 zero matrix, this means we want to find the set of all $ax^2 + bx + c$ so that

$$f(ax^2 + bx + c) = \begin{bmatrix} 0 & 0 \\ 0 & 0 \end{bmatrix}.$$

For our function, this means solving for a, b, and c in the equation

$$\begin{bmatrix} a+c & b-2c \\ 0 & a+c \end{bmatrix} = \begin{bmatrix} 0 & 0 \\ 0 & 0 \end{bmatrix}.$$

Setting corresponding entries equal gives us $a+c = 0$, $b-2c = 0$, and $a+c = 0$. The first and third equations tell us we must have $a = -c$, and the second equation tells us we need $b = 2c$. This means our kernel is

$$\ker(f) = \{-cx^2 + 2cx + c\}.$$

If we aren't sure this is correct, remember we can always check our work by computing $f(-cx^2 + 2cx + c)$ and making sure we get the 2×2 zero matrix.

The range of f is the set of all 2×2 matrices which are outputs of our function. In other words, it is the set of all matrices $\begin{bmatrix} v & w \\ y & z \end{bmatrix}$ where

$$f(ax^2 + bx + c) = \begin{bmatrix} v & w \\ y & z \end{bmatrix}$$

for some polynomial $ax^2 + bx + c$. We can find these matrices by solving for a, b, and c in the equation

$$\begin{bmatrix} a+c & b-2c \\ 0 & a+c \end{bmatrix} = \begin{bmatrix} v & w \\ y & z \end{bmatrix}.$$

(Here we are assuming we know the values of v, w, y, and z.) Setting corresponding entries equal gives us $a + c = v$, $b - 2c = w$, $0 = y$, and $a + c = z$. This immediately tells us that for $\begin{bmatrix} v & w \\ y & z \end{bmatrix}$ to be in the range of f, we must have $y = 0$. What values of v, w, and z can we have? We need there to be a solution to the other three linear equations. These equations have augmented coefficient matrix

$$\begin{bmatrix} 1 & 0 & 1 & v \\ 0 & 1 & -2 & w \\ 1 & 0 & 1 & z \end{bmatrix}.$$

We can answer this question by row reduction, so let's do enough of our algorithm to see where the leading 1s are.

$$\begin{bmatrix} 1 & 0 & 1 & v \\ 0 & 1 & -2 & w \\ 1 & 0 & 1 & z \end{bmatrix} \rightarrow_{r_3 - r_1} \begin{bmatrix} 1 & 0 & 1 & v \\ 0 & 1 & -2 & w \\ 0 & 0 & 0 & z-v \end{bmatrix}$$

It is now clear that our first two rows are not an issue, because they contain leading 1s which aren't in the last column. Our third row is a potential problem though, because its equation is $0 = z - v$ which only has a solution when $z = v$. This means that to have a 2×2 matrix $\begin{bmatrix} v & w \\ y & z \end{bmatrix}$ in the range of f means we need $y = 0$ and $z = v$. Thus

$$\text{range}(f) = \left\{ \begin{bmatrix} v & w \\ 0 & v \end{bmatrix} \right\}.$$

Note: Since the kernel is a subspace of the domain, our kernel must be a set of polynomials, not a set of matrices. On the other hand, the range is a subspace of the codomain, and so must be a set of 2×2 matrices not a set of polynomials. It is often helpful to remind yourself of the types of objects in the kernel and range to check your work.

Exercises 3.2.

1. Let $W = \{ax^2 + bx\}$ where a is even and b is positive. (Both a and b are real numbers.) Show that W is closed under the usual polynomial addition but not under scalar multiplication.

2. Redo Example 2 using the subspace test.

3. Show that $W = \{ax^2\}$ with the usual polynomial addition and scalar multiplication is a vector space.

4. Let $V = \{2ax^2 + 2bx + 2c\}$ be the subset of P_2 where all coefficients are even. Show that with the usual polynomial addition and scalar multiplication, V is a vector space.

5. Show that $1 + x^2$, x, and 1 are a basis for P_2.

6. Show that $x^2 + 6x + 9$, $x^2 - 9$, and $-2x + 1$ are a basis for P_2.

7. Is $\mathcal{B} = \{1 + x^2, 1 - x^2, 4\}$ a basis for P_2?

8. Is $\mathcal{B} = \{-x^3 + x^2 - x + 1, x^3 + x, x^2 - x + 1, 2x^2 + 2\}$ a basis for P_3?

9. The set $\mathcal{B} = \{2x^2, x^2 - x + 1, 3x\}$ is a basis for P_2.

   (a) Find the vector $\vec{v}$ in V whose $\mathcal{B}$-coordinate vector is $\begin{bmatrix} -1 \\ 3 \\ 2 \end{bmatrix}$.

   (b) Find the coordinate vector of $3x^2 + 10x - 1$ with respect to basis $\mathcal{B}$.

10. The set $\mathcal{B} = \{1, 1 + x, 1 + x + x^2\}$ is a basis for P_2.

    (a) Find the polynomial $\vec{v}$ whose coordinate vector with respect to this basis is $[\vec{v}]_\mathcal{B} = \begin{bmatrix} 1 \\ 2 \\ 3 \end{bmatrix}$.

    (b) Find the coordinate vector of $\vec{v} = 4 + 2x + 3x^2$ with respect to this basis.

11. The set $\mathcal{B} = \{x^2 + x + 1, 2x^2 + 2, x^2\}$ is a basis for P_2.

    (a) Find the polynomial whose $\mathcal{B}$-coordinate vector is $\begin{bmatrix} 1 \\ 2 \\ 3 \end{bmatrix}$.

    (b) Find the $\mathcal{B}$-coordinate vector of $x^2 - 4x + 6$.

12. The set $\mathcal{B} = \{x + 2, 3x + 1\}$ is a basis for P_1.

    (a) Find the vector $\vec{v}$ which has $[\vec{v}]_\mathcal{B} = \begin{bmatrix} -1 \\ 2 \end{bmatrix}$.

    (b) Find the $\mathcal{B}$-coordinate vector of $x - 1$.

    (c) Find the codomain, W, of the $\mathcal{B}$-coordinate map $f_\mathcal{B} : P_2 \to W$.

13. Decide whether or not $6x + 4$ is in the span of $x^2 + x + 1$ and $x^2 - 2x - 1$ without using a coordinate map.

14. Decide whether or not $2x^2 - x + 10$ is in the span of $x - 2$ and $x^2 + x + 1$ without using a coordinate map.

15. Use a coordinate map to decide whether or not $6x + 4$ is in the span of $x^2 + x + 1$ and $x^2 - 2x - 1$.

16. Use a coordinate map to decide whether or not $2x^2 - x + 10$ is in the span of $x - 2$ and $x^2 + x + 1$.

17. Redo Example 4 in P_3, i.e., without using a coordinate map.

18. Decide whether $x^2 - x + 1$, $x^2 + 2x + 3$, and $5x^2 - 2x$ are linearly independent or linearly dependent without using a coordinate map.

19. Use a coordinate map to see if $x^3 - x^2 - 2$, $-3x^3 + 2x^2 + x - 1$, and $-x^2 + x - 7$ are linearly independent or linearly dependent.

20. Use a coordinate map to see if $x^2 - x + 1$, $x^2 + 2x + 3$, and $5x^2 - 2x$ are linearly independent or linearly dependent.

21. Let $f\left(\begin{bmatrix} a & b \\ c & d \end{bmatrix}\right) = ax^2 + (b+c)x + a$.

   (a) What is the domain of f?
   (b) What is the codomain of f?

22. Let $f(ax^2 + bx + c) = \begin{bmatrix} a+b & 0 \\ a+c & a+b \end{bmatrix}$.

   (a) What is the domain of f?
   (b) What is the codomain of f?

23. Let $f : M_{23} \to P_4$ by $f\left(\begin{bmatrix} a & b & c \\ d & e & f \end{bmatrix}\right) = ax^4 + (b+c)x^2 + (d-f)$.

   (a) Find the kernel of f.
   (b) Find the range of f.

24. Let $f : P_2 \to M_{22}$ by $f = (ax^2 + bx + c) = \begin{bmatrix} 2a & 0 \\ 0 & b+c \end{bmatrix}$.

   (a) Find the kernel of f.
   (b) Find the range of f.

25. Let $f : P_2 \to M_{22}$ be given by $f(ax^2+bx+c) = \begin{bmatrix} a-b & b+c \\ a+b+2c & 0 \end{bmatrix}$.

   (a) Find the kernel of f.
   (b) Find the range of f.

26. Let $f : P_2 \to M_{33}$ by $f(ax^2 + bx + c) = \begin{bmatrix} a-b+c & 0 & 0 \\ 0 & a-b & 0 \\ 0 & 0 & 2c \end{bmatrix}$.

   (a) Find the kernel of f.
   (b) Find the range of f.

3.3   Other Vector Spaces

Although $\mathbb{R}^n$, M_{mn}, and P_n are our main examples of vector spaces, there are many more examples of familiar mathematical environments which also turn out to be vector spaces. In this section we'll explore two of these: the complex numbers and the set of continuous functions from $\mathbb{R}$ to $\mathbb{R}$.

The complex numbers, $\mathbb{C}$, are an extension of the real numbers formed by including a new number i which is defined to be the square root of -1. They were developed to help solve equations like $x^2 + 4 = 0$ which don't have a solution in $\mathbb{R}$. Such equations may not seem relevant to the real world, but they do come up in practical applications and can help solve real problems. The way we include i into $\mathbb{R}$ is the same way we include any other real number. It is part of the set, but also can be added to and multiplied by other numbers. This means we can have things like $7 + i$ and $-3i$. We can also multiply i by itself, but since $i^2 = -1$ any power of i can be reduced to $\pm i$. We'll formally define $\mathbb{C}$ as follows.

Definition.   The *complex numbers* are $\mathbb{C} = \{a + bi \,|\, a, b \text{ in } \mathbb{R}\}$.

The first term a is typically called the real part of a complex number, while the second term is called the imaginary part. (For a more detailed discussion of $\mathbb{C}$, see Appendix A.1.)

Geometrically, we can think of $\mathbb{C}$ as a plane by plotting the complex number $a + bi$ as the point (a, b). For this reason, many people call the x-axis the real axis and the y-axis the imaginary axis. To illustrate this, look at the plot of $4 + 2i$ in Figure 3.1.

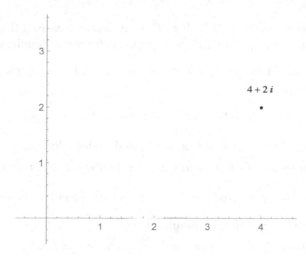

Figure 3.1: A geometric representation of a complex number

Complex numbers are added and multiplied by scalars as if they were polynomials in i. In other words,

$$(a + bi) + (x + yi) = (a + x) + (b + y)i$$

and

$$r(a + bi) = (ra) + (rb)i.$$

Keep in mind that our scalar r is from $\mathbb{R}$ not $\mathbb{C}$. (To learn about the rule for multiplying a complex number by another complex number, see Appendix A.1.)

Example 1. Compute $(2 + 4i) + (-1 + 7i)$.

Following our rule above, to add these two complex numbers, we add their real and imaginary parts separately. This gives us

$$(2 + 4i) + (-1 + 7i) = (2 - 1) + (4 + 7)i = -1 + 11i.$$

Example 2. Compute $-10(3 - 2i)$.

To multiply this complex number by -10, we multiply both the real and imaginary parts by -10. This gives us

$$-10(3 - 2i) = -10(3) - 10(-2)i = -30 + 20i.$$

Theorem 1. $\mathbb{C}$ is a vector space using complex addition as its $+$ and multiplication of a complex number by a scalar as its $\cdot$.

As with our check in 3.2 that P is a vector space, I'll number my explanations to correspond with the conditions from the definition.

1. If we have two complex numbers $a + bi$ and $x + yi$, their sum is defined to be

    $$(a + bi) + (x + yi) = (a + x) + (b + y)i$$

    which is clearly in $\mathbb{C}$. Thus $\mathbb{C}$ is closed under addition.

2. The order in which we add complex numbers doesn't matter since

    $$(a+bi)+(x+yi) = (a+x)+(b+y)i = (x+a)+(y+b)i = (x+yi)+(a+bi)$$

    so complex addition is commutative.

3. If we have three complex numbers $a + bi$, $x + yi$, and $w + vi$, which

pair we add first doesn't matter since

$$[(a + bi) + (x + yi)] + (w + vi) = [(a + x) + (b + y)i] + (w + vi)$$
$$= (a + x + w) + (b + y + v)i$$
$$= (a + bi) + [(x + w) + (y + v)i]$$
$$= (a + bi) + [(x + yi) + (w + vi)].$$

Thus complex addition is associative.

4. Our zero vector in $\mathbb{C}$ will be a complex number $a + bi$ so that

$$(a + bi) + (x + yi) = x + yi$$

for all $x + yi$. We can do this by simply setting $a = b = 0$, so our $\vec{0}$ in $\mathbb{C}$ is simply 0. (Alternately, you can think of $\vec{0}_\mathbb{C}$ as $0 + 0i$.)

5. Since complex numbers add component-wise and $\vec{0}_\mathbb{C} = 0$, the additive inverse of a complex number will be one whose values for a and b are the opposites of our original complex number. This means the additive inverse of $a + bi$ is $-a - bi$.

6. If we have a complex number $a + bi$ and a scalar r, their product is defined to be
$$r(a + bi) = (ra) + (rb)i$$

which is clearly in $\mathbb{C}$, so $\mathbb{C}$ is closed under scalar multiplication.

7. Whether we multiply a complex number $a + bi$ by a scalar s and then another scalar r or simply by the product of the two scalars rs doesn't matter since

$$r(s(a + bi)) = r((sa) + (sb)i) = (rsa) + (rsb)i = (rs)(a + bi)$$

so scalar multiplication is associative.

8. Multiplying a complex number $a + bi$ by 1 just means multiplying a and b by 1, so $1(a + bi) = a + bi$.

9. Multiplying a complex number by a sum of two scalars distributes since

$$(r + s)(a + bi) = ((r + s)a) + ((r + s)b)i = (ra + sa) + (rb + sb)i$$
$$= ((ra) + (rb)i) + ((sa) + (sb)i) = r(a + bi) + s(a + bi).$$

10. Multiplying a sum of complex numbers by a scalar also distributes since

$$r((a + bi) + (x + yi)) = r((x + a) + (y + b)i) = r(x + a) + r(y + b)i$$
$$= (rx + ra) + (ry + rb)i = r(a + bi) + r(x + yi).$$

Therefore $\mathbb{C}$ is a vector space as claimed.

Now that we know $\mathbb{C}$ is a vector space, we can explore all the properties we looked at in $\mathbb{R}^n$, M_{mn}, and P_n. Let's start with subspaces.

Example 3. Show $W = \{bi \mid b \text{ in } \mathbb{R}\}$ is a subspace of $\mathbb{C}$.

Before we run our subspace test, notice that W can be thought of as all complex numbers $a + bi$ where $a = 0$. Since these complex numbers have no real part, they are often called purely imaginary numbers. Geometrically, this means W is the set pictured below.

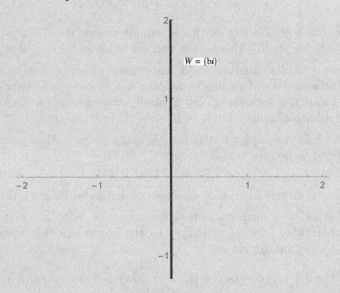

The zero vector of $\mathbb{C}$ is 0, which we can see is in W by letting $b = 0$ to get $0i$. If we have two elements bi and ci in W, their sum is

$$bi + ci = (b + c)i$$

which is also in W. Thus W is closed under addition. If we have an element bi in W and a scalar r, then

$$r(bi) = (rb)i$$

which is in W. Thus W is closed under scalar multiplication. Since it satisfies all three conditions of the subspace test, W is a subspace of $\mathbb{C}$.

Next let's figure out the dimension of $\mathbb{C}$ by finding a basis.

Example 4. Show 1 and i are a basis for $\mathbb{C}$.

From our definition of $\mathbb{C}$ it is clear that every complex number is a linear combination of 1 and i, so $\mathbb{C} = \text{Span}\{1, i\}$. Since there are only two spanning elements, we can check linear independence by seeing if one of them is in the span of the other. There is no real number a which will make ai a multiple of 1, so 1 and i are linearly independent. Therefore 1 and i are a basis for $\mathbb{C}$.

In fact, 1 and i are the standard basis for $\mathbb{C}$, although their ordering isn't standard so be careful to specify whether you're using the basis $1, i$ or the basis $i, 1$.

Since $\mathbb{C}$ has a basis with two elements, we know:

Theorem 2. $\dim(\mathbb{C}) = 2$

Now that we have access to a coordinate map via our standard basis, let's explore linear independence in $\mathbb{C}$.

Example 5. Are $1 + i$, $-2 + 14i$, and $10 - 0i$ linearly independent?

We don't actually need to do any computations to answer this question. The dimension of $\mathbb{C}$ is 2, so no matter which basis we pick for $\mathbb{C}$, our coordinate map is $f_B : \mathbb{C} \to \mathbb{R}^2$. This means if we use our coordinate map to translate this question into $\mathbb{R}^2$, we're asking if three vectors in $\mathbb{R}^2$ are linearly independent. This is clearly impossible, because we can't get leading 1s in all three columns of a 2×3 matrix. Therefore $1 + i$, $-2 + 14i$, and $10 - 6i$ are linearly dependent.

Example 6. Are $-2 + 14i$ and $10 - 6i$ linearly independent?

Now that we're only talking about two complex numbers, it is possible for them to be linearly independent. Let's check using the coordinate map with respect to the standard basis $\mathcal{B} = \{1, i\}$. Coordinate vectors with respect to this basis have first entry equal to the real part of the complex number (the part without the i) and second entry equal to the coefficient on i. This means our two coordinate vectors are

$$[-2 + 14i]_\mathcal{B} = \begin{bmatrix} -2 \\ 14 \end{bmatrix} \text{ and } [10 - 6i]_\mathcal{B} = \begin{bmatrix} 10 \\ -6 \end{bmatrix}.$$

The matrix with these two columns is

$$\begin{bmatrix} -2 & 10 \\ 14 & -6 \end{bmatrix},$$

which has reduced echelon form I_2. Therefore, by the Invertible Matrix Theorem, its columns are linearly independent, which means $-2 + 14i$ and $10 - 6i$ are linearly independent as well.

As with the other vector spaces we've seen, we can create linear maps with $\mathbb{C}$ as their domain or codomain.

Example 7. Find the kernel of the map $f : \mathbb{C} \to M_{22}$ given by

$$f(a + bi) = \begin{bmatrix} a - 2b & 0 \\ 0 & -2a + 4b \end{bmatrix}.$$

As usual, we find the kernel of f by setting f's output equal to the zero vector of the codomain, which in this case is M_{22}. This means we want to solve

$$\begin{bmatrix} a - 2b & 0 \\ 0 & -2a + 4b \end{bmatrix} = \begin{bmatrix} 0 & 0 \\ 0 & 0 \end{bmatrix}$$

so we must have $a - 2b = 0$ and $-2a + 4b = 0$. Both of these equations give us $a = 2b$, so we get

$$\ker(f) = \{2b + bi\}.$$

Using our geometric identification of $\mathbb{C}$ with the plane, we can draw the kernel by rewriting our defining equation of the kernel as $b = \frac{1}{2}a$. This means $\ker(f)$ can be thought of as the line pictured below.

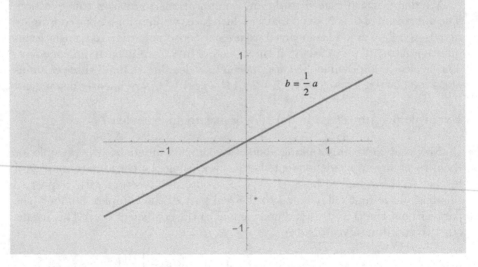

Now that we've explored our new vector space $\mathbb{C}$, let's turn our attention to another, more familiar, collection of mathematical objects: the set of continuous functions from $\mathbb{R}$ to $\mathbb{R}$, which we will call $\mathscr{C}$. Notice that this set includes many functions we haven't talked about in linear algebra so far, like e^x and $2x + \sin(x)$. These may not seem like things that belong in our linear world, but as we did with polynomials, we can use them as the mathematical objects, i.e., vectors in a new vector space. Note that as with polynomials, we mean the "objects" part of this idea seriously. We don't get to treat e^x as a function if we want to use it here, so as with P, resist the urge to plug things into your continuous functions, solve for x, etc.

The other important building block of any vector space is its two operations $+$ and $\cdot$. For our operations, we'll use the usual notion of adding and scaling functions you worked with in calculus.

Theorem 3. $\mathscr{C}$ is a vector space using addition of functions as its $+$ and multiplication of a function by a scalar as its $\cdot$.

As above, I'll number my explanations to correspond with the conditions from the definition.

1. If we have two continuous functions f and g from $\mathbb{R}$ to $\mathbb{R}$, their sum $f + g$ is clearly also a function from $\mathbb{R}$ to $\mathbb{R}$. As you saw in calculus, the fact that limits split up over sums implies that $f + g$ is continuous. Therefore $f + g$ is in $\mathscr{C}$, so $\mathscr{C}$ is closed under addition.

2. In calculus, you also saw that $f + g = g + f$, so function addition is commutative.

3. Again, in calculus you saw that $(f + g) + h = f + (g + h)$, so function addition is associative.

4. To find our additive identity, we need a function $\vec{0}$ in $\mathscr{C}$ so that $f + 0 = f$. Since constant functions are always continuous, we can let $\vec{0}_{\mathscr{C}}$ be the zero function $z(x) = 0$.

5. To find the additive inverse of a continuous function f, we need to find a function which cancels out each output of f so that their sum is the zero function. We can do this by simply multiplying f by -1. This will be continuous since f is continuous.

6. As with addition, the fact from calculus that scalar multiplication can be pulled out of a limit tells us that if $f(x)$ is continuous then $rf(x)$ is also continuous. Therefore $\mathscr{C}$ is closed under scalar multiplication.

7. Since multiplying a function by a scalar really means multiplying the function's outputs by that scalar, the fact that multiplication in $\mathbb{R}$ is associative gives us $r(sf(x)) = (rs)f(x)$. Thus scalar multiplication in $\mathscr{C}$ is associative.

8. Multiplying a function by 1 doesn't change its outputs, so $1f(x) = f(x)$.

9. From working with continuous functions in calculus you've already seen that $(r + s)f(x) = rf(x) + sf(x)$, so scalar multiplication distributes over addition of scalars.

10. Similarly, $r(f(x) + g(x)) = rf(x) + rg(x)$, so scalar multiplication distributes over function addition.

Therefore $\mathscr{C}$ with our usual operations on functions is a vector space.

As with the complex numbers, we can now explore some of our vector space ideas. Again, let's start with subspaces.

Example 8. Show P is a subspace of $\mathscr{C}$.

Since every polynomial in one variable is a continuous function from $\mathbb{R}$ to $\mathbb{R}$, we know P is a subset of $\mathscr{C}$. In fact, our notions of function addition and scalar multiplication are exactly the same as those of polynomial addition and scalar multiplication in the case where our function is a polynomial. Since we've already shown that P is a vector space, we can now conclude that P is a subspace of $\mathscr{C}$.

However, this exposes a problem with finding a basis for $\mathscr{C}$ and using its coordinate map to tackle spans and linear independence. From 3.1 we know that $\dim(P) \leq \dim(\mathscr{C})$ and $\dim(P) = \infty$, so we must have the following.

Theorem 4. $\dim(\mathscr{C}) = \infty$

Since $\mathscr{C}$ is infinite dimensional, it won't have a nice coordinate map to $\mathbb{R}^n$. In fact, although it can be proved that a basis for $\mathscr{C}$ exists, actually constructing a basis for $\mathscr{C}$ is difficult enough to be beyond the scope of this book.

Exercises 3.3.

1. Show $\mathbb{R}$ is a subspace of $\mathbb{C}$.

2. Is $W = \{1 + bi\}$ a subspace of $\mathbb{C}$?

3. Show that $1 + i$ and $1 - i$ form a basis for $\mathbb{C}$.

4. Show that $4 - 2i$ and $1 - 4i$ form a basis for $\mathbb{C}$.

5. Is $\{2 - i, 4 - 2i\}$ a basis for $\mathbb{C}$?

6. Is $\{2 + 2i, -3 + 3i\}$ a basis for $\mathbb{C}$?

7. Let $\mathcal{B} = \{1 + i, i\}$ be a basis for $\mathbb{C}$.

   (a) Find the complex number $\vec{v}$ with $[\vec{v}]_{\mathcal{B}} = \begin{bmatrix} -3 \\ 5 \end{bmatrix}$.

   (b) Find the $\mathcal{B}$-coordinate vector of $2 - 6i$.

8. Let $\mathcal{B} = \{1 + i, 1 - i\}$ be a basis for $\mathbb{C}$.

   (a) Find the complex number $\vec{v}$ with $[\vec{v}]_{\mathcal{B}} = \begin{bmatrix} 1 \\ 7 \end{bmatrix}$.

   (b) Find the $\mathcal{B}$-coordinate vector of $8 + 3i$.

9. Let $\mathcal{B} = \{4 - 2i, 1 - 4i\}$ be a basis for $\mathbb{C}$.

   (a) Find the complex number $\vec{v}$ with $[\vec{v}]_{\mathcal{B}} = \begin{bmatrix} -3 \\ -1 \end{bmatrix}$.

   (b) Find the $\mathcal{B}$-coordinate vector of $12 + 4i$.

10. Let $\mathcal{B} = \{-1 + 2i, 2 + 5i\}$ be a basis for $\mathbb{C}$.

    (a) Find the complex number $\vec{v}$ with $[\vec{v}]_{\mathcal{B}} = \begin{bmatrix} 4 \\ 1 \end{bmatrix}$.

    (b) Find the $\mathcal{B}$-coordinate vector of $10 + i$.

11. Decide whether $6 - 9i$ and $12 - 18i$ are linearly independent or linearly dependent without using a coordinate map.

12. Decide whether $5 + 4i$ and $4 + 5i$ are linearly independent or linearly dependent without using a coordinate map.

13. Use a coordinate map to decide whether $6 - 9i$ and $12 - 18i$ are linearly independent or linearly dependent.

14. Use a coordinate map to decide whether $5 + 4i$ and $4 + 5i$ are linearly independent or linearly dependent.

15. Decide whether or not $4 - 2i$ and $10 + i$ span $\mathbb{C}$ without using a coordinate map.

16. Decide whether or not $20 + 8i$ and $15 + 6i$ span $\mathbb{C}$ without using a coordinate map.

17. Use a coordinate map to decide whether or not $4 - 2i$ and $10 + i$ span $\mathbb{C}$.

18. Use a coordinate map to decide whether or not $20 + 8i$ and $15 + 6i$ span $\mathbb{C}$.

19. Let $f(a + bi) = \begin{bmatrix} -a & a \\ 0 & b \end{bmatrix}$.

    (a) Find the domain of f.

    (b) Find the codomain of f.

20. Let $f\left(\begin{bmatrix} x_1 \\ x_2 \\ x_3 \end{bmatrix} \right) = (x_1 + x_2) + (x_1 - x_3)i$.

    (a) Find the domain of f.

    (b) Find the codomain of f.

21. Let $f(a + bi) = \begin{bmatrix} -a & a \\ 0 & b \end{bmatrix}$.

    (a) Find the kernel of f.

    (b) Find the range of f.

22. Let $f\left(\begin{bmatrix} x_1 \\ x_2 \\ x_3 \end{bmatrix}\right) = (x_1 + x_2) + (x_1 - x_3)i.$

    (a) Find the kernel of f.
    (b) Find the range of f.

23. Let $f\left(\begin{bmatrix} a & b \\ c & d \end{bmatrix}\right) = (a - 2b) + ci.$

    (a) Find the kernel of f.
    (b) Find the range of f.

24. Let $f(a + bi) = -bi.$

    (a) Find the kernel of f.
    (b) Find the range of f.

25. Let $W = \{A\sin(x) + B \mid A, B \text{ in } \mathbb{R}\}$, so W is the set of sine functions with all possible amplitudes and midlines. (Adjusting the amplitude and midline of a sine function can be very helpful in modeling periodic behavior.) Show W is a subspace of $\mathscr{C}$.

26. Show that the integers, $\mathbb{Z}$, do not form a vector space with their usual addition and scalar multiplication.

27. Show that the set of all differentiable functions, $\mathscr{D}$, from $\mathbb{R} \to \mathbb{R}$ with our usual addition and scalar multiplication of functions is a vector space.

28. Let V be the set of positive real numbers. Define "addition" to be our usual multiplication, so $a\text{"+"}b = ab$, and "scalar multiplication" by r to be the rth power, so $r\text{"}\cdot\text{"}a = a^r$. Is V a vector space? Show that your answer is correct.

29. Let $V = \mathbb{R}$ with the usual addition and multiplication of real numbers, and $W = R_{>0}$ be the vector space of positive real numbers whose "addition" is multiplication and whose "scalar multiplication" raises the real number to that scalar power (see the previous exercise). Define $f : V \to W$ by $f(x) = 2^x$. Either show that f is a linear map, or show that it is not.

4

Diagonalization

4.1 Eigenvalues and Eigenvectors

In Example 5 from 2.2, we developed a function to predict population sizes in different life stages of the smooth coneflower from one year to the next. Suppose we aren't interested in just predicting the next year's population, but instead want to know what the population will look like in 10 years or 20 years. We can even ask whether the population will survive in the long term or whether it is dying out. To do this, we can repeatedly apply our linear function representing the population change over each year to our initial population vector. This means we are repeatedly multiplying by a fixed matrix.

This need to multiply many times by the same matrix also comes up in game theory, when we study movement around the board of games like Monopoly. Here our vector entries represent the likelihood of ending up on the various spaces on the board, and the matrix we're multiplying by contains the probabilities of moving from one space to another.

In this chapter, we'll start developing a method which can help us simplify problems with repeated matrix multiplication. As with scalar multiplication, we can rewrite repeated matrix multiplication as multiplication by the powers of a matrix. For example, if f has matrix A, then $f(f(\vec{x})) = A(A(\vec{x})) = A^2\vec{x}$. In general, these problems are complicated whether we are repeatedly applying a linear map, repeatedly multiplying by a matrix, or taking powers of a matrix. However, there is a case where matrix operations behave much better because some strategically chosen entries are zero.

Suppose we are multiplying matrices $A = \begin{bmatrix} a & b \\ c & d \end{bmatrix}$ and $Z = \begin{bmatrix} x & y \\ z & w \end{bmatrix}$. According to our usual rules of matrix multiplication, we'd get

$$AZ = \begin{bmatrix} ax + bz & ay + bw \\ cx + dz & cy + dw \end{bmatrix}.$$

One way to make this less complicated is to set enough entries of A equal to zero so that each entry of their product contains only one term rather than being a sum of two terms. We can eliminate one term from the sum in each matrix entry by making b and c equal zero. This would make $A = \begin{bmatrix} a & 0 \\ 0 & d \end{bmatrix}$

and $AZ = \begin{bmatrix} ax & ay \\ dz & dw \end{bmatrix}$. We could have set other entries of A equal to zero, but this gives the nice additional result that each row of Z is being scaled by the corresponding entry of A.

We could also go back and focus on the entries of Z, but instead let's generalize the type of matrix we created with A.

Definition. An $n \times n$ matrix is *diagonal* if its entries a_{ij} are zero unless $i = j$.

(Remember that our notation is that a_{ij} is the entry of A in the ith row and jth column.)

Example 1. $A = \begin{bmatrix} 2 & 0 & 0 \\ 0 & 0 & 0 \\ 0 & 0 & -3 \end{bmatrix}$ is a diagonal matrix.

All entries off the diagonal are zero, so this matrix is diagonal. (Notice that one of our diagonal entries is also zero, which is fine because diagonal entries are allowed to be any real number.)

Matrix multiplication is much easier if we stick to diagonal matrices. To multiply two $n \times n$ diagonal matrices, simply multiply their corresponding diagonal entries.

Example 2. Compute $\begin{bmatrix} -2 & 0 & 0 \\ 0 & 4 & 0 \\ 0 & 0 & 7 \end{bmatrix} \begin{bmatrix} 1 & 0 & 0 \\ 0 & 3 & 0 \\ 0 & 0 & -2 \end{bmatrix}$.

If we do this using our regular definition of matrix multiplication, we get

$$\begin{bmatrix} -2 & 0 & 0 \\ 0 & 4 & 0 \\ 0 & 0 & 7 \end{bmatrix} \begin{bmatrix} 1 & 0 & 0 \\ 0 & 3 & 0 \\ 0 & 0 & -2 \end{bmatrix} = \begin{bmatrix} -2(1) & 0 & 0 \\ 0 & 4(3) & 0 \\ 0 & 0 & 7(-2) \end{bmatrix} = \begin{bmatrix} -2 & 0 & 0 \\ 0 & 12 & 0 \\ 0 & 0 & -14 \end{bmatrix}.$$

Notice that the product above does indeed work out to be a diagonal matrix whose entries are the products of the diagonal entries of our two diagonal matrices.

Another way to view multiplication by a diagonal matrix is to decompose our diagonal matrix into a product of elementary matrices. Each elementary matrix will correspond to a row operation where we scale some row by a constant. This allows us to compute the matrix product AB where A is a diagonal matrix by scaling each row of B by A's diagonal entry in that row. (This is the bonus nice property we noticed in our discussion before the definition of a diagonal matrix.)

Example 3. Use the multiplication technique described above to compute
AB where $A = \begin{bmatrix} 1 & 0 & 0 \\ 0 & 3 & 0 \\ 0 & 0 & -2 \end{bmatrix}$ and $B = \begin{bmatrix} 2 & 7 & 4 \\ -1 & 5 & 2 \\ 6 & -3 & 0 \end{bmatrix}$.

We could do this using our usual method for matrix multiplication, but it's easier to simply multiply each row of B by the corresponding diagonal entry of A. For our A, this means multiplying B's first row by 1, its second row by 3, and its third row by -2. This gives us

$$AB = \begin{bmatrix} 2(1) & 7(1) & 4(1) \\ -1(3) & 5(3) & 2(3) \\ 6(-2) & -3(-2) & 0(-2) \end{bmatrix} = \begin{bmatrix} 2 & 7 & 4 \\ -3 & 15 & 6 \\ -12 & 6 & 0 \end{bmatrix}.$$

Be careful to remember that this method doesn't work for computing AB if B is diagonal, since matrix multiplication doesn't commute. It turns out that there is a related shortcut in that case, which you'll explore in the exercises.

We can also think about multiplication by a diagonal matrix geometrically. A diagonal matrix can be interpreted as a transformation which scales each axis of $\mathbb{R}^n$ by the corresponding diagonal entry. In Example 1 above, if we think of the three axes of $\mathbb{R}^3$ in order as the x-axis, y-axis, and z-axis, our matrix scales the x-axis by 2, the y-axis by 0, and the z-axis by 3.

Now that we understand diagonal matrix multiplication, let's shift our attention back to taking powers of a matrix. If our matrix is diagonal, the shortcut used in Example 3 says we are repeatedly scaling each row of our matrix by its own diagonal entries.

Example 4. Compute A^2 and A^3 where $A = \begin{bmatrix} 3 & 0 \\ 0 & -2 \end{bmatrix}$.

Let's start with $A^2 = A \cdot A$. We saw in Example 2 that we can compute the product of two diagonal matrices by multiplying their diagonal entries together and using those products as the diagonal entries of the matrix product. Since here we are multiplying A by itself, the diagonal entries will be the products of the old diagonal entries by themselves, i.e., the squares of the diagonal entries. Computationally, this looks like

$$\begin{bmatrix} 3 & 0 \\ 0 & -2 \end{bmatrix}^2 = \begin{bmatrix} 3 & 0 \\ 0 & -2 \end{bmatrix}\begin{bmatrix} 3 & 0 \\ 0 & -2 \end{bmatrix} = \begin{bmatrix} 3(3) & 0 \\ 0 & -2(-2) \end{bmatrix} = \begin{bmatrix} 3^2 & 0 \\ 0 & (-2)^2 \end{bmatrix} = \begin{bmatrix} 9 & 0 \\ 0 & 4 \end{bmatrix}.$$

Finding $A^3 = A \cdot A \cdot A$ means doing this process three times, so we get

$$\begin{bmatrix} 3 & 0 \\ 0 & -2 \end{bmatrix}\begin{bmatrix} 3 & 0 \\ 0 & -2 \end{bmatrix}\begin{bmatrix} 3 & 0 \\ 0 & -2 \end{bmatrix} = \begin{bmatrix} 3(3)(3) & 0 \\ 0 & -2(-2)(-2) \end{bmatrix}$$

$$= \begin{bmatrix} 3^3 & 0 \\ 0 & (-2)^3 \end{bmatrix} = \begin{bmatrix} 27 & 0 \\ 0 & -8 \end{bmatrix}.$$

Notice that our cubed matrix's diagonal entries are the cubes of the original matrix's diagonal entries.

We can generalize what happened here to say that we can take the kth power of a diagonal matrix by simply taking the kth power of each of the matrix's diagonal entries.

Example 5. Compute $\begin{bmatrix} -1 & 0 & 0 & 0 \\ 0 & 3 & 0 & 0 \\ 0 & 0 & 2 & 0 \\ 0 & 0 & 0 & -2 \end{bmatrix}^5$.

This would be annoying to do without our shortcut, but it's easy to do by taking the 5th power of each diagonal entry. Then our computation is just

$$\begin{bmatrix} -1 & 0 & 0 & 0 \\ 0 & 3 & 0 & 0 \\ 0 & 0 & 2 & 0 \\ 0 & 0 & 0 & -2 \end{bmatrix}^5 = \begin{bmatrix} (-1)^5 & 0 & 0 & 0 \\ 0 & (3)^5 & 0 & 0 \\ 0 & 0 & (2)^5 & 0 \\ 0 & 0 & 0 & (-2)^5 \end{bmatrix}$$

$$= \begin{bmatrix} -1 & 0 & 0 & 0 \\ 0 & 243 & 0 & 0 \\ 0 & 0 & 32 & 0 \\ 0 & 0 & 0 & -32 \end{bmatrix}.$$

Diagonal matrices also make multiplying a matrix and a vector easier. If we think about our geometric interpretation of scaling each axis in $\mathbb{R}^n$ by its corresponding diagonal entry of the matrix, it should make sense that we can multiply a vector by a diagonal matrix by multiplying each vector entry by the matrix's corresponding diagonal entry.

Example 6. Compute $\begin{bmatrix} 2 & 0 & 0 \\ 0 & 0 & 0 \\ 0 & 0 & -3 \end{bmatrix} \begin{bmatrix} 5 \\ 11 \\ -2 \end{bmatrix}$.

We can compute this matrix vector product by scaling the first entry of our vector by 2, the second by 0, and the third by -3 to give us

$$\begin{bmatrix} 2 & 0 & 0 \\ 0 & 0 & 0 \\ 0 & 0 & -3 \end{bmatrix} \begin{bmatrix} 5 \\ 11 \\ -2 \end{bmatrix} = \begin{bmatrix} 2(5) \\ 0(11) \\ -3(-2) \end{bmatrix} = \begin{bmatrix} 10 \\ 0 \\ 6 \end{bmatrix}.$$

Combining this with our previous discussion of powers of diagonal matrices, we see that repeated multiplication of a vector by a diagonal matrix is not terribly hard.

Example 7. Compute $\begin{bmatrix} 2 & 0 & 0 \\ 0 & 0 & 0 \\ 0 & 0 & -3 \end{bmatrix}^4 \begin{bmatrix} 5 \\ 11 \\ -2 \end{bmatrix}$.

We can either compute $\begin{bmatrix} 2 & 0 & 0 \\ 0 & 0 & 0 \\ 0 & 0 & -3 \end{bmatrix}^4$ and then multiply it by $\begin{bmatrix} 5 \\ 11 \\ -2 \end{bmatrix}$, or think of repeating the multiplication in Example 6 four times in a row. In either case, we'll get

$$\begin{bmatrix} 2 & 0 & 0 \\ 0 & 0 & 0 \\ 0 & 0 & -3 \end{bmatrix}^4 \begin{bmatrix} 5 \\ 11 \\ -2 \end{bmatrix} = \begin{bmatrix} 2^4(5) \\ 0^4(11) \\ (-3)^4(-2) \end{bmatrix} = \begin{bmatrix} 16(5) \\ 0(11) \\ 81(-2) \end{bmatrix} = \begin{bmatrix} 80 \\ 0 \\ -162 \end{bmatrix}.$$

The only way this could be easier is if all diagonal entries of our matrix were equal. Then we could simply multiply all entries of our vector by the same number. This is possible, and sometimes happens even when our matrix A isn't diagonal. For example, consider $A = \begin{bmatrix} 1 & 2 \\ 4 & 3 \end{bmatrix}$ and $\vec{x} = \begin{bmatrix} 1 \\ 2 \end{bmatrix}$. Their product is $A\vec{x} = \begin{bmatrix} 1 & 2 \\ 4 & 3 \end{bmatrix} \begin{bmatrix} 1 \\ 2 \end{bmatrix} = \begin{bmatrix} 5 \\ 10 \end{bmatrix}$. If we compare this with $\vec{x}$, we see that $A\vec{x} = 5\vec{x}$ so multiplication by A scales each entry of x by 5 even though A is not diagonal! This is a very special situation that we'll spend some time exploring. We start by giving the vectors that have this special sort of relationship with A a name.

Definition. A nonzero vector $\vec{x}$ is an *eigenvector* of a matrix A with *eigenvalue* λ if $A\vec{x} = \lambda\vec{x}$.

In other words, $\vec{x}$ is an eigenvector if multiplying $\vec{x}$ by A is the same as multiplying $\vec{x}$ by some scalar λ. Geometrically, this means that A scales its eigenvector by λ along each axis of $\mathbb{R}^n$. Notice that having $A\vec{x} = \lambda\vec{x}$ means every eigenvector has a unique eigenvalue. (We'll see later on that each eigenvalue has many different eigenvectors.)

Example 8. Show $\vec{y} = \begin{bmatrix} -1 \\ 1 \end{bmatrix}$ is an eigenvector of $A = \begin{bmatrix} 1 & 2 \\ 4 & 3 \end{bmatrix}$ with eigenvalue -1.

To see this, we need to check that $A\vec{y} = (-1)\vec{y}$. Doing this computation gives us

$$\begin{bmatrix} 1 & 2 \\ 4 & 3 \end{bmatrix} \begin{bmatrix} -1 \\ 1 \end{bmatrix} = \begin{bmatrix} -1+2 \\ -4+3 \end{bmatrix} = \begin{bmatrix} 1 \\ -1 \end{bmatrix} = \begin{bmatrix} -1(-1) \\ (-1)1 \end{bmatrix} = (-1) \begin{bmatrix} -1 \\ 1 \end{bmatrix}.$$

This shows $\vec{y}$ is an eigenvector of A with eigenvalue -1 as claimed.

Since this was the same matrix we discussed before the definition of an eigenvector, we know $\vec{x} = \begin{bmatrix} 1 \\ 2 \end{bmatrix}$ is also an eigenvector of A with eigenvalue 5. This shows us that a given matrix can have many eigenvectors with different eigenvalues.

We can use eigenvalues to answer the question of whether a population grows larger or dies out over time. (A population dies out if eventually all entries of its population vector go to the zero, i.e., its population vector goes to the zero vector.)

Example 9. Suppose a population has demographic matrix A so that if $\vec{x}$ is the population vector for one year, then $A\vec{x}$ is the population vector for the next year. If $\vec{x}$ is an eigenvector of A with eigenvalue $\lambda = 0.75$, does this population grow or die out in the long run?

To figure out what happens to our population in the long run, we want to multiply our population vector $\vec{x}$ by increasingly large powers of our demographic matrix A. In other words, we want to look at the limit of $A^k \vec{x}$ as $k \to \infty$.

Since $A\vec{x} = 0.75\vec{x}$, we know $A^2\vec{x} = (0.75)^2\vec{x}$. Multiplying by A again gives us

$$A^3\vec{x} = A(0.75)^2\vec{x} = (0.75)^2 A\vec{x} = (0.75)^2(0.75)\vec{x} = (0.75)^3\vec{x}.$$

Continuing this pattern gives us $A^k\vec{x} = (0.75)^k\vec{x}$.

From calculus we know that $\lim\limits_{k \to \infty} (0.75)^k = 0$. This means that as $k \to \infty$, the limit of $A^k\vec{x} = (0.75)^k\vec{x}$ is $0\vec{x} = \vec{0}$.

Since our population vector had a limit of $\vec{0}$, our population dies out in the long run.

Because of the impact of these eigenvalues on population growth, biologists often refer to a demographic matrix's largest eigenvalue as its "population growth rate."

Now that we've seen how easy eigenvalues and eigenvectors are to work with, the natural next question is how to find them. Given an $n \times n$ matrix A, we can find its eigenvectors and eigenvalues by solving $A\vec{x} = \lambda\vec{x}$ for $\vec{x}$ and λ. As it stands, this doesn't have a very familiar format, so we'll do a bit of algebraic manipulation to try and massage it back into the shape of our usual matrix equation $A\vec{x} = \vec{b}$.

We can start by moving both terms to the left-hand side of the equation, which gives us $A\vec{x} - \lambda\vec{x} = \vec{0}$. At this point, it's tempting to try to factor $\vec{x}$ out of both terms on the left-hand side. However, that would leave a factor of $A - \lambda$, which doesn't make sense because A is a matrix and λ is a scalar. We can fix this by inserting an extra factor of the $n \times n$ identity matrix I_n between λ and $\vec{x}$. (Recall that we did something similar in 2.7's Example 2.) Since

$I_n \vec{x} = \vec{x}$, we must have $\lambda I_n \vec{x} = \lambda \vec{x}$. This makes our equation $A\vec{x} - \lambda I_n \vec{x} = \vec{0}$. Now we can factor $\vec{x}$ out of the left-hand side, leaving $A - \lambda I_n$. Since both A and λI_n are $n \times n$ matrices, this now makes sense. Our factored equation is $(A - \lambda I_n)\vec{x} = \vec{0}$ which has the familiar format of a matrix equation.

Unfortunately, we are still at a bit of a loss. The matrix in our matrix equation above contains the unknown variable λ. This means we are trying to solve for two unknown quantities, $\vec{x}$ and λ, simultaneously within the same equation. Additionally, while we usually invoke row reduction to help solve matrix equations, it is less effective with a variable in the matrix's entries. We need a new tool which we'll develop in the next section. In 4.3, we'll come back to the problem of finding eigenvectors and eigenvalues from the equation $(A - \lambda I_n)\vec{x} = \vec{0}$ armed with the idea of the determinant.

Exercises 4.1.

1. Compute $\begin{bmatrix} -9 & 0 & 0 \\ 0 & 4 & 0 \\ 0 & 0 & 7 \end{bmatrix}^2$.

2. Compute $\begin{bmatrix} 2 & 0 & 0 \\ 0 & -3 & 0 \\ 0 & 0 & 5 \end{bmatrix}^3$.

3. Compute $\begin{bmatrix} -1 & 0 & 0 \\ 0 & 0 & 0 \\ 0 & 0 & 2 \end{bmatrix}^6$.

4. Compute $\begin{bmatrix} 1 & 0 & 0 \\ 0 & -5 & 0 \\ 0 & 0 & 6 \end{bmatrix}^3$.

5. Let $B = \begin{bmatrix} 2 & 0 & 0 \\ 0 & -3 & 0 \\ 0 & 0 & 5 \end{bmatrix}$.

   (a) Compute AB for $A = \begin{bmatrix} 3 & -1 & 6 \\ 2 & 1 & -4 \\ 0 & 7 & -2 \end{bmatrix}$.

   (b) Come up with a shortcut for multiplication on the right by a diagonal matrix which is similar to the one used in Example 3 for multiplication on the left.

6. (a) Find the inverse of $\begin{bmatrix} 10 & 0 & 0 \\ 0 & -5 & 0 \\ 0 & 0 & 6 \end{bmatrix}$ or show it is not invertible.

   (b) Find the inverse of $\begin{bmatrix} -1 & 0 & 0 \\ 0 & 0 & 0 \\ 0 & 0 & 2 \end{bmatrix}$ or show it is not invertible.

(c) Give a rule which lets you easily see when a diagonal matrix is invertible.

(d) Give a formula for the inverse of a diagonal matrix. (Your formula should not involve row reduction.)

7. Let W be the set of $n \times n$ diagonal matrices. Show W is a subspace of M_{nn}.

8. Let $\vec{v} = \begin{bmatrix} -3 \\ 0 \\ -6 \end{bmatrix}$ and $A = \begin{bmatrix} 2 & -2 & 0 \\ -2 & 4 & 1 \\ 0 & 1 & 2 \end{bmatrix}$. Decide whether or not $\vec{v}$ is an eigenvector of A. If it is, give its eigenvalue λ.

9. Let $\vec{v} = \begin{bmatrix} 2 \\ 1 \\ -1 \end{bmatrix}$ and $A = \begin{bmatrix} 2 & -2 & 0 \\ -2 & 4 & 1 \\ 0 & 1 & 2 \end{bmatrix}$. Decide whether or not $\vec{v}$ is an eigenvector of A. If it is, give its eigenvalue λ.

10. Let $\vec{v} = \begin{bmatrix} 5 \\ 0 \\ -10 \end{bmatrix}$ and $A = \begin{bmatrix} 1 & 2 & 1 \\ 0 & -4 & 0 \\ -2 & 1 & -2 \end{bmatrix}$. Decide whether or not $\vec{v}$ is an eigenvector of A. If it is, give its eigenvalue λ.

11. Let $\vec{v} = \begin{bmatrix} 11 \\ -2 \\ 4 \end{bmatrix}$ and $A = \begin{bmatrix} 3 & 0 & 0 \\ 0 & 1 & -1 \\ 2 & -1 & -3 \end{bmatrix}$. Decide whether or not $\vec{v}$ is an eigenvector of A. If it is, give its eigenvalue λ.

12. Let $\vec{v} = \begin{bmatrix} 1 \\ 2 \\ 1 \end{bmatrix}$ be an eigenvector of $A = \begin{bmatrix} 3 & -2 & 4 \\ 0 & 4 & -2 \\ 4 & 0 & 0 \end{bmatrix}$ with eigenvalue $\lambda = 3$. Compute $A^4 \vec{v}$.

13. Let $\vec{v} = \begin{bmatrix} 1 \\ 17 \\ 4 \end{bmatrix}$ be an eigenvector of $A = \begin{bmatrix} 1 & 0 & -1 \\ 7 & -2 & -6 \\ 4 & 0 & -4 \end{bmatrix}$ with eigenvalue $\lambda = -3$. Compute $A^2 \vec{v}$.

14. Let $\vec{v} = \begin{bmatrix} 2 \\ 1 \\ 0 \end{bmatrix}$ be an eigenvector of $A = \begin{bmatrix} 6 & 2 & 8 \\ 4 & -1 & -9 \\ 0 & 0 & 5 \end{bmatrix}$ with eigenvalue $\lambda = 7$. Compute $A^2 \vec{v}$.

15. Let $\vec{v} = \begin{bmatrix} -4 \\ 2 \\ 2 \end{bmatrix}$ be an eigenvector of $A = \begin{bmatrix} 3 & -2 & 4 \\ 0 & 4 & -2 \\ -1 & 0 & 0 \end{bmatrix}$ with eigenvalue $\lambda = 2$. Compute $A^5 \vec{v}$.

16. Let $f\left(\begin{bmatrix} x_1 \\ x_2 \\ x_3 \end{bmatrix} \right) = \begin{bmatrix} 6x_1 + 2x_2 + 8x_3 \\ 4x_1 - x_2 - 9x_3 \\ 5x_3 \end{bmatrix}$. The vector $\vec{v} = \begin{bmatrix} -30 \\ -41 \\ 14 \end{bmatrix}$ is an eigenvector of f's matrix with eigenvalue $\lambda = 5$. Use this to compute $f(\vec{v})$.

17. Let $f\left(\begin{bmatrix} x_1 \\ x_2 \\ x_3 \end{bmatrix}\right) = \begin{bmatrix} -6x_1 + 5x_2 - 4x_3 \\ 10x_2 + 7x_3 \\ -x_2 + 2x_3 \end{bmatrix}$. The vector $\vec{v} = \begin{bmatrix} -13 \\ -35 \\ 5 \end{bmatrix}$ is an eigenvector of f's matrix with eigenvalue $\lambda = 9$. Use this to compute $f(\vec{v})$.

18. Let $f\left(\begin{bmatrix} x_1 \\ x_2 \\ x_3 \end{bmatrix}\right) = \begin{bmatrix} x_1 - x_3 \\ 7x_1 - 2x_2 - 6x_3 \\ 4x_1 - 4x_3 \end{bmatrix}$. The vector $\vec{v} = \begin{bmatrix} 0 \\ -5 \\ 0 \end{bmatrix}$ is an eigenvector of f's matrix with eigenvalue $\lambda = -2$. Use this to compute $f(f(\vec{v}))$.

19. Let $f\left(\begin{bmatrix} x_1 \\ x_2 \\ x_3 \end{bmatrix}\right) = \begin{bmatrix} 3x_1 - 2x_2 + 4x_3 \\ 4x_2 - 2x_3 \\ -x_1 \end{bmatrix}$. The vector $\vec{v} = \begin{bmatrix} -4 \\ 2 \\ 1 \end{bmatrix}$ is an eigenvector of f's matrix with eigenvalue $\lambda = 3$. Use this to compute $f(f(f(\vec{v})))$.

20. In Example 9, we saw that starting with a population eigenvector of a demographic matrix with eigenvalue $\lambda = 0.75$ meant the population would die out in the long run. What if the population eigenvector had eigenvalue $\lambda = 1.6$?

21. Put together Example 9 and Exercise 20 to come up with a condition on λ so that a population whose population eigenvector with eigenvalue λ dies out in the long run. Find another condition on λ so that the population grows in the long run.

4.2 Determinants

In this section we'll develop a computational tool called the determinant which assigns a number to each $n \times n$ matrix. Then we'll discuss what that number tells us about the matrix. We'll start in the smallest interesting case where $n = 2$. Here our definition of the determinant is a byproduct of computing of the inverse of a general 2×2 matrix. This will get a bit messy in the middle, but hang in until the computational dust settles and you'll see something interesting happen.

Let $A = \begin{bmatrix} a & b \\ c & d \end{bmatrix}$. To find the inverse, we need to row reduce $[A \,|\, I_2]$. This looks like

$$\begin{bmatrix} a & b & 1 & 0 \\ c & d & 0 & 1 \end{bmatrix} \to (\frac{1}{a} \cdot r_1) \begin{bmatrix} 1 & \frac{b}{a} & \frac{1}{a} & 0 \\ c & d & 0 & 1 \end{bmatrix} \to (r_2 - c \cdot r_1) \begin{bmatrix} 1 & \frac{b}{a} & \frac{1}{a} & 0 \\ 0 & d - \frac{bc}{a} & \frac{c}{a} & 1 \end{bmatrix}.$$

Since $d - \dfrac{bc}{a} = \dfrac{ad - bc}{a}$, the last matrix can be rewritten as

$$\begin{bmatrix} 1 & \frac{b}{a} & \frac{1}{a} & 0 \\ 0 & \frac{ad-bc}{a} & \frac{c}{a} & 1 \end{bmatrix}.$$

Continuing our row reduction from here we get

$$\begin{bmatrix} 1 & \frac{b}{a} & \frac{1}{a} & 0 \\ 0 & \frac{ad-bc}{a} & -\frac{c}{a} & 1 \end{bmatrix} \to (\frac{a}{ad-bc} \cdot r_2) \begin{bmatrix} 1 & \frac{b}{a} & \frac{1}{a} & 0 \\ 0 & 1 & -\frac{c}{ad-bc} & \frac{a}{ad-bc} \end{bmatrix}$$

$$\to (r_1 - \frac{b}{a} \cdot r_2) \begin{bmatrix} 1 & 0 & \bigstar & -\frac{b}{ad-bc} \\ 0 & 1 & -\frac{c}{ad-bc} & \frac{a}{ad-bc} \end{bmatrix}$$

where

$$\bigstar = \frac{1}{a} - \frac{b}{a}\left(-\frac{c}{ad-bc}\right) = \frac{ad-bc}{a(ad-bc)} + \frac{bc}{a(ad-bc)} = \frac{ad-bc+bc}{a(ad-bc)}$$

$$= \frac{ad}{a(ad-bc)} = \frac{d}{ad-bc}.$$

Substituting this back into our matrix gives us

$$\begin{bmatrix} 1 & 0 & \frac{d}{ad-bc} & -\frac{b}{ad-bc} \\ 0 & 1 & -\frac{c}{ad-bc} & \frac{a}{ad-bc} \end{bmatrix}.$$

Since A row reduced to I_2, we get

$$A^{-1} = \begin{bmatrix} \frac{d}{ad-bc} & -\frac{b}{ad-bc} \\ -\frac{c}{ad-bc} & \frac{a}{ad-bc} \end{bmatrix}.$$

We can make this more appealing looking by factoring out $\frac{1}{ad-bc}$ from each entry to get

$$\begin{bmatrix} a & b \\ c & d \end{bmatrix}^{-1} = \frac{1}{ad-bc} \begin{bmatrix} d & -b \\ -c & a \end{bmatrix}.$$

This formula for the inverse of a 2×2 matrix makes sense as long as our denominator $ad - bc$ isn't zero. This means the number $ad - bc$ tells us something interesting about our matrix A, which prompts the following definition.

Definition. The *determinant* of $A = \begin{bmatrix} a & b \\ c & d \end{bmatrix}$ is $\det(A) = ad - bc$.

Notice that ad is the product of the two diagonal entries, and bc is the product of the two non-diagonal entries. Therefore one way to remember this formula is to think of it as "the product of the diagonal entries minus the product of the non-diagonal entries".

Example 1. Compute the determinant of $A = \begin{bmatrix} 3 & 2 \\ 6 & 5 \end{bmatrix}$.

Here $a = 3$, $b = 2$, $c = 6$ and $d = 5$. Plugging those values into the formula above, we get

$$\det(A) = 3(5) - 2(6) = 3.$$

We can restate the conclusion from our inverse computation at the beginning of this section as follows: a 2×2 matrix A is invertible exactly when $\det(A) \neq 0$. This means the matrix from Example 1 is invertible, because its determinant was nonzero. The determinant also tells us more about A than just whether or not it is invertible, which should make sense. After all, the determinant's output gives us more information than just zero or nonzero.

Thinking of a 2×2 matrix A as a map from $\mathbb{R}^2$ to itself, we can ask what the determinant of the matrix tells us about its map's effect on the plane. It turns out that the determinant tells us two things about the geometric effects of the map: how it affects areas and whether or not it flips the plane over.

Let's start with the question of whether A's map flips the plane over. Just to make sure we are all interpreting this idea the same way, think of the maps f and g where f reflects $\mathbb{R}^2$ about the line $y = x$ and g rotates $\mathbb{R}^2$ counterclockwise by $90°$. If we imagine $\mathbb{R}^2$ as an infinite sheet of paper, we can identify one side of the paper as the top side and the other as the bottom side. After applying the map f, our top and bottom sides have switched places, i.e., f has flipped the plane over. The map g on the other hand leaves the top and bottom sides in their original positions, so it does not flip over the plane. If we're looking at a map which isn't defined geometrically, it can be more complicated to figure out whether that map flips the plane over or not.

However, we can use the sign of the determinant of A to quickly determine whether or not A's map flips over the plane. If $\det(A)$ is positive, then it does not flip $\mathbb{R}^2$. Alternately, if $\det(A)$ is negative, then it does flip $\mathbb{R}^2$. To justify this relationship, let's compute the determinants of our two example maps discussed above.

Example 2. Find the signs of the determinants for the maps f and g discussed above.

Before we can compute these two determinants, we need to find the matrix of each map. Let's call f's matrix A and g's matrix B to avoid any confusion. We saw in 2.2 that we can find the matrix of a map from a geometric description of its action on $\mathbb{R}^2$ by putting the image of $\begin{bmatrix} 1 \\ 0 \end{bmatrix}$ in the first column and the image of $\begin{bmatrix} 0 \\ 1 \end{bmatrix}$ in the second column.

Our map f is reflection about the line $y = x$, so geometrically it sends $\begin{bmatrix} 1 \\ 0 \end{bmatrix}$ to $\begin{bmatrix} 0 \\ 1 \end{bmatrix}$ and $\begin{bmatrix} 0 \\ 1 \end{bmatrix}$ to $\begin{bmatrix} 1 \\ 0 \end{bmatrix}$.

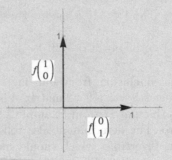

This means f's matrix has first column $\begin{bmatrix} 0 \\ 1 \end{bmatrix}$ and second column $\begin{bmatrix} 1 \\ 0 \end{bmatrix}$, so $A = \begin{bmatrix} 0 & 1 \\ 1 & 0 \end{bmatrix}$.

The other map g is counterclockwise rotation by $90°$, so it sends $\begin{bmatrix} 1 \\ 0 \end{bmatrix}$ to $\begin{bmatrix} 0 \\ 1 \end{bmatrix}$ and $\begin{bmatrix} 0 \\ 1 \end{bmatrix}$ to $\begin{bmatrix} -1 \\ 0 \end{bmatrix}$.

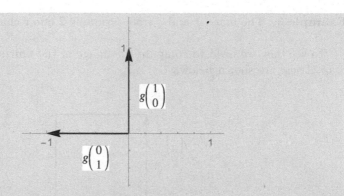

This means f's matrix has first column $\begin{bmatrix} 0 \\ 1 \end{bmatrix}$ and second column $\begin{bmatrix} -1 \\ 0 \end{bmatrix}$, so

$B = \begin{bmatrix} 0 & -1 \\ 1 & 0 \end{bmatrix}$.

Computing our two determinants, we get $\det(A) = 0(0) - 1(1) = -1$ and $\det(B) = 0(0) - (-1)(1) = 1$. Thus, as expected, the map that flips the plane has a negative determinant and the map that doesn't flip the plan has a positive determinant.

Next let's see what determinants can tell us about the effect that a matrix's map has on area. We'll explore this by looking at the effect of the map on the unit square, i.e., the area enclosed by the unit vectors along the x and y axes in Figure 4.1.

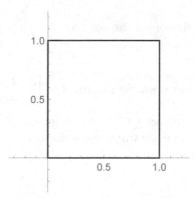

Figure 4.1: The unit square

This square has area 1, so we can compare the area of the unit square's image under various matrix maps to 1 to see what effect those maps have on areas.

Example 3. The maps f and g from Example 2 don't change areas.

To see this, we need to compute the image of the unit square under each map. These are shown below.

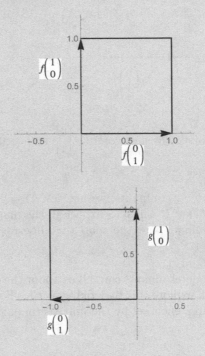

From the pictures above, we can see that both of these images still have area 1.

Example 4. What effect does the matrix $A = \begin{bmatrix} 2 & 1 \\ 1 & 3 \end{bmatrix}$ have on areas?

To figure this out, we need to find the image of the unit square after multiplication by A. This is shown in the following picture.

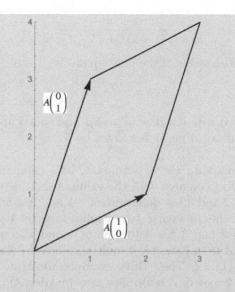

This image is a parallelogram, so we can compute its area using the formula *Area* = *base* · *height* with base b and height h are shown below.

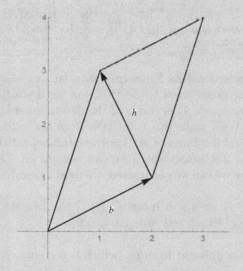

We can compute the lengths of both b and h using the formula for the distance between two points in the plane:

$$d = \sqrt{(x_1 - x_2)^2 + (y_1 - y_2)^2}.$$

Plugging in the points at the ends of b and h gives us

$$b = \sqrt{(2-0)^2 + (1-0)^2} = \sqrt{5}$$

and

$$h = \sqrt{(1-2)^2 + (3-1)^2} = \sqrt{5}.$$

Thus the area of the image of the unit square is

$$b \cdot h = \sqrt{5} \cdot \sqrt{5} = 5.$$

Since we started with a square of area 1 and ended up with an image that has area 5, we can say A multiplies areas by 5.

To connect area with the determinant, let's compare each of our three example maps' effects on area with the value of their determinants. The map f from Examples 2 and 3 has determinant -1 and has no effect on area. (We can also think of this as saying f multiplies area by 1.) Similarly, the map g from Examples 2 and 3 has determinant 1 and also multiplies area by 1. Finally, the map A from Example 4 multiplies area by 5 and has determinant $\det(A) = 2(3) - 1(1) = 5$. These three examples illustrate the general pattern: the map of a 2×2 matrix A multiplies area by $|\det(A)|$.

We can summarize our geometric exploration of the determinant of a 2×2 matrix as follows.

Theorem 1. Let $f : \mathbb{R}^2 \to \mathbb{R}^2$ have matrix A. If $\det(A)$ is negative, then f flips the plane over, while if $\det(A)$ is positive, then f does not. Also, f multiplies areas by $|\det(A)|$.

Now that we understand 2×2 determinants, let's extend this idea to $n \times n$ matrices. There are many ways to do this, but we'll start with an iterative process which rewrites $\det(A)$ in terms of the determinants of $(n-1) \times (n-1)$ submatrices of A. We'll make this transition from $n \times n$ to $(n-1) \times (n-1)$ by deleting a row and a column of A. Applying this repeatedly will eventually reduce down to the 2×2 case where we can use our $ad - bc$ formula for the determinant. Before we can we get started, we need to establish some notation.

Definition. If A is an $n \times n$ matrix, A_{ij} is the $(n-1) \times (n-1)$ matrix formed by removing A's ith row and jth column.

(Note that this is different from a_{ij} which is the entry of A in the ith row and jth column.)

Example 5. Find A_{24} and A_{31} for $A = \begin{bmatrix} 2 & 0 & -1 & 5 \\ 3 & 1 & 0 & -4 \\ -1 & 2 & 1 & 1 \\ 7 & -3 & 0 & 1 \end{bmatrix}$.

We can find the first submatrix A_{24} by removing A's 2nd row and 4th column. Crossing out this row and column looks like

$$\begin{bmatrix} 2 & 0 & -1 & 5 \\ 3 & 1 & 0 & 4 \\ -1 & 2 & 1 & \\ 7 & -3 & 0 & \end{bmatrix}$$

which means

$$A_{24} = \begin{bmatrix} 2 & 0 & -1 \\ -1 & 2 & 1 \\ 7 & -3 & 0 \end{bmatrix}.$$

Similarly, we can find the second submatrix A_{31} by removing A's 3rd row and 1st column. Crossing out this row and column looks like

$$\begin{bmatrix} 2 & 0 & -1 & 5 \\ 3 & 1 & 0 & -4 \\ -1 & 2 & 1 & 1 \\ 7 & -3 & 0 & 1 \end{bmatrix}$$

which means

$$A_{31} = \begin{bmatrix} 0 & -1 & 5 \\ 1 & 0 & -4 \\ -3 & 0 & 1 \end{bmatrix}.$$

Now that we've established our notation, we can state the following formulas which serve as the $n \times n$ to $(n-1) \times (n-1)$ reduction step at the heart of our algorithm for computing an $n \times n$ determinant. The first formula is usually called expansion along a row, because it contains one term for each entry along some particular row of A.

Theorem 2. Let A be an $n \times n$ matrix and fix i with $1 \leq i \leq n$. Then

$$\det(A) = (-1)^{i+1}a_{i1}\det(A_{i1}) + (-1)^{i+2}a_{i2}\det(A_{i2}) + \cdots + (-1)^{i+n}a_{in}\det(A_{in}).$$

Note that there are n terms, each of which contains three parts: a sign (plus or minus), an entry of A, and the determinant of an $(n-1) \times (n-1)$ submatrix of A.

Example 6. Implement this formula for the determinant along the second row of $A = \begin{bmatrix} 2 & 0 & -1 & 5 \\ 3 & 1 & 0 & -4 \\ -1 & 2 & 1 & 1 \\ 7 & -3 & 0 & 1 \end{bmatrix}.$

Since we're expanding along the second row of A, we have $i = 2$. Plugging this into our formula gives us

$$\det(A) = (-1)^{2+1}a_{21}\det(A_{21}) + (-1)^{2+2}a_{22}\det(A_{22})$$
$$+ (-1)^{2+3}a_{23}\det(A_{23}) + (-1)^{2+4}a_{24}\det(A_{24}).$$

The a_{2j} are the entries (in order) along the second row of A, so we have $a_{21} = 3$, $a_{22} = 1$, $a_{23} = 0$, and $a_{24} = -4$. The A_{2j} are the 3×3 submatrices of A where we've deleted the second row and jth column, so we have

$$A_{21} = \begin{bmatrix} 0 & -1 & 5 \\ 2 & 1 & 1 \\ -3 & 0 & 1 \end{bmatrix}, \ A_{22} = \begin{bmatrix} 2 & -1 & 5 \\ -1 & 1 & 1 \\ 7 & 0 & 1 \end{bmatrix},$$

$$A_{23} = \begin{bmatrix} 2 & 0 & 5 \\ -1 & 2 & 1 \\ 7 & -3 & 1 \end{bmatrix}, \ A_{24} = \begin{bmatrix} 2 & 0 & -1 \\ -1 & 2 & 1 \\ 7 & -3 & 0 \end{bmatrix}.$$

Plugging these back into our formula gives us

$$\det(A) = (-1)^3 3\det\left(\begin{bmatrix} 0 & -1 & 5 \\ 2 & 1 & 1 \\ -3 & 0 & 1 \end{bmatrix}\right) + (-1)^4 1\det\left(\begin{bmatrix} 2 & -1 & 5 \\ -1 & 1 & 1 \\ 7 & 0 & 1 \end{bmatrix}\right)$$

$$+ (-1)^5 0\det\left(\begin{bmatrix} 2 & 0 & 5 \\ -1 & 2 & 1 \\ 7 & -3 & 1 \end{bmatrix}\right) + (-1)^6 (-4)\det\left(\begin{bmatrix} 2 & 0 & -1 \\ -1 & 2 & 1 \\ 7 & -3 & 0 \end{bmatrix}\right).$$

Simplifying the powers of -1 gives us

$$\det(A) = -3\det\left(\begin{bmatrix} 0 & -1 & 5 \\ 2 & 1 & 1 \\ -3 & 0 & 1 \end{bmatrix}\right) + 1\det\left(\begin{bmatrix} 2 & -1 & 5 \\ -1 & 1 & 1 \\ 7 & 0 & 1 \end{bmatrix}\right)$$

$$- 0\det\left(\begin{bmatrix} 2 & 0 & 5 \\ -1 & 2 & 1 \\ 7 & -3 & 1 \end{bmatrix}\right) + (-4)\det\left(\begin{bmatrix} 2 & 0 & -1 \\ -1 & 2 & 1 \\ 7 & -3 & 0 \end{bmatrix}\right).$$

Alternately, we can use the formula below which is often called expansion down a column, because it contains one term for each entry down a particular column of A.

Theorem 3. Let A be an $n \times n$ matrix and fix j with $1 \le j \le n$. Then

$$\det(A) = (-1)^{1+j}a_{1j}\det(A_{1j}) + (-1)^{2+j}a_{2j}\det(A_{2j}) + \cdots + (-1)^{n+j}a_{nj}\det(A_{nj}).$$

As with expansion along a row, we have n terms each with the same three parts: a sign, a matrix entry, and the determinant of a submatrix.

Example 7. Implement this formula for the determinant down the first

column of $A = \begin{bmatrix} 2 & 0 & -1 & 5 \\ 3 & 1 & 0 & -4 \\ -1 & 2 & 1 & 1 \\ 7 & -3 & 0 & 1 \end{bmatrix}$.

Since we're expanding down the first column, we have $j = 1$. Plugging this into our formula above gives us

$$\det(A) = (-1)^{1+1}a_{11}\det(A_{11}) + (-1)^{2+1}a_{21}\det(A_{21})$$
$$+ (-1)^{3+1}a_{31}\det(A_{31}) + (-1)^{4+1}a_{41}\det(A_{41}).$$

The a_{i1} are the entries (in order) down the first column, so we have $a_{11} = 2$, $a_{21} = 3$, $a_{31} = -1$, and $a_{41} = 7$. The A_{i1} are the 3×3 submatrices of A we get by deleting the ith row and first column, so we have

$$A_{11} = \begin{bmatrix} 1 & 0 & -4 \\ 2 & 1 & 1 \\ -3 & 0 & 1 \end{bmatrix}, A_{21} = \begin{bmatrix} 0 & -1 & 5 \\ 2 & 1 & 1 \\ -3 & 0 & 1 \end{bmatrix},$$

$$A_{31} = \begin{bmatrix} 0 & -1 & 5 \\ 1 & 0 & -4 \\ -3 & 0 & 1 \end{bmatrix}, A_{41} = \begin{bmatrix} 0 & -1 & 5 \\ 1 & 0 & -4 \\ 2 & 1 & 1 \end{bmatrix}.$$

Plugging these into our formula gives us

$$\det(A) = (-1)^2 2 \det\left(\begin{bmatrix} 1 & 0 & -4 \\ 2 & 1 & 1 \\ -3 & 0 & 1 \end{bmatrix}\right) + (-1)^3 3 \det\left(\begin{bmatrix} 0 & -1 & 5 \\ 2 & 1 & 1 \\ -3 & 0 & 1 \end{bmatrix}\right)$$
$$+ (-1)^4(-1)\det\left(\begin{bmatrix} 0 & -1 & 5 \\ 1 & 0 & -4 \\ -3 & 0 & 1 \end{bmatrix}\right) + (-1)^5 7 \det\left(\begin{bmatrix} 0 & -1 & 5 \\ 1 & 0 & -4 \\ 2 & 1 & 1 \end{bmatrix}\right).$$

The signs on our four terms simplify to

$$\det(A) = 2 \det\left(\begin{bmatrix} 1 & 0 & -4 \\ 2 & 1 & 1 \\ -3 & 0 & 1 \end{bmatrix}\right) - 3 \det\left(\begin{bmatrix} 0 & -1 & 5 \\ 2 & 1 & 1 \\ -3 & 0 & 1 \end{bmatrix}\right)$$
$$+ (-1)\det\left(\begin{bmatrix} 0 & -1 & 5 \\ 1 & 0 & -4 \\ -3 & 0 & 1 \end{bmatrix}\right) - 7 \det\left(\begin{bmatrix} 0 & -1 & 5 \\ 1 & 0 & -4 \\ 2 & 1 & 1 \end{bmatrix}\right).$$

Note that together these formulas give us $2n$ different choices on how to tackle $\det(A)$, because A has n rows and n columns. Miraculously, it doesn't matter which row or column we pick, so let's make the best, i.e., easiest, choice possible. Usually this means picking the row or column with the most zero entries, because if $a_{ij} = 0$ its whole term is multiplied by zero and can be ignored.

Example 8. Which row(s) or column(s) make computing the determinant

of $A = \begin{bmatrix} 2 & 0 & -1 & 5 \\ 3 & 1 & 0 & -4 \\ -1 & 2 & 1 & 1 \\ 7 & -3 & 0 & 1 \end{bmatrix}$ easiest?

This question can be rephrased as "Which row(s) or column(s) of A have the most 0 entries?" Stated in this format, the answer is clearly the second column of A, which contains two 0s. If for some reason we didn't want to expand down the second column, the first, second, and third rows and second column would be our next best choices since they each contain one 0 entry.

While we can continue to use these two formulas directly as in Examples 6 and 7, there is a commonly used way to find the sign and submatrix for a given term based on the position of that term's matrix entry. In every term of our formula for $\det(A)$, the subscripts on our matrix entry a_{ij} and our $(n-1) \times (n-1)$ submatrix A_{ij} match. This means that if we're looking at the term with a_{ij}, we're removing the ith row and jth column of A to get A_{ij}. In other words, we can find the submatrix which goes with a_{ij} by removing the row and column of A where a_{ij} appears as shown below.

$$\begin{bmatrix} a_{11} & a_{12} & \cdots & a_{1j} & \cdots & a_{1n} \\ a_{21} & a_{22} & \cdots & a_{2j} & \cdots & a_{2n} \\ \vdots & & \ddots & & & \vdots \\ \overline{a_{i1}} & \overline{a_{i2}} & & \overline{a_{ij}} & & \overline{a_{in}} \\ \vdots & & & & \ddots & \vdots \\ a_{m1} & a_{m2} & \cdots & a_{nj} & \cdots & a_{mn} \end{bmatrix}$$

We can also link a_{ij}'s position to the sign associated with its term in our sum. The term containing a_{ij} has sign $(-1)^{i+j}$ which is positive if $i + j$ is even and negative if $i + j$ is odd. The top left corner of any matrix is a_{11}. Since $1 + 1 = 2$ is even, the sign piece of a_{11}'s term is positive. If we travel along a single row, we see that the signs of the terms alternate, because $i + j$ is followed by $i + (j+1) = (i+j) + 1$, which will have the opposite even/odd parity. Similarly, as we travel down a single column, the signs alternate, since $i + j$ is followed by $(i+1) + j = (i+j) + 1$ which again has the opposite even/odd parity. We can use this to fill in a matrix with plus and minus signs corresponding to where $(-1)^{i+j}$ is positive and negative respectively as follows

$$\begin{bmatrix} + & - & + & - & \cdots \\ - & + & - & + & \cdots \\ + & - & + & & \\ - & + & & & \ddots \\ \vdots & \vdots & & & \end{bmatrix}.$$

We can read off the sign of a_{ij}'s term of our determinant sum by checking which sign is in the ijth spot in the matrix above.

Example 9. What are the signs of the terms if we compute the determinant of $A = \begin{bmatrix} 2 & 0 & -1 & 5 \\ 3 & 1 & 0 & -4 \\ -1 & 2 & 1 & 1 \\ 7 & -3 & 0 & 1 \end{bmatrix}$ along the fourth row?

In this case, our matrix of pluses and minuses above is the 4×4 matrix

$$\begin{bmatrix} + & - & + & - \\ - & + & - & + \\ + & - & + & - \\ - & + & - & + \end{bmatrix}.$$

Therefore if we compute the determinant of A by expanding along the fourth row, our signs will be $- + - +$.

If you like these ideas for finding $(-1)^{i+j}$ and A_{ij} from the position of a_{ij}, feel free to use them. If not, feel free to use the original formulas for expansion along a row or down a column.

Now that we understand how to rewrite the determinant of an $n \times n$ matrix A in terms of determinants of $(n-1) \times (n-1)$ matrices, we can give the procedure for computing an $n \times n$ determinant. Pick a row or column of A, and use one of the formulas above to rewrite $\det(A)$ in terms of smaller determinants. Pick a row or column of each smaller matrix and repeat this process. Eventually the smaller matrices will be 2×2, where we can compute their determinants using our $ad - bc$ formula. This process is fairly tedious for large matrices, but is easily implemented on a computer. We'll usually practice with $n \leq 4$.

Example 10. Compute the determinant of $A = \begin{bmatrix} 3 & 0 & -5 \\ 1 & 1 & 2 \\ 4 & -2 & -1 \end{bmatrix}$ using expansion along a row.

Since we are allowed to pick whichever row we like, we should pick the 1st row since it is the only one which has a zero entry. This means we'll have $i = 1$, so

$$\det(A) = (-1)^{1+1}a_{11}\det(A_{11}) + (-1)^{1+2}a_{12}\det(A_{12}) + (-1)^{1+3}a_{13}\det(A_{13}).$$

Plugging in our matrix entries a_{1j} along the first row and our submatrices A_{1j} with the first row and jth column deleted gives us

$$\det(A) = (-1)^2 3 \det\left(\begin{bmatrix} 1 & 2 \\ -2 & -1 \end{bmatrix}\right) + (-1)^3 0 \det\left(\begin{bmatrix} 1 & 2 \\ 4 & -1 \end{bmatrix}\right)$$
$$+ (-1)^4 (-5) \det\left(\begin{bmatrix} 1 & 1 \\ 4 & -2 \end{bmatrix}\right).$$

Our middle term is 0, so we can remove it and simplify our signs to get

$$\det(A) = 3 \det\left(\begin{bmatrix} 1 & 2 \\ -2 & -1 \end{bmatrix}\right) - 5 \det\left(\begin{bmatrix} 1 & 1 \\ 4 & -2 \end{bmatrix}\right).$$

Our smaller submatrices are now 2×2, so we can use our $ad - bc$ formula to compute each of their determinants. This gives us

$$\det\left(\begin{bmatrix} 1 & 2 \\ -2 & -1 \end{bmatrix}\right) = 1(-1) - 2(-2) = -1 + 4 = 3$$

and

$$\det\left(\begin{bmatrix} 1 & 1 \\ 4 & -2 \end{bmatrix}\right) = 1(-2) - 1(4) = -2 - 4 = -6.$$

Plugging these 2×2 determinants back into our formula for $\det(A)$ gives us

$$\det(A) = 3(3) - 5(-6) = 39.$$

To check that it doesn't matter which method we use to compute the determinant, let's recompute $\det(A)$ using expansion down a column instead of along a row.

Example 11. Compute the determinant of $A = \begin{bmatrix} 3 & 0 & -5 \\ 1 & 1 & 2 \\ 4 & -2 & -1 \end{bmatrix}$ using expansion down a column.

Again, we can pick whichever column we like, so let's pick the second column since it has a zero entry. Plugging $j = 2$ into our formula for the determinant gives us

$$\det(A) = (-1)^{1+2}a_{12}\det(A_{12}) + (-1)^{2+2}a_{22}\det(A_{22}) + (-1)^{3+2}a_{32}\det(A_{32}).$$

Next we can plug in our matrix entries a_{i2} down the second column and our submatrices A_{i2} with the ith row and second column removed to get

$$\det(A) = (-1)^3 0 \det\left(\begin{bmatrix} 1 & 2 \\ 4 & -1 \end{bmatrix}\right) + (-1)^4 1 \det\left(\begin{bmatrix} 3 & -5 \\ 4 & -1 \end{bmatrix}\right)$$
$$+ (-1)^5 (-2) \det\left(\begin{bmatrix} 3 & -5 \\ 1 & 2 \end{bmatrix}\right).$$

Our first term is 0, so this simplifies to

$$\det(A) = 1 \det\left(\begin{bmatrix} 3 & -5 \\ 4 & -1 \end{bmatrix}\right) + 2 \det\left(\begin{bmatrix} 3 & -5 \\ 1 & 2 \end{bmatrix}\right).$$

As in our previous example, we can now use our 2×2 determinant formula to get

$$\det\left(\begin{bmatrix} 3 & -5 \\ 4 & -1 \end{bmatrix}\right) = 3(-1) - (-5)(4) = -3 + 20 = 17$$

and

$$\det\left(\begin{bmatrix} 3 & -5 \\ 1 & 2 \end{bmatrix}\right) = 3(2) - (-5)(1) = 6 + 5 = 11.$$

Plugging these back into our formula for $\det(A)$ gives us

$$\det(A) = 1(17) + 2(11) = 39.$$

(Notice that this is the same answer we got by expanding along the 1st row!)

There are two classes of matrices where computing the determinant is easy even when they are very large: lower triangular matrices and upper triangular matrices. (These were discussed in 2.9 as tools for solving $A\vec{x} = \vec{b}$ with an extremely large A.) If A is lower triangular, then all its entries above the diagonal are zeros, i.e., $a_{ij} = 0$ if $i < j$. In particular, this means that the first row of A looks like $a_{11} \, 0 \ldots 0$. If we start computing $\det(A)$ by expanding along this top row, we'll get

$$\det(A) = +a_{11} \det(A_{11}).$$

The matrix A_{11} is also lower triangular as shown by the picture below.

$$\begin{bmatrix} a_{11} & 0 & 0 & & 0 \\ a_{21} & a_{22} & 0 & \cdots & 0 \\ & & \ddots & & \vdots \\ & & & \ddots & 0 \\ a_{n1} & a_{n2} & & \cdots & a_{nn} \end{bmatrix}$$

The first row of A_{11} is $a_{22}\ 0 \ldots 0$, so expanding along this row gives

$$\det(A_{11}) = +a_{22}\det(B)$$

where B is A_{11} with its first row and first column removed. Plugging this back into our formula for $\det(A)$ gives us

$$\det(A) = a_{11}a_{22}\det(B)$$

where B is A with its top 2 rows and leftmost 2 columns removed. We can keep repeating this process until we get down to the 2×2 case where our submatrix is

$$\begin{bmatrix} a_{(n-1)(n-1)} & 0 \\ a_{n(n-1)} & a_{nn} \end{bmatrix}$$

which has determinant $a_{(n-1)(n-1)}a_{nn}$. This gives us the following fact.

Theorem 4. An $n \times n$ lower triangular matrix has $\det(A) = a_{11}a_{22}\cdots a_{nn}$.

Example 12. Find the determinant of $A = \begin{bmatrix} -3 & 0 & 0 & 0 \\ 7 & 2 & 0 & 0 \\ 0 & 1 & 8 & 0 \\ 4 & -3 & 5 & -1 \end{bmatrix}$.

Since A is lower triangular, its determinant is the product of its diagonal entries. Thus

$$\det(A) = -3(2)(8)(-1) = 48.$$

On the other hand, if A is upper triangular, then all its entries below the diagonal are zeros, i.e., $a_{ij} = 0$ if $i > j$. This means that the nth row of A has the form $0 \ldots 0\ a_{nn}$. If we start computing $\det(A)$ by expanding along this bottom row, we'll get

$$\det(A) = +a_{nn}\det(A_{nn}).$$

The matrix A_{nn} is also upper triangular as shown by the picture below.

$$\begin{bmatrix} a_{11} & \cdots & \cdots & a_{1(n-1)} & a_{1n} \\ 0 & \ddots & & & \\ \vdots & & \ddots & & \\ 0 & \cdots & 0 & a_{(n-1)(n-1)} & a_{(n-1)n} \\ 0 & & 0 & 0 & a_{nn} \end{bmatrix}$$

The last row of A_{nn} is $0 \ldots 0\ a_{(n-1)(n-1)}$, so expanding along this row gives

$$\det(A_{nn}) = +a_{(n-1)(n-1)}\det(B)$$

where B is A_{nn} with its last row and last column removed. Plugging this back into our formula for $\det(A)$ gives us

$$\det(A) = a_{nn}a_{(n-1)(n-1)}\det(B)$$

where B is A with its bottom 2 rows and rightmost 2 columns removed. We can keep repeating this process until we get down to the 2×2 case where our submatrix is

$$\begin{bmatrix} a_{11} & a_{12} \\ 0 & a_{22} \end{bmatrix}$$

which has determinant $a_{11}a_{22}$. This gives us the following fact.

Theorem 5. An $n \times n$ upper triangular matrix has $\det(A) = a_{11}a_{22}\cdots a_{nn}$.

Example 13. Compute the determinant of $A = \begin{bmatrix} 7 & 0 & -1 & 9 \\ 0 & 3 & 4 & 0 \\ 0 & 0 & 1 & -1 \\ 0 & 0 & 0 & 4 \end{bmatrix}$.

Since A is upper triangular, its determinant is the product of its diagonal entries. This means we have

$$\det(A) = 7(3)(1)(4) = 84.$$

Theorem 5 is intriguing, because the first half of our row reduction algorithm transforms a general matrix A into an upper triangular form U. This means we can link $\det(A)$, which is harder to compute, with $\det(U)$ which is just the product of U's diagonal entries. In order to link these determinants, we'll need to understand what each of our row operations do to the determinant.

Suppose we multiply the ith row of our matrix A by some scalar s, to create a new matrix B. (This is how we create leading 1s during row reduction.) We can choose how to compute the determinant of B, so let's expand along the ith row. This gives us

$$\det(B) = (-1)^{i+1}b_{i1}\det(B_{i1}) + \cdots + (-1)^{i+n}b_{in}\det(B_{in}).$$

However, all rows of A and B are exactly the same except for the ith row, so $B_{ij} = A_{ij}$ since both have the ith row removed. The ith row of B is s times the ith row of A, which means $b_{ij} = sa_{ij}$. Plugging this into our formula for B's determinant gives us

$$\det(B) = (-1)^{i+1}sa_{i1}\det(A_{i1}) + \cdots + (-1)^{i+n}sa_{in}\det(A_{in}).$$

Factoring an s out of every term gives us

$$\det(B) = s((-1)^{i+1}a_{i1}\det(A_{i1}) + \cdots + (-1)^{i+n}a_{in}\det(A_{in}))$$
$$= s\det(A).$$

Next let's consider the effect of swapping two rows of A. (This is how we put a nonzero entry into the top left corner during row reduction.) Suppose we swap the ith and $i+1$st rows of A to create B. If we expand along the ith row of B we get

$$\det(B) = (-1)^{i+1} b_{i1} \det(B_{i1}) + \cdots + (-1)^{i+n} b_{in} \det(B_{in}).$$

However, the ith row of B is the $i+1$st row of A, so $b_{ij} = a_{(i+1)j}$ and $B_{ij} = A_{(i+1)j}$. Plugging this back into our computation of B's determinant, we get

$$\det(B) = (-1)^{i+1} a_{(i+1)1} \det(A_{(i+1)1}) + \cdots + (-1)^{i+n} a_{(i+1)n} \det(A_{(i+1)n}).$$

This is almost the expansion of $\det(A)$ along the $i+1$st row, but the signs have changed. Instead of $(-1)^{(i+1)+j}$ we have $(-1)^{i+j}$. Changing the power on -1 by 1 means we've switched the sign of each term, so $\det(B) = -\det(A)$.

Since we don't always swap two adjacent rows of A, we'll also need to consider the more general case of swapping the ith and jth rows. For the ease of explanation, I'll suppose $i < j$ and rewrite j as $i+k$ for some positive k. I'll label the rows of A as r_1 through r_n. After swapping the ith and $i+k$th rows of A, the rows are in the following order from top to bottom (with the swapped rows in bold):

$$r_1, \ldots, r_{i-1}, \mathbf{r_{i+k}}, r_{i+1}, \ldots, r_{i+(k-1)}, \mathbf{r_i}, r_{i+(k+1)}, \ldots r_n.$$

Since swapping two consecutive rows multiplies the determinant by -1, we'll perform consecutive swaps on the rows of A until we've gotten them into the order listed above. If the total number of consecutive swaps needed is odd, the net effect will be to multiply the determinant by -1. If it is even, the determinant will be multiplied by 1 and so remain unchanged.

We start with the rows in their original order $r_1, \ldots, r_n$. First we'll perform consecutive swaps of r_i with the rows below it until it is directly below r_{i+k}. Again I'll put the two rows being swapped in bold. The first consecutive swap switches r_i and r_{i+1} to give us

$$r_1, \ldots, r_{i-1}, \mathbf{r_{i+1}}, \mathbf{r_i}, r_{i+2} \ldots, r_n.$$

The second switches r_i and r_{i+2} to give us

$$r_1, \ldots, r_{i-1}, r_{i+1}, \mathbf{r_{i+2}}, \mathbf{r_i}, r_{i+3}, \ldots, r_n.$$

Continuing in the same fashion, the kth consecutive swap switches r_i and r_{i+k} to give us

$$r_1, \ldots, r_{i-1}, r_{i+1}, \ldots, \mathbf{r_{i+k}}, \mathbf{r_i}, r_{i+(k+1)}, \ldots, r_n.$$

Now we'll perform consecutive swaps of r_{i+k} with the rows above it until it is between r_{i-1} and r_{i+1} where r_i was in our original matrix. Again, the two rows being switched are in bold. The first of these consecutive swaps switches r_{i+k} with $r_{i+(k-1)}$ to give us

$$r_1, \ldots, r_{i-1}, r_{i+1}, \ldots, r_{i+(k-2)}, \mathbf{r_{i+k}}, \mathbf{r_{i+(k-1)}}, r_i, r_{i+(k+1)}, \ldots, r_n.$$

The second of these consecutive swaps switches r_{i+k} with $r_{i+(k-2)}$ to give us

$$r_1, \ldots, r_{i-1}, r_{i+1}, \ldots, r_{i+(k-3)}, \mathbf{r_{i+k}}, \mathbf{r_{i+(k-2)}}, r_{i+(k-1)}, r_i, r_{i+(k+1)}, \ldots, r_n.$$

The $k - 1$st of these switches r_{i+k} with r_{i+1} to give us

$$r_1, \ldots, r_{i-1}, \mathbf{r_{i+k}}, \mathbf{r_{i+1}}, \ldots, r_{i+(k-1)}, r_i, r_{i+(k+1)}, \ldots, r_n.$$

This is the same as simply swapping the ith and $i + k$th rows directly, and we got there via $k + (k - 1)$ consecutive swaps. Since $k + k - 1 = 2k - 1$ is always odd, this means that in swapping any two rows of a matrix multiplies the determinant by -1.

Our final row operation is adding a multiple of one row to another row. (We use this to create the needed zeros below each leading 1 during row reduction.) To make this precise, suppose that we add s times the kth row of A to the ith row of A to get a new matrix B. If we compute B's determinant along the ith row, we get

$$\det(B) = (-1)^{i+1} b_{i1} \det(B_{i1}) + \cdots + (-1)^{i+n} b_{in} \det(B_{in}).$$

In our new matrix B we have $b_{ij} = a_{ij} + sa_{kj}$ and $B_{ij} = A_{ij}$, so our determinant is really

$$\det(B) = (-1)^{i+1} (a_{i1} + sa_{k1}) \det(A_{i1}) + \cdots + (-1)^{i+n} (a_{in} + sa_{kn}) \det(A_{in}).$$

We can split this sum up as

$$\det(B) = \left[(-1)^{i+1} a_{i1} \det(A_{i1}) + \cdots + (-1)^{i+n} a_{in} \det(A_{in}) \right]$$
$$+ \left[(-1)^{i+1} sa_{k1} \det(A_{i1}) + \cdots + (-1)^{i+n} sa_{kn} \det(A_{in}) \right].$$

The sum inside the first set of brackets is just $\det(A)$, and we can factor an s out of the second set of brackets to get

$$\det(B) = \det(A) + s \left[(-1)^{i+1} a_{k1} \det(A_{i1}) + \cdots + (-1)^{i+n} a_{kn} \det(A_{in}) \right].$$

Now the sum inside the remaining set of brackets is the determinant of the matrix C we'd get by replacing the ith row of A by its kth row. This means C has identical ith and kth rows, so swapping these two rows would leave the determinant unchanged. However, we just saw that swapping two rows changes the sign of the determinant. This means $\det(C) = -\det(C)$, so we must have $\det(C) = 0$. Plugging this back into our equation for $\det(B)$ gives us

$$\det(B) = \det(A) + s \det(C) = \det(A).$$

Therefore adding a multiple of one row to another row doesn't change the determinant.

We summarize these results in the following theorem.

Theorem 6. Let A be an $n \times n$ matrix. Multiplying a row of A by the scalar s multiplies $\det(A)$ by s, swapping two rows of A multiplies $\det(A)$ by -1, and adding a multiple of one row to another doesn't change $\det(A)$.

Now that we understand how row operations change the determinant and how to compute the determinant of an upper triangular matrix, we have an alternative to our expansion method for finding $\det(A)$. The plan here is to use row operations to transform A into an upper triangular matrix U. As we go we'll keep track of each time we swapped rows or multiplied a row by a constant. (We don't need to keep track of adding a multiple of one row to another since that doesn't have any effect on the determinant.) Then we can compute $\det(U)$ and then undo each of the changes our row operations made to the determinant to get $\det(A)$. This is illustrated in the example below.

Example 14. Compute the determinant of $A = \begin{bmatrix} 3 & 0 & -5 \\ 1 & 1 & 2 \\ 4 & -2 & -1 \end{bmatrix}$ by using row operations to link A to an upper triangular matrix.

Our first job here is to use row operations to transform A into an upper triangular matrix. I'll use our usual notation for row operations since we need to keep track of when we've swapped two rows or scaled a row by a constant.

$$\begin{bmatrix} 3 & 0 & -5 \\ 1 & 1 & 2 \\ 4 & -2 & -1 \end{bmatrix} \rightarrow_{r_1 \leftrightarrow r_2} \begin{bmatrix} 1 & 1 & 2 \\ 3 & 0 & -5 \\ 4 & -2 & -1 \end{bmatrix} \rightarrow_{r_2 - 3r_1} \begin{bmatrix} 1 & 1 & 2 \\ 0 & -3 & -11 \\ 4 & -2 & -1 \end{bmatrix}$$

$$\rightarrow_{r_3 - 4r_1} \begin{bmatrix} 1 & 1 & 2 \\ 0 & -3 & -11 \\ 0 & -6 & -9 \end{bmatrix} \rightarrow_{r_2 \leftrightarrow r_3} \begin{bmatrix} 1 & 1 & 2 \\ 0 & -6 & -9 \\ 0 & -3 & -11 \end{bmatrix}$$

$$\rightarrow_{-\frac{1}{6}r_2} \begin{bmatrix} 1 & 1 & 2 \\ 0 & 1 & \frac{3}{2} \\ 0 & -3 & -11 \end{bmatrix} \rightarrow_{r_3 + 3r_2} \begin{bmatrix} 1 & 1 & 2 \\ 0 & 1 & \frac{3}{2} \\ 0 & 0 & -\frac{13}{2} \end{bmatrix}$$

This last matrix is upper triangular, so let's call it U. We can compute $\det(U) = -\frac{13}{2}$ by taking the product of U's diagonal entries. However, we also have a link between $\det(U)$ and $\det(A)$. To figure out what that relationship is, we need to look back at the row operations we used to get from A to U. We can ignore all the places we added a multiple of one row to another, since they don't affect the determinant. We swapped two rows twice during the process and scaled a row by $-\frac{1}{6}$ once. This means to get $\det(U)$ we multiplied A's determinant by -1 twice (once for each row swap) and by $-\frac{1}{6}$ once (when we scaled a row by $-\frac{1}{6}$). Thus

$$\det(U) = (-1)(-1)\left(-\frac{1}{6}\right)\det(A)$$

or

$$\det(U) = \left(-\frac{1}{6}\right)\det(A).$$

Plugging in $\det(U)$ gives us

$$-\frac{13}{2} = \left(-\frac{1}{6}\right)\det(A).$$

Solving for $\det(A)$ now gives

$$\det(A) = (-6)\left(-\frac{13}{2}\right) = 39.$$

(Notice that this matrix is the same one we used for Examples 6 and 7 where we also got $\det(A) = 39$.)

We now have $2n + 1$ choices when computing the determinant of an $n \times n$ matrix: the $2n$ options for expansion along a row or down a column method plus this new way via row reduction. Sometimes we'll want to use one of them, sometimes another. If A has a row or column which is mostly zeros, I'd personally prefer to use expansion. If A is close to triangular already, I'll use the method outlined in the example above. You may choose to use another method from the one I or your friends would use, but the beauty of this mathematical choice is that we'll all end up with the same value for the determinant.

Now that we understand how to compute the determinant of any $n \times n$ matrix, let's go back to our inspiration for the 2×2 determinant: that a 2×2 matrix is invertible if and only if its determinant is nonzero. Is this also true in the $n \times n$ case? We know that a square matrix is invertible if and only if its reduced echelon form is the identity matrix, so we'd like to link that to the determinant. If we think for a minute about how row operations change the determinant, none of them change it from zero to nonzero or vice versa unless we multiply a row of our matrix by the constant 0. Since we aren't allowed to multiply a row by 0 as we move from A to its reduced echelon form, A and its reduced echelon form either both have a nonzero determinant or both have determinant equal to zero. The reduced echelon form of A is always an upper triangular matrix, so its determinant is simply the product of its diagonal entries. If A is invertible, then its reduced echelon form is I_n. Since $\det(I_n) = 1$, we must have $\det(A) \neq 0$. If A isn't invertible, then its reduced echelon form must have at least one zero on the diagonal. This would mean that the determinant of the reduced echelon form of A, and hence also that of A itself, would be 0. Therefore we get the following addition to the Invertible Matrix Theorem.

Theorem 7. An $n \times n$ matrix A is invertible if and only if $\det(A) \neq 0$.

We will rely on this fact when we return to finding eigenvectors and eigenvalues in the next section.

In addition to our link between invertibility and determinants, we can use our understanding of row operations' effects on the determinant to show that determinants interact nicely with matrix multiplication. To do this, recall the elementary matrices we discussed in 2.9 which link our row operations to matrix multiplication in the following way. Suppose E is the elementary matrix of some row operation and A is any $n \times n$ matrix. Then EA is the matrix we get by doing that row operation to A. In other words, multiplication on the left by the elementary matrix of a row operation does that row operation to the matrix we multiplied. Let's explore the determinant of each type of elementary matrix.

If E is the elementary matrix of a row operation which adds a multiple of one row to another row, then its row operation has no effect on the determinant, so $\det(EA) = \det(A)$. Since we got E by doing this row operation to I_n, we must have $\det(E) = \det(I_n) = 1$. This means that for any $n \times n$ matrix A we have

$$\det(EA) = \det(A) = \det(E)\det(A).$$

If E is the elementary matrix of a row operation which swaps two rows, then its row operation changes the sign the determinant, which means $\det(EA) = -\det(A)$. Since we got E by doing this row operation to I_n, we must have $\det(E) = -\det(I_n) = -1$. This means that for any $n \times n$ matrix A we have

$$\det(EA) = -\det(A) = \det(E)\det(A).$$

If E is the elementary matrix of a row operation which multiplies a row by a constant s, then its row operation multiplies the determinant by s, so $\det(EA) = s\det(A)$. Since we got E by doing this row operation to I_n, we must have $\det(E) = s\det(I_n) = s$. This means that for any $n \times n$ matrix A we have

$$\det(EA) = s\det(A) = \det(E)\det(A).$$

Looking at our three types of row operations, we can see that in all three cases we get

$$\det(EA) = \det(E)\det(A).$$

We would like to extend this idea that the determinant splits up over matrix products to any two $n \times n$ matrices A and B.

If A or B isn't invertible, we know from 2.11's Exercises 10 and 11 that AB also isn't invertible. This means that $\det(AB) = 0$. Since either $\det(A) = 0$ or $\det(B) = 0$, we have $\det(AB) = \det(A)\det(B)$.

If both A and B are invertible, they both have reduced echelon form I_n, i.e., they each have a series of row operations which takes them to I_n. Since each row operation is reversible by another row operation, we can also go from I_n to A or B via row operations. This means we have elementary matrices $E_1, \ldots E_k$ and $F_1, \ldots F_\ell$ so that $A = E_1 \cdots E_k I_n$ and $B = F_1 \cdots F_\ell I_n$. Repeatedly using the fact that $\det(EA) = \det(E)\det(A)$, we get

$$\det(A) = \det(E_1)\cdots\det(E_k)\det(I_n)$$

and

$$\det(B) = \det(F_1)\cdots\det(F_\ell)\det(I_n).$$

Since $\det(I_n) = 1$, this means

$$\det(A)\det(B) = \det(E_1)\cdots\det(E_k)\det(F_1)\cdots\det(F_\ell).$$

Their product is

$$AB = E_1\cdots E_k I_n F_1 \cdots F_\ell I_n = E_1 \cdots E_k F_1 \cdots F_\ell.$$

Taking the determinant and applying the fact that $\det(EA) = \det(E)\det(A)$ repeatedly, we get

$$\det(AB) = \det(E_1)\cdots\det(E_k)\det(F_1)\cdots\det(F_\ell)$$

which is equal to our computation of $\det(A)\det(B)$ above.

We can summarize this as follows:

Theorem 8. For any $n \times n$ matrices A and B, $\det(AB) = \det(A)\det(B)$.

If A is invertible, this theorem gives us a nice fact about the determinant of A^{-1}.

Theorem 9. Let A be an invertible matrix. Then $\det(A^{-1}) = \dfrac{1}{\det(A)}$.

To see why this is true, remember $AA^{-1} = I_n$, so we must have

$$\det(AA^{-1}) = \det(I_n) = 1.$$

But the determinant on the left-hand side of the equation can be split up into the product of the determinants of A and A^{-1} to give us

$$\det(A)\det(A^{-1}) = 1.$$

Since $\det(A) \neq 0$, we can divide both sides by $\det(A)$ to get

$$\det(A^{-1}) = \frac{1}{\det(A)}.$$

Exercises 4.2.

1. Compute $\det \left(\begin{bmatrix} -5 & 3 \\ 1 & 6 \end{bmatrix} \right)$.

2. Compute $\det \left(\begin{bmatrix} 2 & 7 \\ -1 & -8 \end{bmatrix} \right)$.

3. Compute $\det \left(\begin{bmatrix} 1 & 4 \\ 3 & -2 \end{bmatrix} \right)$.

4. Compute $\det \left(\begin{bmatrix} 9 & 0 \\ -6 & 7 \end{bmatrix} \right)$.

5. Use the determinant to describe the geometric effect $\begin{bmatrix} -1 & 2 \\ -2 & 1 \end{bmatrix}$'s map has on the plane?

6. Use the determinant to describe the geometric effect $\begin{bmatrix} 4 & 0 \\ 2 & 3 \end{bmatrix}$'s map has on the plane?

7. Use the determinant to describe the geometric effect $\begin{bmatrix} 1 & 3 \\ 0 & -1 \end{bmatrix}$'s map has on the plane?

8. Use the determinant to describe the geometric effect $\begin{bmatrix} 3 & 5 \\ 6 & 2 \end{bmatrix}$'s map has on the plane?

9. Compute the determinant of $\begin{bmatrix} 5 & 0 & 0 \\ 6 & -2 & 0 \\ 1 & 7 & 4 \end{bmatrix}$.

10. Compute the determinant of $\begin{bmatrix} -1 & 0 & 0 \\ 4 & 3 & 0 \\ 9 & -2 & 6 \end{bmatrix}$.

11. Compute the determinant of $\begin{bmatrix} 2 & 0 & 7 \\ 0 & 8 & -3 \\ 0 & 0 & 1 \end{bmatrix}$.

12. Compute the determinant of $\begin{bmatrix} -4 & 1 & -6 \\ 0 & -3 & -5 \\ 0 & 0 & -2 \end{bmatrix}$.

13. Find A_{23} where $A = \begin{bmatrix} 2 & -1 & 0 \\ 4 & -5 & 2 \\ 1 & 0 & -2 \end{bmatrix}$.

14. Find A_{12} where $A = \begin{bmatrix} 1 & 0 & 8 \\ 6 & -3 & 1 \\ 3 & -5 & 0 \end{bmatrix}$.

15. Find A_{31} where $A = \begin{bmatrix} 0 & -5 & 2 \\ 8 & 3 & 1 \\ 7 & -4 & -1 \end{bmatrix}$.

16. Find A_{22} where $A = \begin{bmatrix} -3 & 1 & 4 \\ 2 & 0 & -1 \\ 9 & 3 & 6 \end{bmatrix}$.

17. What effect does swapping rows 1 and 2 have on the determinant?

18. What effect does adding 3 times row 2 to row 1 have on the determinant?

19. What effect does multiplying row 1 by -4 have on the determinant?

20. What effect does adding -2 times row 3 to row 1 have on the determinant?

21. Compute the determinant of $A = \begin{bmatrix} 1 & 3 & 4 \\ 2 & 0 & -2 \\ -1 & 1 & -1 \end{bmatrix}$ three times: once by expanding along a row, once by expanding down a column, and once by using row operations to reduce it to an upper triangular matrix.

22. Compute the determinant of $A = \begin{bmatrix} 1 & 0 & 6 \\ 2 & 2 & -9 \\ 1 & 1 & -3 \end{bmatrix}$ three times: once by expanding along a row, once by expanding down a column, and once by using row operations to reduce it to an upper triangular matrix.

23. Compute the determinant of $A = \begin{bmatrix} 4 & 8 & 10 \\ 1 & -1 & 3 \\ -4 & 0 & 2 \end{bmatrix}$ three times: once by expanding along a row, once by expanding down a column, and once by using row operations to reduce it to an upper triangular matrix.

24. Compute the determinant of $A = \begin{bmatrix} 2 & 3 & -1 \\ 0 & 5 & 3 \\ -4 & -6 & 2 \end{bmatrix}$ three times: once by expanding along a row, once by expanding down a column, and once by using row operations to reduce it to an upper triangular matrix.

25. Use the determinant of $A = \begin{bmatrix} -5 & 2 & 7 \\ 3 & 0 & -2 \\ 4 & 0 & 1 \end{bmatrix}$ to decide whether or not A is invertible.

26. Use the determinant of $A = \begin{bmatrix} -7 & 3 & 0 & 1 \\ 4 & 0 & 0 & 0 \\ 5 & 2 & 1 & 6 \\ -8 & -2 & 1 & 3 \end{bmatrix}$ to decide whether

    or not A is invertible.

27. Use the determinant of $A = \begin{bmatrix} 2 & 0 & 0 & 4 \\ 2 & 0 & 3 & 0 \\ -7 & 3 & 10 & -5 \\ -1 & 0 & 6 & 0 \end{bmatrix}$ to decide whether

    or not A is invertible.

28. Use the determinant of $A = \begin{bmatrix} -1 & 3 & 0 & 2 \\ 0 & 1 & 0 & -4 \\ 7 & -4 & 2 & 9 \\ -2 & 0 & 0 & 1 \end{bmatrix}$ to decide whether

    or not A is invertible.

29. Let A and B be $n \times n$ matrices with $\det(A) = 3$ and $\det(B) = -6$.

    (a) Compute $\det(AB)$.
    (b) Compute $\det(A^{-1})$.

30. Let A and B be 2×2 matrices with $\det(A) = 10$ and $\det(B) = -1$.

    (a) Compute $\det(AB)$.
    (b) Compute $\det(B^{-1})$.

31. Let A and B be 4×4 matrices with $\det(A) = 12$ and $\det(B) = -2$.

    (a) Compute $\det(AB)$.
    (b) Compute $\det(A^{-1})$.

32. Let A and B be 3×3 matrices with $\det(A) = -2$ and $\det(B) = 4$.

    (a) Find $\det(A^{-1})$.
    (b) Find $\det(AB)$.

33. Let A be a 3×3 matrix with $\det(A) = 6$. What is $\det(2 \cdot A)$?

34. Let A be an $n \times n$ matrix with $\det(A) = 6$. What is $\det(2 \cdot A)$?

35. If A is a 1×1 matrix, we can define its determinant to be its only entry a_{11}. Use this definition to show that the determinant of $\begin{bmatrix} a & b \\ c & d \end{bmatrix}$ is $ad - bc$ if we compute it using our formulas for expansion along a row and down a column.

36. Let $V = \{A \in M_{nn} \mid A \text{ is diagonal and } \det(A) \neq 0\}$ with operations $A \text{ "} + \text{" } B = AB$ and $r \text{ "} \cdot \text{" } A = A^r$. Show V is a vector space.

4.3   Eigenspaces

After our interlude developing determinants in 4.2, let's get back to finding the eigenvalues and eigenvectors of an $n \times n$ matrix A. Recall from the end of 4.1 that we'd reduced the problem of finding the eigenvalues, λ, and the eigenvectors, $\vec{x}$, of a matrix A to solving the equation $(A - \lambda I_n)\vec{x} = \vec{0}$. Our issue back then was that our equation had two unknowns and we didn't know how to solve for both of them at once. In this section we'll use the determinant to solve for λ first, and then use our value of λ to solve for $\vec{x}$ later.

To isolate solving for λ and solving for $\vec{x}$, remember that the Invertible Matrix Theorem from 2.11 tells us that we have a nonzero solution to $A\vec{x} = \vec{0}$ exactly when A isn't invertible. Since eigenvectors are defined to be nonzero, we can apply this to $(A - \lambda I_n)\vec{x} = \vec{0}$ and get that we'll have eigenvectors for A precisely when $A - \lambda I_n$ isn't invertible. Usually we check invertibility of matrices by row reducing to see if their reduced echelon form is I_n, however the presence of the variable λ makes this unappealing. Instead we'll rely on our newest addition to the Invertible Matrix Theorem: the fact that a matrix is invertible exactly when its determinant is nonzero. In other words, whenever $\det(A - \lambda I_n) = 0$ we'll know $A - \lambda I_n$ isn't invertible so we'll get eigenvectors for A and λ will be an eigenvector of A.

Example 1. Find all the eigenvalues of $A = \begin{bmatrix} 1 & 2 \\ 4 & 3 \end{bmatrix}$.

(Note that this is the matrix from 4.1's Example 8.)

We want to find all values of λ for which $\det(A - \lambda I_n) = 0$. Our matrix is 2×2, so $n = 2$. We'll start by computing

$$A - \lambda I_2 = \begin{bmatrix} 1 & 2 \\ 4 & 3 \end{bmatrix} - \lambda \begin{bmatrix} 1 & 0 \\ 0 & 1 \end{bmatrix} = \begin{bmatrix} 1 & 2 \\ 4 & 3 \end{bmatrix} - \begin{bmatrix} \lambda & 0 \\ 0 & \lambda \end{bmatrix} = \begin{bmatrix} 1 - \lambda & 2 \\ 4 & 3 - \lambda \end{bmatrix}.$$

Taking the determinant gives us

$$\det(A - \lambda I_2) = (1 - \lambda)(3 - \lambda) - (2)(4) = 3 - 4\lambda + \lambda^2 - 8 = \lambda^2 - 4\lambda - 5.$$

This means our eigenvalues are the solutions to

$$\lambda^2 - 4\lambda - 5 = 0.$$

We can factor our polynomial in λ to get

$$(\lambda - 5)(\lambda + 1) = 0$$

which has solutions $\lambda = 5$ and $\lambda = -1$. Therefore A has two eigenvalues: 5 and -1.

Example 2. Find all the eigenvalues of $A = \begin{bmatrix} 1 & 0 & 7 \\ -2 & 2 & 14 \\ 1 & 0 & -5 \end{bmatrix}$.

As in the previous example, we start by computing $\det(A - \lambda I_n)$. Here $n = 3$, so we get

$$A - \lambda I_3 = \begin{bmatrix} 1 & 0 & 7 \\ -2 & 2 & 14 \\ 1 & 0 & -5 \end{bmatrix} - \lambda \begin{bmatrix} 1 & 0 & 0 \\ 0 & 1 & 0 \\ 0 & 0 & 1 \end{bmatrix} = \begin{bmatrix} 1-\lambda & 0 & 7 \\ -2 & 2-\lambda & 14 \\ 1 & 0 & -5-\lambda \end{bmatrix}.$$

The best choice for computing the determinant is to expand down the 2nd column since it has two zero entries. This gives us

$$\det\left(\begin{bmatrix} 1-\lambda & 0 & 7 \\ -2 & 2-\lambda & 14 \\ 1 & 0 & -5-\lambda \end{bmatrix}\right) = (-1)^{2+2}(2-\lambda)\det\left(\begin{bmatrix} 1-\lambda & 7 \\ 1 & -5-\lambda \end{bmatrix}\right)$$

$$= (2-\lambda)((1-\lambda)(-5-\lambda) - 1(7)).$$

Since we want to set this equal to zero and solve for λ, I'll leave the factor of $(2-\lambda)$ in the front and try to expand and factor the rest of this polynomial. This gives us

$$(2-\lambda)(\lambda^2 - 4\lambda - 5 - 7) = (2-\lambda)(\lambda^2 + 4\lambda - 12) = (2-\lambda)(\lambda - 2)(\lambda + 6).$$

Now we can find the eigenvalues of A by setting $\det(A - \lambda I_3) = 0$. Solving

$$(2-\lambda)(\lambda - 2)(\lambda + 6) = 0$$

gives us $\lambda = 2$, $\lambda = 2$, and $\lambda = -6$, so 2 and -6 are the eigenvalues of A.

Example 3. Find the population growth rate of a population which has demographic matrix $A = \begin{bmatrix} 0 & 0.25 & 0.25 \\ 0.5 & 0.25 & 0.5 \\ 0.5 & 0.5 & 0.25 \end{bmatrix}$.

Recall from 4.1 that the population growth rate is the largest eigenvalue of a demographic matrix, so we need to find the eigenvalues of A and select the biggest one. A is 3×3, so we start by computing

$$A - \lambda I_3 = \begin{bmatrix} 0-\lambda & 0.25 & 0.25 \\ 0.5 & 0.25-\lambda & 0.5 \\ 0.5 & 0.5 & 0.25-\lambda \end{bmatrix}.$$

Computing the determinant by expanding down the first column gives us

$$\det(A - \lambda I_3) = -\lambda \det \begin{bmatrix} 0.25 - \lambda & 0.5 \\ 0.5 & 0.25 - \lambda \end{bmatrix} - 0.5 \det \begin{bmatrix} 0.25 & 0.25 \\ 0.5 & 0.25 - \lambda \end{bmatrix}$$
$$+ 0.5 \det \begin{bmatrix} 0.25 & 0.25 \\ 0.25 - \lambda & 0.5 \end{bmatrix}.$$

Expanding each of our 2×2 determinants gives us

$$\det(A - \lambda I_3) = -\lambda((0.25 - \lambda)^2 - (0.5)^2) - 0.5((0.25)(0.25 - \lambda) - (0.25)(0.5))$$
$$+ 0.5((0.25)(0.5) - (0.25)(0.25 - \lambda)).$$

This simplifies to

$$-\lambda^3 + 0.5\lambda^2 + 0.4375\lambda + 0.0625 = -(\lambda - 1)(\lambda + 0.25)^2$$

so our eigenvalues are the solutions to

$$-(\lambda - 1)(\lambda + 0.25)^2 = 0,$$

namely $\lambda = 1$ and $\lambda = -0.25$.

The largest of these eigenvalues is $\lambda = 1$, which means the population growth rate is 1.

The examples above show us if A is an $n \times n$ matrix, $\det(A - \lambda I_n)$ is always a polynomial in λ of degree n. This tells us a few things. First of all, an $n \times n$ matrix can have at most n different eigenvalues, although it can have fewer than n as in our Example 2. Second, it is possible to have a matrix with no eigenvalues among the real numbers, because not every polynomial has real roots. (Think of $x^2 + 4$.) We have chosen to work with vector spaces where our scalars are real numbers, but many people choose to work with vector spaces whose scalars are complex numbers so that they can be assured of always having eigenvalues. If you're interested in learning more about this, see Appendix A.1. For the rest of this book, I'll make sure to choose matrices for examples and exercises whose eigenvalues are real numbers.

In chemistry, eigenvalues are used as part of the Hückel method to compute molecular orbitals for planar organic molecules like hydrocarbons. These in turn allow chemists to calculate things like charge densities, molecular reactivity, and even the color of a molecule! To do this, chemists compute the eigenvalues of a matrix H with one row and column for each carbon atom where $H_{ii} = \alpha$ and $H_{ij} = \beta$ if atoms i and j share an atomic bond and $H_{ij} = 0$ if the ith and jth atoms are not bonded. (Here α and β are different relative energy levels of electrons.) To find eigenvalues of H we need to solve $\det(A) = 0$ where $A = H - \lambda I_n$ has diagonal entries $\alpha - \lambda$ and the same off-diagonal entries as H. If we divide all entries by β, then our diagonal entries are $\dfrac{\alpha - \lambda}{\beta}$, which we will call x, and our off-diagonal entries are 1 for bonded atoms and 0 otherwise. (This is similar to the adjacency vectors discussed in Example 7 from Chapter 0.) Solving $\det(A) = 0$ for x and using the values of x, α, and β then allows us to find λ.

Example 4. Find the matrix A and solve $\det(A) = 0$ for x for the molecule 1,3-butadiene.

Since our matrix A depends on the number and bonds between the carbon atoms in a molecule of 1,3-butadiene, let's start by looking at a picture of its molecular structure. (Here Hs are hydrogen atoms and Cs are carbon.)

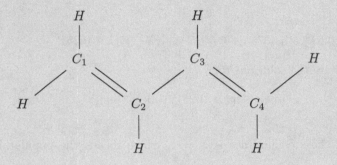

There are 4 carbon atoms which I've helpfully labeled C_1 through C_4, so our matrix A will be 4×4. Remember that A's diagonal entries are all x, and A_{ij} is 1 if there is a bond between C_i and C_j and 0 otherwise. Looking at the picture above, we can see that there are bonds between C_1 and C_2, C_2 and C_3, and C_3 and C_4. Therefore our matrix is $A = \begin{bmatrix} x & 1 & 0 & 0 \\ 1 & x & 1 & 0 \\ 0 & 1 & x & 1 \\ 0 & 0 & 1 & x \end{bmatrix}$.

We want to solve $\det(A) = 0$ for x, so let's take the determinant of A by expanding along the first row. This gives us

$$\det(A) = x \det\left(\begin{bmatrix} x & 1 & 0 \\ 1 & x & 1 \\ 0 & 1 & x \end{bmatrix}\right) - 1 \det\left(\begin{bmatrix} 1 & 1 & 0 \\ 0 & x & 1 \\ 0 & 1 & x \end{bmatrix}\right).$$

There are choices of how to expand each of our 3×3 subdeterminants, so let's expand down the first column of each. This gives us

$$\det\left(\begin{bmatrix} x & 1 & 0 \\ 1 & x & 1 \\ 0 & 1 & x \end{bmatrix}\right) = x \det\left(\begin{bmatrix} x & 1 \\ 1 & x \end{bmatrix}\right) - 1 \det\left(\begin{bmatrix} 1 & 0 \\ 1 & x \end{bmatrix}\right)$$

$$= x(x^2 - 1) - 1(x - 0)$$

and

$$\det\left(\begin{bmatrix} 1 & 1 & 0 \\ 0 & x & 1 \\ 0 & 1 & x \end{bmatrix}\right) = 1\det\left(\begin{bmatrix} x & 1 \\ 1 & x \end{bmatrix}\right) = 1(x^2 - 1).$$

Plugging these subdeterminants back into our formula for $\det(A)$ gives us

$$\det(A) = x(x(x^2 - 1) - 1(x - 0)) - 1(1(x^2 - 1)) = x^4 - 3x^2 + 1.$$

This doesn't factor nicely over the integers, but we can solve $\det(A) = 0$ using technological assistance to get the solutions $x = \frac{1}{2}(1 + \sqrt{5})$, $x = \frac{1}{2}(1 - \sqrt{5})$, $x = \frac{1}{2}(-1 + \sqrt{5})$, and $x = \frac{1}{2}(-1 - \sqrt{5})$.

Now that we know how to find the eigenvalues of a matrix A, we can go back to our original equation $(A - \lambda I_n)\vec{x} = \vec{0}$ to solve for the eigenvectors for each eigenvalue. Recall that a matrix equation has either no solutions, one solution, or infinitely many solutions. Clearly the equation $(A - \lambda I_n)\vec{x} = \vec{0}$ always has at least the solution $\vec{x} = \vec{0}$, and if we have chosen λ to be an eigenvalue, then there will also be at least one nonzero eigenvector solution as well. This means that when λ is an eigenvalue of A the equation $(A - \lambda I_n)\vec{x} = \vec{0}$ has infinitely many solutions which are eigenvectors of A with eigenvalue λ.

Definition. Let λ be an eigenvalue of a matrix A. The *eigenspace* of λ is $E_\lambda = \{\vec{x} \mid A\vec{x} = \lambda\vec{x}\}$.

In other words, the eigenspace of an eigenvalue λ is the set of all eigenvectors which have λ as their eigenvalue.

Example 5. Find the eigenspace of $A = \begin{bmatrix} 1 & 2 \\ 4 & 3 \end{bmatrix}$ for $\lambda = 5$.

We can find all eigenvectors of A whose eigenvalue is 5 using the equation $(A - \lambda I_2)\vec{x} = \vec{0}$. Plugging in our A and λ gives us

$$\left(\begin{bmatrix} 1 & 2 \\ 4 & 3 \end{bmatrix} - 5\begin{bmatrix} 1 & 0 \\ 0 & 1 \end{bmatrix}\right)\vec{x} = \vec{0}$$

which can be simplified to

$$\begin{bmatrix} -4 & 2 \\ 4 & -2 \end{bmatrix}\vec{x} = \vec{0}.$$

At this point, we are finding the null space of

$$\begin{bmatrix} -4 & 2 \\ 4 & -2 \end{bmatrix}$$

just as we did in 2.5 and 2.7. The reduced echelon form of this matrix is

$$\begin{bmatrix} 1 & -\frac{1}{2} \\ 0 & 0 \end{bmatrix}$$

so $x_1 = \frac{1}{2}x_2$ and x_2 is free. As in 2.8, we can write this solution set as a span. Since our solution vector is $\begin{bmatrix} \frac{1}{2}x_2 \\ x_2 \end{bmatrix}$, our null space is $\text{Span}\left\{ \begin{bmatrix} \frac{1}{2} \\ 1 \end{bmatrix} \right\}$. We saw in 3.1 that this spanning set is also a basis for the null space.

This means that

$$E_5 = \text{Span}\left\{ \begin{bmatrix} \frac{1}{2} \\ 1 \end{bmatrix} \right\}$$

and $\begin{bmatrix} \frac{1}{2} \\ 1 \end{bmatrix}$ is a basis for E_5.

Example 6. Find the eigenspace of $A = \begin{bmatrix} 1 & 0 & 7 \\ -2 & 2 & 14 \\ 1 & 0 & -5 \end{bmatrix}$ for $\lambda = 2$.

Mirroring the previous example, we need to find the null space of

$$A - 2I_3 = \begin{bmatrix} -1 & 0 & 7 \\ -2 & 0 & 14 \\ 1 & 0 & -7 \end{bmatrix}.$$

This matrix row reduces to

$$\begin{bmatrix} 1 & 0 & -7 \\ 0 & 0 & 0 \\ 0 & 0 & 0 \end{bmatrix}$$

so $x_1 = 7x_3$ and x_2 and x_3 are free. Thus our eigenspace E_2 is the set of vectors of the form $\begin{bmatrix} 7x_3 \\ x_2 \\ x_3 \end{bmatrix}$.

As we did above, we can write this as

$$E_2 = \text{Span}\left\{ \begin{bmatrix} 0 \\ 1 \\ 0 \end{bmatrix}, \begin{bmatrix} 7 \\ 0 \\ 1 \end{bmatrix} \right\}$$

or say E_2 has basis $\begin{bmatrix} 0 \\ 1 \\ 0 \end{bmatrix}, \begin{bmatrix} 7 \\ 0 \\ 1 \end{bmatrix}$.

Biologists call the eigenvector of length 1 with the largest eigenvector (aka the population growth rate) the "stable stage distribution". We'll learn how to compute the length of a vector and find vectors of given lengths within a span in Chapter 5, but for now we can at least find a basis for the eigenspace containing the stable state distribution.

Example 7. Find the eigenspace containing the stable stage distribution for the population discussed in Example 3.

In Example 3 we found that the population growth rate was $\lambda = 1$, so we want to find the eigenspace E_1 of our demographic matrix

$$A = \begin{bmatrix} 0 & 0.25 & 0.25 \\ 0.5 & 0.25 & 0.5 \\ 0.5 & 0.5 & 0.25 \end{bmatrix}.$$

This means finding the null space of

$$A - 1I_3 = \begin{bmatrix} -1 & 0.25 & 0.25 \\ 0.5 & -0.75 & 0.5 \\ 0.5 & 0.5 & -0.75 \end{bmatrix}$$

which row reduces to

$$\begin{bmatrix} 1 & 0 & -0.5 \\ 0 & 1 & -1 \\ 0 & 0 & 0 \end{bmatrix}$$

so $x_1 = 0.5x_3$, $x_2 = x_3$, and x_3 is free. This means E_1 is all vectors of the form $\begin{bmatrix} 0.5x_3 \\ x_3 \\ x_3 \end{bmatrix}$.

This tells us our stable stage distribution is (the vector of length 1) in the span of $\begin{bmatrix} 0.5 \\ 1 \\ 1 \end{bmatrix}$.

As we've seen in the examples above, there is always at least one free variable when we solve for $\vec{x}$. If there were no free variables, we'd know the only solution was $\vec{x} = \vec{0}$, which would mean our matrix was invertible and λ wasn't an eigenvalue. For me, this usually means I've messed up my determinant somehow. If we have more than one free variable, as in Example 6, that just means our eigenspace has dimension greater than 1.

We can get an upper bound on the dimension of an eigenspace E_λ by looking at how its eigenvalue solves the polynomial $\det(A - \lambda I_n)$. Some roots of a polynomial occur only once, while some occur multiple times. We saw this in Example 2, where $\det(A - \lambda I_n)$ was $(\lambda - 2)^2(\lambda + 6)$. In this case, -6 is a single root, because only one copy of $(\lambda + 6)$ factors out of $\det(A - \lambda I_n)$. However, two copies of $(\lambda - 2)$ factor out of $\det(A - \lambda I_n)$, so 2 is a double root. In general, the dimension of an eigenspace is bounded above by the number of times it is a root of $\det(A - \lambda I_n)$. Applying this to Example 2, we get that $\dim(E_{-6}) \leq 1$ while $\dim(E_2) \leq 2$.

Exercises 4.3.

1. Suppose $\det(A - \lambda I_4) = (4 - \lambda)(-1 - \lambda)^2(\frac{1}{2} - \lambda)$. Find all the eigenvalues of A, and say how many times each eigenvalue is a root of $\det(A - \lambda I_4)$.

2. Suppose $\det(A - \lambda I_4) = (7 - \lambda)^3(-10 - \lambda)$. Find all the eigenvalues of A, and say how many times each eigenvalue is a root of $\det(A - \lambda I_4)$.

3. Suppose $\det(A - \lambda I_3) = \lambda^3 - \lambda^2 - 6\lambda$. Find all the eigenvalues of A, and say how many times each eigenvalue is a root of $\det(A - \lambda I_3)$.

4. Suppose $\det(A - \lambda I_3) = \lambda^3 + 8\lambda^2 + 7\lambda$. Find all the eigenvalues of A, and say how many times each eigenvalue is a root of $\det(A - \lambda I_3)$.

5. Find all the eigenvalues of $A = \begin{bmatrix} 1 & 2 & 1 \\ 0 & -4 & 0 \\ -2 & 1 & -2 \end{bmatrix}$.

6. Find all the eigenvalues of $A = \begin{bmatrix} -1 & 0 & 8 \\ 6 & 2 & -5 \\ 0 & 0 & 7 \end{bmatrix}$.

7. Find all the eigenvalues of $A = \begin{bmatrix} 0 & 9 & -1 \\ 0 & 3 & 6 \\ 0 & 8 & -5 \end{bmatrix}$.

8. Find all the eigenvalues of $A = \begin{bmatrix} 1 & 0 & -2 \\ -4 & 0 & 8 \\ 0 & 2 & 1 \end{bmatrix}$.

9. Find the eigenspace of $A = \begin{bmatrix} 1 & 0 & -1 \\ 7 & -2 & -6 \\ 4 & 0 & -4 \end{bmatrix}$ for eigenvalue $\lambda = -2$.

10. Find the eigenspace of $A = \begin{bmatrix} 6 & 2 & 8 \\ 4 & -1 & -9 \\ 0 & 0 & 5 \end{bmatrix}$ for eigenvalue $\lambda = 5$.

11. Find the eigenspace of $A = \begin{bmatrix} 3 & -2 & 4 \\ 0 & 4 & -2 \\ -1 & 0 & 0 \end{bmatrix}$ for eigenvalue $\lambda = 3$.

12. Find the eigenspace of $A = \begin{bmatrix} 0 & -2 & 0 \\ 1 & 6 & 2 \\ 2 & 3 & 4 \end{bmatrix}$ for eigenvalue $\lambda = 0$.

13. Compute all the eigenvalues and eigenspaces of $A = \begin{bmatrix} 1 & 2 & -11 \\ 2 & -2 & 6 \\ 0 & 0 & -9 \end{bmatrix}$.

14. Compute all the eigenvalues and eigenspaces of $A = \begin{bmatrix} 0 & -7 & 0 \\ 1 & -6 & 2 \\ 2 & -9 & 4 \end{bmatrix}$.

15. Compute all the eigenvalues and eigenspaces of $A = \begin{bmatrix} -6 & 5 & -4 \\ 0 & 10 & 7 \\ 0 & -1 & 2 \end{bmatrix}$.

16. Compute all the eigenvalues and eigenspaces of $A = \begin{bmatrix} -2 & 0 & 0 \\ 7 & 4 & 2 \\ 1 & 5 & 1 \end{bmatrix}$.

17. As we did in Example 4, set up the matrix A and solve for x for ethylene (pictured below).

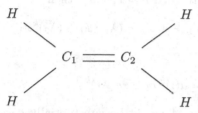

18. As we did in Example 4, set up the matrix A and solve for x for benzene (pictured below).

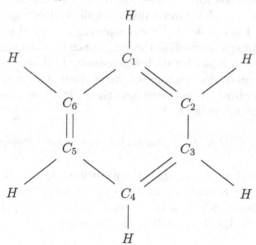

19. Find the growth rate of a population whose demographic matrix is $A = \begin{bmatrix} 0 & 1 & 6 \\ 0.7 & 0.6 & 0 \\ 0.3 & 0.4 & 0 \end{bmatrix}$. (As with many realistic examples, the numbers here do not come out clean. You will probably want to use some technology to help, see Appendix 2 for help using Mathematica.)

20. For the demographic matrix in the previous problem, find a basis for the eigenspace which contains the stable stage distribution.

4.4 Diagonalization

Remember that we started investigating eigenvectors because we were looking for an easier way to repeatedly multiply a vector by the same matrix. Obviously this is very easy if our vector is an eigenvector, but in our applications we won't always have that luxury. For example, our population vector is unlikely to magically turn out to be an eigenvector of our demographic matrix for every population. In this section we'll explore a way to use eigenvectors to make things easier for general vectors.

To start, let's suppose $\vec{x}_1$ and $\vec{x}_2$ are eigenvectors of an $n \times n$ matrix A with eigenvectors λ_1 and λ_2 respectively. If $\vec{v} = \vec{x}_1 + \vec{x}_2$, then

$$A^k \vec{v} = A^k(\vec{x}_1 + \vec{x}_2) = A^k \vec{x}_1 + A^k \vec{x}_2 = (\lambda_1)^k \vec{x}_1 + (\lambda_2)^k \vec{x}_2.$$

Similarly, if $\vec{w} = a\vec{x}_1$, then

$$A^k \vec{v} = A^k(a\vec{x}_1) = a(A^k \vec{x}_1) = a(\lambda_1)^k \vec{x}_1.$$

This suggests that multiplication by A would be substantially easier if we could write any vector as a linear combination of eigenvectors. To make this possible, we'll try to find a basis for $\mathbb{R}^n$ made up of eigenvectors of A. Since $\dim(\mathbb{R}^n) = n$, we know it is enough to find n linearly independent eigenvectors of A. Each eigenvector of A is in one of A's eigenspaces, so we'll start there.

In the last section, we practiced finding the eigenspace for each eigenvalue of A, and ended up finding a basis for each eigenspace. This means by the time we've finished finding all of A's eigenspaces, we've created several linearly independent sets of eigenvectors. To help combine these eigenspace bases into a basis for $\mathbb{R}^n$, we'll use the following fact.

Theorem 1. Eigenvectors with different eigenvalues are linearly independent.

Suppose we have eigenvectors $\vec{v}_1, \ldots, \vec{v}_k$ with eigenvalues $\lambda_1, \ldots, \lambda_k$. If they are linearly dependent, there is some $\vec{v}_i$ which is in the span of the others. Even stronger, there is some $\vec{v}_i$ which is in the span of $\vec{v}_1, \ldots, \vec{v}_{i-1}$. Pick the first place that this happens, i.e., the smallest such i. Then

$$\vec{v}_i = a_1 \vec{v}_1 + \cdots + a_{i-1} \vec{v}_{i-1}$$

and $\vec{v}_1, \ldots, \vec{v}_{i-1}$ are linearly independent. Substituting $a_1 \vec{v}_1 + \cdots + a_{i-1} \vec{v}_{i-1}$ for $\vec{v}_i$ in the equation $A\vec{v}_i = \lambda_i \vec{v}_i$ gives us

$$A(a_1 \vec{v}_1 + \cdots + a_{i-1} \vec{v}_{i-1}) = \lambda_i(a_1 \vec{v}_1 + \cdots + a_{i-1} \vec{v}_{i-1}).$$

Expanding both sides gives

$$A(a_1 \vec{v}_1) + \cdots + A(a_{i-1} \vec{v}_{i-1}) = \lambda_i(a_1 \vec{v}_1) + \cdots + \lambda_i(a_{i-1} \vec{v}_{i-1}).$$

Since $A(a_1\vec{v}_j) = \lambda_j a_1 \vec{v}_j$, we get

$$(\lambda_1 a_1)\vec{v}_1 + \cdots + (\lambda_{i-1} a_{i-1})\vec{v}_{i-1} = (\lambda_i a_1)\vec{v}_1 + \cdots + (\lambda_i a_{i-1})\vec{v}_{i-1}.$$

Moving all terms to the right-hand side of the equation gives us

$$\vec{0} = (\lambda_i a_1)\vec{v}_1 + \cdots + (\lambda_i a_{i-1})\vec{v}_{i-1} - (\lambda_1 a_1)\vec{v}_1 - \cdots - (\lambda_{i-1} a_{i-1})\vec{v}_{i-1}$$

which simplifies to

$$\vec{0} = (\lambda_i - \lambda_1)a_1\vec{v}_1 + \cdots (\lambda_i - \lambda_{i-1})a_{i-1}\vec{v}_{i-1}.$$

Since $\vec{v}_1, \ldots, \vec{v}_{i-1}$ are linearly independent, we must have $(\lambda_i - \lambda_j)a_j = 0$ for $j = 1, \ldots, i-1$. Since the eigenvalues were different, we must have $\lambda_i - \lambda_j \neq 0$ for all j, which means $a_j = 0$ for all j. Thus it wasn't possible to have any $\vec{v}_i$ in the span of the other $\vec{v}$s, so these eigenvectors are linearly independent.

Since eigenvectors with different eigenvalues are linearly independent, we can combine our bases from all the eigenspaces of A to create a set of linearly independent eigenvectors. In fact, this is the largest linearly independent collection of eigenvectors of A. We know from 3.1's Theorem 3 that if a linearly independent set contains $\dim(V)$ vectors, it is a basis for V. This means that if we've collected n vectors from the bases of A's eigenspaces, we've found a basis for $\mathbb{R}^n$ composed of eigenvectors of A. However, if there are fewer than n vectors in our collection, then we cannot create a basis of eigenvectors for $\mathbb{R}^n$. Because the number of vectors in a basis equals the dimension, another way to state this is as follows.

Theorem 2. There is a basis for $\mathbb{R}^n$ made up of eigenvectors of an $n \times n$ matrix A exactly when the sum of the dimensions of A's eigenspaces is n.

Below we'll see examples of both cases.

Example 1. Find a basis for $\mathbb{R}^3$ which is made up of eigenvectors of

$$A = \begin{bmatrix} 1 & 0 & 7 \\ -2 & 2 & 14 \\ 1 & 0 & -5 \end{bmatrix}.$$

This is the matrix from Examples 2 and 6 in 4.3, so we already know that A has eigenvalues 2 and -6 and that E_2 has basis $\begin{bmatrix} 0 \\ 1 \\ 0 \end{bmatrix}, \begin{bmatrix} 7 \\ 0 \\ 1 \end{bmatrix}$. This means we've already got two out of three basis vectors for $\mathbb{R}^3$.

To complete our basis for $\mathbb{R}^3$, we need to find another basis vector from E_{-6}. We can do this by solving $(A - (-6)I_3)\vec{x} = \vec{0}$. Plugging in A and simplifying gives us

$$\begin{bmatrix} 7 & 0 & 7 \\ -2 & 8 & 14 \\ 1 & 0 & 1 \end{bmatrix} \vec{x} = \vec{0}.$$

This new matrix has reduced echelon form

$$\begin{bmatrix} 1 & 0 & 1 \\ 0 & 1 & 2 \\ 0 & 0 & 0 \end{bmatrix}$$

so $x_1 = -x_3$, $x_2 = -2x_3$, and x_3 is free. This means

$$E_{-6} = \left\{ \begin{bmatrix} -x_3 \\ -2x_3 \\ x_3 \end{bmatrix} \right\}$$

which has basis $\begin{bmatrix} -1 \\ -2 \\ 1 \end{bmatrix}$.

Putting this newest basis vector for E_{-6} together with our two basis vectors for E_2 gives us the basis

$$\begin{bmatrix} -1 \\ -2 \\ 1 \end{bmatrix}, \begin{bmatrix} 0 \\ 1 \\ 0 \end{bmatrix}, \begin{bmatrix} 7 \\ 0 \\ 1 \end{bmatrix}$$

for $\mathbb{R}^3$ made up of eigenvectors of A.

Example 2. Show that we can't find a basis for $\mathbb{R}^3$ made up of eigenvectors of $A = \begin{bmatrix} -6 & -3 & 5 \\ 3 & 0 & -2 \\ 0 & 0 & 4 \end{bmatrix}$.

This isn't a matrix we've discussed before, so we'll have to take it from the top: find the eigenvalues, find a basis for each eigenspace, and notice that we don't get three basis vectors from that process.

To find the eigenvalues, we need to compute

$$\det \left(\begin{bmatrix} -6-\lambda & -3 & 5 \\ 3 & 0-\lambda & -2 \\ 0 & 0 & 4-\lambda \end{bmatrix} \right).$$

Let's do that by expanding along the 3rd row since it has two zero entries. This gives us

$$\det \left(\begin{bmatrix} -6-\lambda & -3 & 5 \\ 3 & -\lambda & -2 \\ 0 & 0 & 4-\lambda \end{bmatrix} \right) = (-1)^{3+3}(4-\lambda) \det \left(\begin{bmatrix} -6-\lambda & -3 \\ 3 & -\lambda \end{bmatrix} \right)$$

$$= (4-\lambda)((-6-\lambda)(-\lambda) - (-3)(3)).$$

Factoring our determinant and setting it equal to zero gives us

$$(4 - \lambda)((-6 - \lambda)(-\lambda) - (-3)(3)) = (4 - \lambda)(\lambda^2 + 6\lambda + 9)$$
$$= (4 - \lambda)(\lambda + 3)^2 = 0$$

which means A's eigenvalues are 4 and -3.

Next we need to find a basis for E_4 and E_{-3}. Let's start with E_4. This eigenspace is the null space of

$$A - 4I_3 = \begin{bmatrix} -6 - 4 & 3 & 5 \\ 3 & 0 - 4 & -2 \\ 0 & 0 & 4 - 4 \end{bmatrix} = \begin{bmatrix} -10 & -3 & 5 \\ 3 & -4 & -2 \\ 0 & 0 & 0 \end{bmatrix}$$

which has reduced echelon form

$$\begin{bmatrix} 1 & 0 & -\frac{26}{49} \\ 0 & 1 & \frac{5}{49} \\ 0 & 0 & 0 \end{bmatrix}.$$

This means

$$E_4 = \left\{ \begin{bmatrix} \frac{26}{49} x_3 \\ -\frac{5}{49} x_3 \\ x_3 \end{bmatrix} \right\}$$

which has basis $\begin{bmatrix} \frac{26}{49} \\ -\frac{5}{49} \\ 1 \end{bmatrix}$. Our other eigenspace E_{-3} is the null space of

$$A - (-3)I_3 = \begin{bmatrix} -6 - (-3) & -3 & 5 \\ 3 & 0 - (-3) & -2 \\ 0 & 0 & 4 - (-3) \end{bmatrix} = \begin{bmatrix} -3 & -3 & 5 \\ 3 & 3 & -2 \\ 0 & 0 & 7 \end{bmatrix}$$

which has reduced echelon form

$$\begin{bmatrix} 1 & 1 & 0 \\ 0 & 0 & 1 \\ 0 & 0 & 0 \end{bmatrix}.$$

Here $x_1 = -x_2$ and $x_3 = 0$, so

$$E_{-3} = \left\{ \begin{bmatrix} -x_2 \\ x_2 \\ 0 \end{bmatrix} \right\}$$

which has basis $\begin{bmatrix} -1 \\ 1 \\ 0 \end{bmatrix}$.

Notice that between our two eigenspaces, we only have two basis vectors. Since we're looking for a basis for $\mathbb{R}^3$, this isn't enough. Therefore we can't get a basis for $\mathbb{R}^3$ made up of eigenvectors of A.

From the discussion at the end of 4.3, we know that the dimension of any eigenspace satisfies $1 \leq \dim(E_\lambda) \leq k$ where k is the number of times λ is a solution to the polynomial $\det(A - \lambda I_n) = 0$. (This is often called the *multiplicity* of λ.) If λ is a so-called single root, i.e., $k = 1$, then clearly $\dim(E_\lambda) = 1$. For larger values of k, we've seen that it's possible to have $\dim(E_\lambda) < k$ as in Example 2 for $\lambda = -3$. If we list each repeated root the number of times it solves a polynomial, the degree of any polynomial equals the number of its roots. Thus our degree n polynomial, $\det(A - \lambda I_n)$, has n roots in total, so the sum of the value of k over all eigenvalues $\lambda_1, \ldots, \lambda_\ell$ is n. As a formula, this looks like $n = \sum_{i=1}^{\ell} k_i$. Since k_i is the upper bound on the number of basis vectors for E_{λ_i}, to get the sum of the dimensions of our eigenspaces equal to n we must have $\dim(E_{\lambda_i}) = k_i$ for every eigenvalue of A. We can see that this is true in Example 1 but not in Example 2. Practically speaking, this means if you come across any eigenspace whose dimension is smaller than the multiplicity of its eigenvalue, you already know it is impossible to find a basis of eigenvectors for $\mathbb{R}^n$, which may save you some work. As a consequence of this discussion we also get the following fact.

Theorem 3. An $n \times n$ matrix with n different eigenvalues always has a basis of eigenvectors for $\mathbb{R}^n$.

For example, the matrix in Example 1 of 4.3 was 2×2 and had eigenvalues 5 and -1, so we know there is a basis for $\mathbb{R}^2$ made up of its eigenvectors.

Suppose A has enough linearly independent eigenvectors to form a basis $\mathcal{B}$ for $\mathbb{R}^n$. We found $\mathcal{B}$ to simplify multiplication by A for vectors which aren't eigenvectors, which we can do by rewriting those vectors in terms of our eigenvector basis $\mathcal{B}$. In other words, we want to replace vectors in $\mathbb{R}^n$ by their $\mathcal{B}$-coordinate vectors. We're already working in $\mathbb{R}^n$, so our coordinate vector function $f_\mathcal{B}$ is a map from $\mathbb{R}^n$ to itself. We'll explore this special case of coordinate functions in the next section. For now, let's see what we've gained by computing $A[\vec{v}]_\mathcal{B}$ instead of $A\vec{v}$.

Let $\mathcal{B} = \{\vec{b}_1, \ldots, \vec{b}_n\}$ be a basis for $\mathbb{R}^n$ where $\vec{b}_i$ is an eigenvector of A with eigenvalue λ_i. Let $\vec{v}$ be any vector in $\mathbb{R}^n$. We can write $\vec{v} = a_1 \vec{b}_1 + \cdots + a_n \vec{b}_n$ so

$$[\vec{v}]_\mathcal{B} = \begin{bmatrix} a_1 \\ \vdots \\ a_n \end{bmatrix}.$$

Now

$$A\vec{v} = A(a_1 \vec{b}_1 + \cdots + a_n \vec{b}_n) = a_1(A\vec{b}_1) + \cdots + a_n(A\vec{b}_n)$$
$$= a_1(\lambda_1 \vec{b}_1) + \cdots + a_n(\lambda_n \vec{b}_n) = (a_1 \lambda_1)\vec{b}_1 + \cdots + (a_n \lambda_n)\vec{b}_n.$$

This may not look immediately better, but if we think in terms of $\mathcal{B}$-coordinate vectors, this means

$$[A\vec{v}]_B = \begin{bmatrix} a_1\lambda_1 \\ \vdots \\ a_n\lambda_n \end{bmatrix}.$$

We can summarize this relationship as follows.

Theorem 4. Let A be an $n \times n$ matrix and $\mathcal{B} = \{\vec{b}_1, \ldots, \vec{b}_n\}$ be a basis for $\mathbb{R}^n$ of A's eigenvectors with eigenvalues $\lambda_1, \ldots, \lambda_n$ respectively. If we let

$$D = \begin{bmatrix} \lambda_1 & 0 & \cdots & 0 \\ 0 & \ddots & & \vdots \\ \vdots & & \ddots & 0 \\ 0 & \cdots & 0 & \lambda_n \end{bmatrix}, \text{ then } [A\vec{v}]_B = D[\vec{v}]_B \text{ and } [A^k\vec{v}]_B = D^k[\vec{v}]_B.$$

This means that if we are willing to work in terms of a basis of eigenvectors we can make A behave like a diagonal matrix, which makes repeated multiplication much easier.

Example 3. Let $\mathcal{B} = \left\{ \begin{bmatrix} 1 \\ 2 \end{bmatrix}, \begin{bmatrix} -1 \\ 1 \end{bmatrix} \right\}$ be a basis for $\mathbb{R}^2$ made up of eigenvectors of $A = \begin{bmatrix} 1 & 2 \\ 4 & 3 \end{bmatrix}$. Use the diagonal form of A which is $D = \begin{bmatrix} 5 & 0 \\ 0 & -1 \end{bmatrix}$ to show that we have $[A\vec{v}]_B = D[\vec{v}]_B$ for $\vec{v} = \begin{bmatrix} 5 \\ 4 \end{bmatrix}$.

We saw in 4.1 that our two basis vectors are eigenvectors of A with eigenvalues 5 and -1 respectively, which means $D = \begin{bmatrix} 5 & 0 \\ 0 & -1 \end{bmatrix}$ is the diagonal matrix whose diagonal entries are the eigenvalues of our basis vectors (in order).

To check the rest of this claim, let's compute both $[A\vec{v}]_B$ and $D[\vec{v}]_B$ to show they're the same.

$$A\vec{v} = \begin{bmatrix} 1 & 2 \\ 4 & 3 \end{bmatrix} \begin{bmatrix} 5 \\ 4 \end{bmatrix} = \begin{bmatrix} 13 \\ 32 \end{bmatrix}.$$

To get $[A\vec{v}]_B$ we need to solve

$$x_1 \begin{bmatrix} 1 \\ 2 \end{bmatrix} + x_2 \begin{bmatrix} -1 \\ 1 \end{bmatrix} = \begin{bmatrix} 13 \\ 32 \end{bmatrix}$$

which has augmented coefficient matrix

$$\begin{bmatrix} 1 & -1 & 13 \\ 2 & 1 & 32 \end{bmatrix}.$$

This row reduces to

$$\begin{bmatrix} 1 & 0 & | & 15 \\ 0 & 1 & | & 2 \end{bmatrix}$$

so

$$[A\vec{v}]_B = \begin{bmatrix} 15 \\ 2 \end{bmatrix}.$$

To find $D[\vec{v}]_B$, we first need to compute $[\vec{v}]_B$. This means we need to solve

$$x_1 \begin{bmatrix} 1 \\ 2 \end{bmatrix} + x_2 \begin{bmatrix} -1 \\ 1 \end{bmatrix} = \begin{bmatrix} 5 \\ 4 \end{bmatrix}$$

which has augmented coefficient matrix

$$\begin{bmatrix} 1 & -1 & | & 5 \\ 2 & 1 & | & 4 \end{bmatrix}.$$

This row reduces to

$$\begin{bmatrix} 1 & 0 & | & 3 \\ 0 & 1 & | & -2 \end{bmatrix}$$

so

$$[\vec{v}]_B = \begin{bmatrix} 3 \\ -2 \end{bmatrix}.$$

Multiplying by D gives us

$$D[\vec{v}]_B = \begin{bmatrix} 5 & 0 \\ 0 & -1 \end{bmatrix} \begin{bmatrix} 3 \\ -2 \end{bmatrix} = \begin{bmatrix} 15 \\ 2 \end{bmatrix}$$

which matches $[A\vec{v}]_B$ as claimed.

Since putting $\mathbb{R}^n$ in terms of a basis of eigenvectors makes A act like a diagonal matrix, we make the following definition.

Definition.   An $n \times n$ matrix A is *diagonalizable* if there is a basis for $\mathbb{R}^n$ made up of eigenvectors of A.

Example 4. The matrix $A = \begin{bmatrix} -6 & -3 & 5 \\ 3 & 0 & -2 \\ 0 & 0 & 4 \end{bmatrix}$ is diagonalizable.

This is the matrix from Example 1, which had the basis of eigenvectors $\begin{bmatrix} -1 \\ -2 \\ 1 \end{bmatrix}, \begin{bmatrix} 0 \\ 1 \\ 0 \end{bmatrix}, \begin{bmatrix} 7 \\ 0 \\ 1 \end{bmatrix}$ for $\mathbb{R}^3$.

Example 5. The matrix $A = \begin{bmatrix} 1 & 0 & 7 \\ -2 & 2 & 14 \\ 1 & 0 & -5 \end{bmatrix}$ is not diagonalizable.

This is the matrix from Example 2, which we showed did not have a basis of eigenvectors for $\mathbb{R}^3$.

Exercises 4.4.

1. How many linearly independent eigenvectors does $\begin{bmatrix} 7 & -1 & 0 \\ 2 & 4 & 3 \\ -4 & 5 & 8 \end{bmatrix}$ need to be diagonalizable?

2. How many linearly independent eigenvectors does $\begin{bmatrix} -5 & 0 & 2 & 1 \\ 3 & -4 & 1 & 9 \\ 2 & 0 & 0 & -4 \\ 1 & 6 & -2 & 5 \end{bmatrix}$ need to be diagonalizable?

3. How many linearly independent eigenvectors does $A = \begin{bmatrix} 9 & 0 \\ 4 & 13 \end{bmatrix}$ need to be diagonalizable?

4. Is a 3×3 matrix A with eigenvalues $\lambda = 4$, $\lambda = -2$, and $\lambda = 1$ where $\dim(E_4) = 1$, $\dim(E_{-2}) = 1$, and $\dim(E_1) = 1$ diagonalizable?

5. Is a 3×3 matrix A with eigenvalues $\lambda = 0$ and $\lambda = -5$ where $\dim(E_0) = 1$ and $\dim(E_{-5}) = 1$ diagonalizable?

6. Is a 4×4 matrix A with eigenvalues $\lambda = 2$, $\lambda = -8$, and $\lambda = 3$ where $\dim(E_2) = 1$, $\dim(E_{-8}) = 1$, and $\dim(E_3) = 1$ diagonalizable?

7. Is a 4×4 matrix A with eigenvalues $\lambda = 7$, $\lambda = 0$, and $\lambda = 10$ where $\dim(E_7) = 1$, $\dim(E_0) = 1$, and $\dim(E_{10}) = 2$ A diagonalizable?

8. Is a 3×3 matrix A with eigenspaces $E_0 = \text{Span}\left\{ \begin{bmatrix} -2 \\ -1 \\ 1 \end{bmatrix} \right\}$,

   $E_3 = \text{Span}\left\{ \begin{bmatrix} 0 \\ 4 \\ 1 \end{bmatrix} \right\}$, and $E_{-1} = \text{Span}\left\{ \begin{bmatrix} 1 \\ 0 \\ 1 \end{bmatrix} \right\}$ diagonalizable?

9. Is a 3×3 matrix A with eigenspaces $E_1 = \text{Span}\left\{ \begin{bmatrix} -2 \\ 3 \\ 1 \end{bmatrix} \right\}$ and

   $E_{-11} = \text{Span}\left\{ \begin{bmatrix} 8 \\ 1 \\ 0 \end{bmatrix}, \begin{bmatrix} -3 \\ 0 \\ 1 \end{bmatrix} \right\}$ diagonalizable?

10. Is a 4×4 matrix A with eigenspaces $E_2 = \text{Span} \left\{ \begin{bmatrix} 1 \\ 1 \\ 0 \\ 1 \end{bmatrix} \right\}$,

    $E_{-6} = \text{Span} \left\{ \begin{bmatrix} 12 \\ 0 \\ 1 \\ 0 \end{bmatrix} \right\}$, and $E_5 = \text{Span} \left\{ \begin{bmatrix} 9 \\ 0 \\ -2 \\ 1 \end{bmatrix} \right\}$ diagonalizable?

11. Is a 4×4 matrix A with eigenspaces $E_{-2} = \text{Span} \left\{ \begin{bmatrix} -3 \\ 2 \\ 0 \\ 1 \end{bmatrix} \right\}$, and

    $E_4 = \text{Span} \left\{ \begin{bmatrix} 8 \\ 1 \\ 0 \\ 0 \end{bmatrix}, \begin{bmatrix} -1 \\ 0 \\ 1 \\ 0 \end{bmatrix}, \begin{bmatrix} 5 \\ 0 \\ 0 \\ 1 \end{bmatrix} \right\}$ diagonalizable?

12. Is $A = \begin{bmatrix} 8 & -3 \\ 9 & -4 \end{bmatrix}$ diagonalizable?

13. Is $A = \begin{bmatrix} 3 & 0 & 5 \\ 2 & 1 & 4 \\ 5 & 0 & 3 \end{bmatrix}$ diagonalizable?

14. Is $A = \begin{bmatrix} 2 & 3 & 0 \\ 5 & 4 & 0 \\ 6 & 6 & -1 \end{bmatrix}$ diagonalizable?

15. Is $A = \begin{bmatrix} 2 & 5 & 1 \\ 3 & 4 & 8 \\ 0 & 0 & 7 \end{bmatrix}$ diagonalizable?

16. Explain why a matrix is diagonalizable if and only if the dimension of each eigenspace, E_λ, equals the multiplicity of λ.

4.5 Change of Basis Matrices

In the last section, we saw that while it is usually easiest to use the standard basis for a vector space, there are situations where it is easier to use another basis instead. Our example from 4.4 was using a basis of eigenvectors to simplify matrix multiplication, but you've probably seen echoes of this idea in other areas of mathematics as well – think of using cylindrical or spherical coordinates in 3D modeling. Guidance systems for some machining tools also use matrices to change coordinates. In this section we'll develop a computational technique to make this change of basis go smoothly. In particular, we'll see what is gained when our change of basis happens inside $\mathbb{R}^n$.

Suppose we have an n-dimensional vector space V where we have been working with a basis $\mathcal{B}$. However, we now want to work with V in terms of another basis $\mathcal{C}$. Since $\dim(V) = n$, working with any basis for V means using the coordinate map to create a correspondence between V and $\mathbb{R}^n$. If we're using basis $\mathcal{B}$, this means using $f_{\mathcal{B}}$ and working with $\mathcal{B}$-coordinate vectors. If we're using basis $\mathcal{C}$, it's $f_{\mathcal{C}}$ and $\mathcal{C}$-coordinate vectors. We can think of this visually in Figure 4.2.

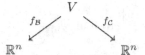

Figure 4.2: Two coordinate maps

If we have a vector $\vec{v}$ from V, this means we have two different coordinate vectors of $\vec{v}$ in $\mathbb{R}^n$: $[\vec{v}]_{\mathcal{B}}$ and $[\vec{v}]_{\mathcal{C}}$. It is certainly possible to get from $[\vec{v}]_{\mathcal{B}}$ to $[\vec{v}]_{\mathcal{C}}$ by using $[\vec{v}]_{\mathcal{B}}$ to find $\vec{v}$ and then using $f_{\mathcal{C}}$ to get $[\vec{v}]_{\mathcal{C}}$. However, tracing that route on the picture above lets us see that we'd be traveling along two legs of the triangle. It would be quicker to just cut across the bottom of the diagram by creating a map directly from $\mathbb{R}^n$ in terms of $\mathcal{B}$-coordinates to $\mathbb{R}^n$ in terms of $\mathcal{C}$-coordinates. This function is appropriately called the change of coordinates map or sometimes the change of basis map. Since it is a map from $\mathbb{R}^n$ to itself, it corresponds to an $n \times n$ matrix as defined below.

Definition. Let $\mathcal{B}$ and $\mathcal{C}$ each be a basis for an n-dimensional vector space V. The *change of coordinates matrix* is the $n \times n$ matrix $P_{\mathcal{C} \leftarrow \mathcal{B}}$ which has $P_{\mathcal{C} \leftarrow \mathcal{B}} [\vec{v}]_{\mathcal{B}} = [\vec{v}]_{\mathcal{C}}$ for every $\vec{v}$ in V.

In other words, multiplication by this change of coordinates matrix changes a vector's $\mathcal{B}$-coordinate vector into its $\mathcal{C}$-coordinate vector. This allows us to fill in our diagram of coordinate maps as in Figure 4.3.

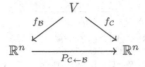

Figure 4.3: The change of coordinates matrix

Pay attention to the notation here! As with function composition, the subscript of $P_{C \leftarrow B}$ is written in the opposite direction from how we usually read. This is because we put $[\vec{v}]_B$ on the right of $P_{C \leftarrow B}$ in our multiplication. I remember this by thinking that the B on the subscript needs to touch the B-coordinate vector.

Example 1. Use basis $B = \left\{ \begin{bmatrix} 1 & 0 \\ 0 & 1 \end{bmatrix}, \begin{bmatrix} 0 & 1 \\ 1 & 0 \end{bmatrix}, \begin{bmatrix} 1 & 0 \\ 0 & -1 \end{bmatrix}, \begin{bmatrix} 0 & -1 \\ 1 & 0 \end{bmatrix} \right\}$ and basis

$C = \left\{ \begin{bmatrix} 1 & 0 \\ 0 & 0 \end{bmatrix}, \begin{bmatrix} 1 & 1 \\ 0 & 0 \end{bmatrix}, \begin{bmatrix} 1 & 1 \\ 1 & 0 \end{bmatrix}, \begin{bmatrix} 1 & 1 \\ 1 & 1 \end{bmatrix} \right\}$ for M_{22}. The matrix $\vec{v} = \begin{bmatrix} 3 & 5 \\ 1 & -5 \end{bmatrix}$

has B-coordinate vector $[\vec{v}]_B = \begin{bmatrix} -1 \\ 3 \\ 4 \\ -2 \end{bmatrix}$. Use the change of coordinates matrix

$P_{C \leftarrow B} = \begin{bmatrix} 1 & -1 & 1 & 1 \\ 0 & 0 & 0 & -2 \\ -1 & 1 & 1 & 1 \\ 1 & 0 & -1 & 0 \end{bmatrix}$ to find $[\vec{v}]_C$.

Plugging our $P_{C \leftarrow B}$ and $[\vec{v}]_B$ into $[\vec{v}]_C = P_{C \leftarrow B}[\vec{v}]_B$, gives us

$$[\vec{v}]_C = \begin{bmatrix} 1 & -1 & 1 & 1 \\ 0 & 0 & 0 & -2 \\ -1 & 1 & 1 & 1 \\ 1 & 0 & -1 & 0 \end{bmatrix} \begin{bmatrix} -1 \\ 3 \\ 4 \\ -2 \end{bmatrix} = \begin{bmatrix} -2 \\ 4 \\ 6 \\ -5 \end{bmatrix}.$$

To check our work, we can make sure that $[\vec{v}]_B$ and $[\vec{v}]_C$ give us the same

matrix $\vec{v}$. Since $[\vec{v}]_B = \begin{bmatrix} -1 \\ 3 \\ 4 \\ -2 \end{bmatrix}$, we get

$$\vec{v} = (-1)\begin{bmatrix} 1 & 0 \\ 0 & 1 \end{bmatrix} + 3\begin{bmatrix} 0 & 1 \\ 1 & 0 \end{bmatrix} + 4\begin{bmatrix} 1 & 0 \\ 0 & -1 \end{bmatrix} - 2\begin{bmatrix} 0 & -1 \\ 1 & 0 \end{bmatrix} = \begin{bmatrix} 3 & 5 \\ 1 & -5 \end{bmatrix}.$$

Similarly, $[\vec{v}]_C$ tells us

$$\vec{v} = (-2)\begin{bmatrix} 1 & 0 \\ 0 & 0 \end{bmatrix} + 4\begin{bmatrix} 1 & 1 \\ 0 & 0 \end{bmatrix} + 6\begin{bmatrix} 1 & 1 \\ 1 & 0 \end{bmatrix} - 5\begin{bmatrix} 1 & 1 \\ 1 & 1 \end{bmatrix} = \begin{bmatrix} 3 & 5 \\ 1 & -5 \end{bmatrix}.$$

Since our matrices agree, our change of coordinates computation checks out.

Given bases $\mathcal{B}$ and $\mathcal{C}$, how do we compute $P_{\mathcal{C} \leftarrow \mathcal{B}}$? We want $P_{\mathcal{C} \leftarrow \mathcal{B}}$ to satisfy $P_{\mathcal{C} \leftarrow \mathcal{B}} [\vec{v}]_{\mathcal{B}} = [\vec{v}]_{\mathcal{C}}$ for every vector $\vec{v}$ in V. Suppose $\mathcal{B} = \{\vec{b}_1, \ldots, \vec{b}_n\}$ and $\vec{v} = x_1 \vec{b}_1 + \cdots + x_n \vec{b}_n$, so

$$[\vec{v}]_{\mathcal{B}} = \begin{bmatrix} x_1 \\ \vdots \\ x_n \end{bmatrix}.$$

This means we really want

$$P_{\mathcal{C} \leftarrow \mathcal{B}} \begin{bmatrix} x_1 \\ \vdots \\ x_n \end{bmatrix} = [\vec{v}]_{\mathcal{C}}.$$

However, we can also compute $[\vec{v}]_{\mathcal{C}}$ as $f_{\mathcal{C}}(\vec{v})$. From this perspective,

$$[\vec{v}]_{\mathcal{C}} = f_{\mathcal{C}}(\vec{v}) = f_{\mathcal{C}}(x_1 \vec{b}_1 + \cdots + x_n \vec{b}_n).$$

Since coordinate functions are linear, this splits up as

$$x_1 f_{\mathcal{C}}(\vec{b}_1) + \cdots + x_n f_{\mathcal{C}}(\vec{b}_n)$$

which can also be written as

$$x_1 \left[\vec{b}_1\right]_{\mathcal{C}} + \cdots + x_n \left[\vec{b}_n\right]_{\mathcal{C}}.$$

Plugging this back into our equation with $P_{\mathcal{C} \leftarrow \mathcal{B}}$ gives us

$$P_{\mathcal{C} \leftarrow \mathcal{B}} \begin{bmatrix} x_1 \\ \vdots \\ x_n \end{bmatrix} = x_1 \left[\vec{b}_1\right]_{\mathcal{C}} + \cdots + x_n \left[\vec{b}_n\right]_{\mathcal{C}}.$$

This equation identifies two different versions of a linear system: on the right as a matrix equation and on the left as a vector equation. This gives us

Theorem 1. If $\mathcal{B}$ and $\mathcal{C}$ are bases for V, then $P_{\mathcal{C} \leftarrow \mathcal{B}} = \begin{bmatrix} \vdots & & \vdots \\ \left[\vec{b}_1\right]_{\mathcal{C}} & \cdots & \left[\vec{b}_n\right]_{\mathcal{C}} \\ \vdots & & \vdots \end{bmatrix}.$

In other words, $P_{\mathcal{C} \leftarrow \mathcal{B}}$ is the matrix whose columns are the $\mathcal{C}$-coordinate vectors of the basis vectors from $\mathcal{B}$. Since this formula isn't symmetric, I remember that I'm transforming vectors from $\mathcal{B}$ into vectors with respect to $\mathcal{C}$ to get $P_{\mathcal{C} \leftarrow \mathcal{B}}$.

Example 2. Using M_{22} with basis $\mathcal{B}$ and $\mathcal{C}$ from Example 1, compute $P_{\mathcal{C} \leftarrow \mathcal{B}}$.

To find $P_{\mathcal{C} \leftarrow \mathcal{B}}$, we need to compute the $\mathcal{C}$-coordinate vector of each matrix from our basis $\mathcal{B}$. Since

$$\mathcal{C} = \left\{ \begin{bmatrix} 1 & 0 \\ 0 & 0 \end{bmatrix}, \begin{bmatrix} 1 & 1 \\ 0 & 0 \end{bmatrix}, \begin{bmatrix} 1 & 1 \\ 1 & 0 \end{bmatrix}, \begin{bmatrix} 1 & 1 \\ 1 & 1 \end{bmatrix} \right\}$$

this means solving the equation

$$x_1 \begin{bmatrix} 1 & 0 \\ 0 & 0 \end{bmatrix} + x_2 \begin{bmatrix} 1 & 1 \\ 0 & 0 \end{bmatrix} + x_3 \begin{bmatrix} 1 & 1 \\ 1 & 0 \end{bmatrix} + x_4 \begin{bmatrix} 1 & 1 \\ 1 & 1 \end{bmatrix} = \begin{bmatrix} b_1 & b_2 \\ b_3 & b_4 \end{bmatrix}$$

for each $\begin{bmatrix} b_1 & b_2 \\ b_3 & b_4 \end{bmatrix}$ in $\mathcal{B}$. We can simplify the left-hand side to get

$$\begin{bmatrix} x_1 + x_2 + x_3 + x_4 & x_2 + x_3 + x_4 \\ x_3 + x_4 & x_4 \end{bmatrix} = \begin{bmatrix} b_1 & b_2 \\ b_3 & b_4 \end{bmatrix}.$$

Setting entries equal gives us the four equations $x_1 + x_2 + x_3 + x_4 = b_1$, $x_2 + x_3 + x_4 = b_2$, $x_3 + x_4 = b_3$, and $x_4 = b_4$. This set of linear equations has augmented coefficient matrix

$$\begin{bmatrix} 1 & 1 & 1 & 1 & b_1 \\ 0 & 1 & 1 & 1 & b_2 \\ 0 & 0 & 1 & 1 & b_3 \\ 0 & 0 & 0 & 1 & b_4 \end{bmatrix}.$$

For $\begin{bmatrix} b_1 & b_2 \\ b_3 & b_4 \end{bmatrix} = \begin{bmatrix} 1 & 0 \\ 0 & 1 \end{bmatrix}$, this means our augmented coefficient matrix is

$$\begin{bmatrix} 1 & 1 & 1 & 1 & 1 \\ 0 & 1 & 1 & 1 & 0 \\ 0 & 0 & 1 & 1 & 0 \\ 0 & 0 & 0 & 1 & 1 \end{bmatrix}$$

which has reduced echelon form

$$\begin{bmatrix} 1 & 0 & 0 & 0 & 1 \\ 0 & 1 & 0 & 0 & 0 \\ 0 & 0 & 1 & 0 & -1 \\ 0 & 0 & 0 & 1 & 1 \end{bmatrix}.$$

Therefore

$$\left[\begin{bmatrix} 1 & 0 \\ 0 & 1 \end{bmatrix} \right]_{\mathcal{C}} = \begin{bmatrix} 1 \\ 0 \\ -1 \\ 1 \end{bmatrix}.$$

Similarly, when $\begin{bmatrix} b_1 & b_2 \\ b_3 & b_4 \end{bmatrix} = \begin{bmatrix} 0 & 1 \\ 1 & 0 \end{bmatrix}$, our augmented coefficient matrix is

$$\left[\begin{array}{cccc|c} 1 & 1 & 1 & 1 & 0 \\ 0 & 1 & 1 & 1 & 1 \\ 0 & 0 & 1 & 1 & 1 \\ 0 & 0 & 0 & 1 & 0 \end{array}\right]$$

which has reduced echelon form

$$\left[\begin{array}{cccc|c} 1 & 0 & 0 & 0 & -1 \\ 0 & 1 & 0 & 0 & 0 \\ 0 & 0 & 1 & 0 & 1 \\ 0 & 0 & 0 & 1 & 0 \end{array}\right].$$

Therefore

$$\left[\begin{bmatrix} 0 & 1 \\ 1 & 0 \end{bmatrix}\right]_c = \begin{bmatrix} -1 \\ 0 \\ 1 \\ 0 \end{bmatrix}.$$

When $\begin{bmatrix} b_1 & b_2 \\ b_3 & b_4 \end{bmatrix} = \begin{bmatrix} 1 & 0 \\ 0 & -1 \end{bmatrix}$, our augmented coefficient matrix is

$$\left[\begin{array}{cccc|c} 1 & 1 & 1 & 1 & 1 \\ 0 & 1 & 1 & 1 & 0 \\ 0 & 0 & 1 & 1 & 0 \\ 0 & 0 & 0 & 1 & -1 \end{array}\right]$$

which has reduced echelon form

$$\left[\begin{array}{cccc|c} 1 & 0 & 0 & 0 & 1 \\ 0 & 1 & 0 & 0 & 0 \\ 0 & 0 & 1 & 0 & 1 \\ 0 & 0 & 0 & 1 & -1 \end{array}\right].$$

Therefore

$$\left[\begin{bmatrix} 1 & 0 \\ 0 & -1 \end{bmatrix}\right]_c = \begin{bmatrix} 1 \\ 0 \\ 1 \\ -1 \end{bmatrix}.$$

Finally, when $\begin{bmatrix} b_1 & b_2 \\ b_3 & b_4 \end{bmatrix} = \begin{bmatrix} 0 & -1 \\ 1 & 0 \end{bmatrix}$, our augmented coefficient matrix is

$$\left[\begin{array}{cccc|c} 1 & 1 & 1 & 1 & 0 \\ 0 & 1 & 1 & 1 & -1 \\ 0 & 0 & 1 & 1 & 1 \\ 0 & 0 & 0 & 1 & 0 \end{array}\right]$$

which has reduced echelon form

$$\left[\begin{array}{cccc|c} 1 & 0 & 0 & 0 & 1 \\ 0 & 1 & 0 & 0 & -2 \\ 0 & 0 & 1 & 0 & 1 \\ 0 & 0 & 0 & 1 & 0 \end{array}\right].$$

Therefore

$$\left[\begin{bmatrix} 0 & -1 \\ 1 & 0 \end{bmatrix}\right]_{\mathcal{C}} = \begin{bmatrix} 1 \\ -2 \\ 1 \\ 0 \end{bmatrix}.$$

Using these four $\mathcal{C}$-coordinate vectors (in order) as the columns of our change of coordinates matrix we get

$$P_{\mathcal{C} \leftarrow \mathcal{B}} = \begin{bmatrix} 1 & -1 & 1 & 1 \\ 0 & 0 & 0 & -2 \\ -1 & 1 & 1 & 1 \\ 1 & 0 & -1 & 0 \end{bmatrix}.$$

Notice in our last example, that we solved four matrix equations whose augmented coefficient matrices were the same except for the augmentation column. This means we could have borrowed the shortcut we use to compute matrix inverses and solved all four equations simultaneously by row reducing the 4×8 matrix formed by putting the common piece of the augmented coefficient matrices on the left and the four augmentation columns on the right. As with matrix inverses, the left-hand side would reduce to I_4 while the right-hand side would become $P_{\mathcal{C} \leftarrow \mathcal{B}}$. If $\mathcal{B}$ and $\mathcal{C}$ are bases for $\mathbb{R}^n$, these augmentation columns are just the basis vectors from $\mathcal{B}$ and the common part of the augmented coefficient matrix has columns that are the basis vectors from $\mathcal{C}$. That means we can find $P_{\mathcal{C} \leftarrow \mathcal{B}}$ as the right half of the reduced echelon form of

$$\begin{bmatrix} \vdots & & \vdots & \vdots & & \vdots \\ \vec{c}_1 & \cdots & \vec{c}_n & \vec{b}_1 & \cdots & \vec{b}_n \\ \vdots & & \vdots & \vdots & & \vdots \end{bmatrix}.$$

Example 3. In $\mathbb{R}^3$, find $P_{\mathcal{C} \leftarrow \mathcal{B}}$ where $\mathcal{B} = \left\{ \begin{bmatrix} 2 \\ 1 \\ -1 \end{bmatrix}, \begin{bmatrix} -3 \\ 0 \\ 2 \end{bmatrix}, \begin{bmatrix} 5 \\ 4 \\ 1 \end{bmatrix} \right\}$ and

$\mathcal{C} = \left\{ \begin{bmatrix} 1 \\ -1 \\ 1 \end{bmatrix}, \begin{bmatrix} 0 \\ 1 \\ -1 \end{bmatrix}, \begin{bmatrix} 0 \\ 0 \\ 1 \end{bmatrix} \right\}$.

Since $\mathcal{B}$ and $\mathcal{C}$ are bases for $\mathbb{R}^3$, we can use the shortcut described above to find $P_{C \leftarrow B}$ by row reducing

$$\begin{bmatrix} 1 & 0 & 0 & 2 & -3 & 5 \\ -1 & 1 & 0 & 1 & 0 & 4 \\ 1 & -1 & 1 & -1 & 2 & 1 \end{bmatrix}$$

to get

$$\begin{bmatrix} 1 & 0 & 0 & 2 & -3 & 5 \\ 0 & 1 & 0 & 3 & -3 & 9 \\ 0 & 0 & 1 & 0 & 2 & 5 \end{bmatrix}.$$

Thus

$$P_{C \leftarrow B} = \begin{bmatrix} 2 & -3 & 5 \\ 3 & -3 & 9 \\ 0 & 2 & 5 \end{bmatrix}.$$

We can extend this trick to a more general vector space V if we are willing to use a coordinate map to translate our problem from V to $\mathbb{R}^n$. You can redo Example 2 by using the standard basis for M_{22} to translate to $\mathbb{R}^4$. (This will actually give you the same augmented coefficient matrix pieces as we computed there.)

Since $\mathcal{B}$ is linearly independent, the only vector with $[\vec{v}]_B = \vec{0}$ is $\vec{0}_V$. This means $P_{C \leftarrow B} \vec{x} = \vec{0}$ has only one solution, $x = \vec{0}$, and therefore by the Invertible Matrix Theorem $P_{C \leftarrow B}$ is invertible. Its inverse, $(P_{C \leftarrow B})^{-1}$, satisfies

$$(P_{C \leftarrow B})^{-1} P_{C \leftarrow B} = I_n$$

so

$$(P_{C \leftarrow B})^{-1} [\vec{v}]_C = (P_{C \leftarrow B})^{-1} (P_{C \leftarrow B} [\vec{v}]_B) = [\vec{v}]_B.$$

This gives us the following:

Theorem 2. $(P_{C \leftarrow B})^{-1} = P_{B \leftarrow C}$

In other words, the inverse of a change of coordinates matrix is the change of coordinates matrix in the other direction. We can visualize this in Figure 4.4.

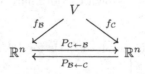

Figure 4.4: The inverse change of coordinates matrix

Example 4. Find $P_{\mathcal{B} \leftarrow \mathcal{C}}$ for M_{22} with the same $\mathcal{B}$ and $\mathcal{C}$ as in Examples 1 and 2.

We could redo our original process for finding change of coordinates matrices as we did in Example 2, by finding the $\mathcal{B}$-coordinate vectors of the matrices from our basis $\mathcal{C}$ and using them as the columns of our matrix. However, it's much shorter to just compute the inverse of our change of coordinates matrix $P_{\mathcal{C} \leftarrow \mathcal{B}}$ in the other direction.

We can find this inverse by row reducing

$$\begin{bmatrix} 1 & -1 & 1 & 1 & 1 & 0 & 0 & 0 \\ 0 & 0 & 0 & -2 & 0 & 1 & 0 & 0 \\ -1 & 1 & 1 & 1 & 0 & 0 & 1 & 0 \\ 1 & 0 & -1 & 0 & 0 & 0 & 0 & 1 \end{bmatrix}$$

to get

$$\begin{bmatrix} 1 & 0 & 0 & 0 & \frac{1}{2} & \frac{1}{2} & \frac{1}{2} & 1 \\ 0 & 1 & 0 & 0 & 0 & \frac{1}{2} & 1 & 1 \\ 0 & 0 & 1 & 0 & \frac{1}{2} & \frac{1}{2} & \frac{1}{2} & 0 \\ 0 & 0 & 0 & 1 & 0 & -\frac{1}{2} & 0 & 0 \end{bmatrix}.$$

This means

$$P_{\mathcal{B} \leftarrow \mathcal{C}} = (P_{\mathcal{C} \leftarrow \mathcal{B}})^{-1} = \begin{bmatrix} \frac{1}{2} & \frac{1}{2} & \frac{1}{2} & 1 \\ 0 & \frac{1}{2} & 1 & 1 \\ \frac{1}{2} & \frac{1}{2} & \frac{1}{2} & 0 \\ 0 & -\frac{1}{2} & 0 & 0 \end{bmatrix}.$$

There is one special case where the process of finding $P_{\mathcal{C} \leftarrow \mathcal{B}}$ is much easier: when $V = \mathbb{R}^n$, and $\mathcal{C}$ is the standard basis for $\mathbb{R}^n$. This may sound too specific to be very useful, but we'll want it when we're working with a basis of eigenvectors.

Suppose $V = \mathbb{R}^n$ and $\mathcal{C}$ is its standard basis. Our process for finding $P_{\mathcal{C} \leftarrow \mathcal{B}}$ is to find the $\mathcal{C}$-coordinate vector of each basis vector from our basis $\mathcal{B}$ and use them as the columns of $P_{\mathcal{C} \leftarrow \mathcal{B}}$. However, the way we write vectors in $\mathbb{R}^n$ is as their coordinate vectors in terms of the standard basis. This means that each basis vector from $\mathcal{B}$ is its own $\mathcal{C}$-coordinate vector, which gives us the following.

Theorem 3. Let $\mathcal{C}$ be the standard basis for $\mathbb{R}^n$ and $\mathcal{B}$ be any other basis.

Then $P_{\mathcal{C} \leftarrow \mathcal{B}} = \begin{bmatrix} \vdots & & \vdots \\ \vec{b}_1 & \cdots & \vec{b}_n \\ \vdots & & \vdots \end{bmatrix}.$

In other words, $P_{C \leftarrow B}$ is just the matrix whose columns are the basis vectors from B.

Example 5. Find $P_{C \leftarrow B}$ where $B = \left\{ \begin{bmatrix} -1 \\ -2 \\ 1 \end{bmatrix}, \begin{bmatrix} 0 \\ 1 \\ 0 \end{bmatrix}, \begin{bmatrix} 7 \\ 0 \\ 1 \end{bmatrix} \right\}$ is the basis of

eigenvectors of $A = \begin{bmatrix} 1 & 0 & 7 \\ -2 & 2 & 14 \\ 1 & 0 & -5 \end{bmatrix}$ from Example 1 in 4.4 and C is the

standard basis for $\mathbb{R}^3$.

Since C is the standard basis for $\mathbb{R}^3$, the basis vectors from B are already written as C-coordinate vectors. This means the columns of $P_{C \leftarrow B}$ are simply the basis vectors from B (in order), which means we have

$$P_{C \leftarrow B} = \begin{bmatrix} -1 & 0 & 7 \\ -2 & 1 & 0 \\ 1 & 0 & 1 \end{bmatrix}.$$

As we saw in the previous example, this links back to our work in the previous section, by letting B be a basis of eigenvectors for an $n \times n$ matrix A. From 4.4, we know that A acts like an $n \times n$ diagonal matrix D when multiplied by B-coordinate vectors. (Recall that D's diagonal entries are the eigenvalues of our basis vectors from B (in order).) With our new change of coordinates matrices, we get an alternate way to compute $A^k \vec{v}$ in three stages: change $\vec{v}$ from standard C-coordinates to B-coordinates using $P_{B \leftarrow C}$, multiply by D^k which is the diagonal version of A^k, and change the result back to standard C-coordinates using $P_{B \leftarrow C}$. We can visualize this as follows, where we start at the top left corner and end up at the top right corner. (Simply multiplying $\vec{v}$ by A^k can be visualized as starting in the same place but simply going straight across the top to the same endpoint.)

$$\begin{array}{ccc} \mathbb{R}^n & \xrightarrow{\ A^k\ } & \mathbb{R}^n \\ \scriptstyle{P_{B \leftarrow C}} \downarrow & & \uparrow \scriptstyle{P_{C \leftarrow B}} \\ \mathbb{R}^n & \xrightarrow{\ D^k\ } & \mathbb{R}^n \end{array}$$

Figure 4.5: Visualizing diagonalization

Since our change of coordinates matrices are in the special case discussed above, we know $P_{C \leftarrow B}$ is just the matrix P whose columns are the eigenvectors in our basis B and $P_{B \leftarrow C} = P^{-1}$. This means we can update Figure 4.5 to get Figure 4.6.

$$\mathbb{R}^n \xrightarrow{\ A^k\ } \mathbb{R}^n$$

$$P^{-1} \Big\downarrow \qquad\qquad \Big\uparrow P$$

$$\mathbb{R}^n \xrightarrow{\ D^k\ } \mathbb{R}^n$$

Figure 4.6: Simplified notation for diagonalization

As in our previous picture, computing $A^k\vec{v}$ directly means going from the top left to the top right via the arrow for the function A^k. Our other computational option is changing to $\mathcal{B}$-coordinates, multiplying by D^k, and changing back to $\mathcal{C}$-coordinates, which means going from the top left to the bottom left, then across to the bottom right, and finally up to the top right via the functions P^{-1}, D^k, and P respectively. Either route gives us the same answer for $A^k\vec{v}$, so we can choose whichever option seems easier for our particular problem. For example, if k is small and we're only planning to do this for one vector $\vec{v}$, it may be easier to just go across the top, i.e., compute $A^k\vec{v}$ directly. On the other hand, if k is large or we're planning to do this many times for many different $\vec{v}$s, it may be easier to find $A^k\vec{v}$ using eigenvectors instead. Many people sum up the relationship between these two paths with one of the following formulas.

Theorem 4. $A^k\vec{v} = \left(PD^kP^{-1}\right)\vec{v}$

or simply

Theorem 5. $A^k = PD^kP^{-1}$

Example 6. Find P, P^{-1}, and D for $A = \begin{bmatrix} 1 & 0 & 7 \\ -2 & 2 & 14 \\ 1 & 0 & -5 \end{bmatrix}$.

This is the matrix we worked with in Example 5 of this section and Example 1 of 4.4. From 4.4's Example 1, we have the basis

$$\mathcal{B} = \left\{ \begin{bmatrix} -1 \\ -2 \\ 1 \end{bmatrix}, \begin{bmatrix} 0 \\ 1 \\ 0 \end{bmatrix}, \begin{bmatrix} 7 \\ 0 \\ 1 \end{bmatrix} \right\}$$

of eigenvectors of A which have eigenvalues -6, 2, and 2 respectively. This means our change of coordinates matrix from $\mathcal{B}$-coordinates to standard coordinates is

$$P = \begin{bmatrix} -1 & 0 & 7 \\ -2 & 1 & 0 \\ 1 & 0 & 1 \end{bmatrix}.$$

The change of coordinates matrix in the other direction is P^{-1}, which we can compute by row reducing

$$\begin{bmatrix} -1 & 0 & 7 & 1 & 0 & 0 \\ -2 & 1 & 0 & 0 & 1 & 0 \\ 1 & 0 & 1 & 0 & 0 & 1 \end{bmatrix}$$

to get

$$\begin{bmatrix} 1 & 0 & 0 & -\frac{1}{8} & 0 & \frac{7}{8} \\ 0 & 1 & 0 & -\frac{1}{4} & 1 & \frac{7}{4} \\ 0 & 0 & 1 & \frac{1}{8} & 0 & \frac{1}{8} \end{bmatrix}$$

so

$$P^{-1} = \begin{bmatrix} -\frac{1}{8} & 0 & \frac{7}{8} \\ -\frac{1}{4} & 1 & \frac{7}{4} \\ \frac{1}{8} & 0 & \frac{1}{8} \end{bmatrix}.$$

Once we change to $\mathcal{B}$-coordinates, A behaves like the diagonal matrix whose diagonal entries are the eigenvalues of the basis vectors from $\mathcal{B}$ (in order), so

$$D = \begin{bmatrix} -6 & 0 & 0 \\ 0 & 2 & 0 \\ 0 & 0 & 2 \end{bmatrix}.$$

Note that if we were interested in dealing with A^k, we'd simply replace D with

$$D^k = \begin{bmatrix} (-6)^k & 0 & 0 \\ 0 & 2^k & 0 \\ 0 & 0 & 2^k \end{bmatrix}.$$

Exercises 4.5.

1. Suppose $P_{\mathcal{C}\leftarrow\mathcal{B}} = \begin{bmatrix} 3 & -4 \\ -1 & 5 \end{bmatrix}$ and $[\vec{v}]_\mathcal{B} = \begin{bmatrix} 2 \\ 6 \end{bmatrix}$. Find $[\vec{v}]_\mathcal{C}$.

2. Suppose $P_{\mathcal{C}\leftarrow\mathcal{B}} = \begin{bmatrix} 2 & 7 \\ -3 & 1 \end{bmatrix}$ and $[\vec{v}]_\mathcal{B} = \begin{bmatrix} -1 \\ 4 \end{bmatrix}$. Find $[\vec{v}]_\mathcal{C}$.

3. Suppose $P_{\mathcal{C}\leftarrow\mathcal{B}} = \begin{bmatrix} 5 & -2 & 1 \\ 0 & 8 & -2 \\ 1 & 5 & 3 \end{bmatrix}$ and $[\vec{v}]_\mathcal{B} = \begin{bmatrix} 2 \\ 0 \\ -1 \end{bmatrix}$. Find $[\vec{v}]_\mathcal{C}$.

4. Suppose $P_{\mathcal{C}\leftarrow\mathcal{B}} = \begin{bmatrix} 4 & 0 & -3 \\ 1 & 6 & 2 \\ -2 & -1 & 0 \end{bmatrix}$ and $[\vec{v}]_\mathcal{B} = \begin{bmatrix} 3 \\ -1 \\ 1 \end{bmatrix}$. Find $[\vec{v}]_\mathcal{C}$.

5. Suppose $P_{\mathcal{C}\leftarrow\mathcal{B}} = \begin{bmatrix} 8 & 3 \\ 2 & 5 \end{bmatrix}$ and $[\vec{v}]_\mathcal{C} = \begin{bmatrix} 3 \\ 1 \end{bmatrix}$. Find $[\vec{v}]_\mathcal{B}$.

6. Suppose $P_{\mathcal{C}\leftarrow\mathcal{B}} = \begin{bmatrix} 4 & 6 \\ 3 & 5 \end{bmatrix}$ and $[\vec{v}]_\mathcal{C} = \begin{bmatrix} 10 \\ 2 \end{bmatrix}$. Find $[\vec{v}]_\mathcal{B}$.

7. Suppose $P_{C \leftarrow B} = \begin{bmatrix} -3 & 2 & 0 \\ 8 & -12 & 4 \\ 1 & -1 & 0 \end{bmatrix}$ and $[\vec{v}]_C = \begin{bmatrix} 3 \\ -4 \\ 2 \end{bmatrix}$. Find $[\vec{v}]_B$.

8. Suppose $P_{C \leftarrow B} = \begin{bmatrix} 1 & 0 & 1 \\ -1 & 2 & 0 \\ 0 & -1 & 1 \end{bmatrix}$ and $[\vec{v}]_C = \begin{bmatrix} -3 \\ 1 \\ 4 \end{bmatrix}$. Find $[\vec{v}]_B$.

9. Let $\mathcal{B} = \left\{ \begin{bmatrix} -11 \\ -5 \\ 2 \end{bmatrix}, \begin{bmatrix} 13 \\ 19 \\ 10 \end{bmatrix}, \begin{bmatrix} 1 \\ 12 \\ 1 \end{bmatrix} \right\}$ and $\mathcal{C} = \left\{ \begin{bmatrix} 2 \\ 4 \\ 2 \end{bmatrix}, \begin{bmatrix} -3 \\ 1 \\ 0 \end{bmatrix}, \begin{bmatrix} 0 \\ -1 \\ 1 \end{bmatrix} \right\}$
   be bases for $\mathbb{R}^3$. Compute $P_{C \leftarrow B}$.

10. Let $\mathcal{B} = \left\{ \begin{bmatrix} -12 \\ 24 \\ -13 \end{bmatrix}, \begin{bmatrix} -2 \\ 8 \\ 11 \end{bmatrix}, \begin{bmatrix} -28 \\ 34 \\ -1 \end{bmatrix} \right\}$ and $\mathcal{C} = \left\{ \begin{bmatrix} -4 \\ 4 \\ 1 \end{bmatrix}, \begin{bmatrix} 2 \\ 0 \\ 3 \end{bmatrix}, \begin{bmatrix} -1 \\ 5 \\ -2 \end{bmatrix} \right\}$
    be bases for $\mathbb{R}^3$. Compute $P_{C \leftarrow B}$.

11. Use the bases $\mathcal{B} = \left\{ \begin{bmatrix} -3 & 3 \\ -3 & -7 \end{bmatrix}, \begin{bmatrix} 6 & 32 \\ 8 & 44 \end{bmatrix}, \begin{bmatrix} 4 & 18 \\ 5 & 26 \end{bmatrix}, \begin{bmatrix} 8 & 8 \\ 0 & 12 \end{bmatrix} \right\}$ and
    $\mathcal{C} = \left\{ \begin{bmatrix} 2 & 4 \\ 2 & 8 \end{bmatrix}, \begin{bmatrix} 1 & 1 \\ -1 & -1 \end{bmatrix}, \begin{bmatrix} 0 & 0 \\ 1 & 0 \end{bmatrix}, \begin{bmatrix} -1 & 3 \\ 0 & 1 \end{bmatrix} \right\}$ for M_{22}. Compute
    $P_{C \leftarrow B}$.

12. Use the bases $\mathcal{B} = \left\{ \begin{bmatrix} -10 & 3 \\ -6 & 12 \end{bmatrix}, \begin{bmatrix} -12 & 3 \\ 0 & 1 \end{bmatrix}, \begin{bmatrix} 0 & 5 \\ 18 & 2 \end{bmatrix}, \begin{bmatrix} 0 & 11 \\ 12 & 15 \end{bmatrix} \right\}$
    and
    $\mathcal{C} = \left\{ \begin{bmatrix} -1 & 2 \\ 0 & 4 \end{bmatrix}, \begin{bmatrix} 2 & 1 \\ 6 & 1 \end{bmatrix}, \begin{bmatrix} 0 & -1 \\ 0 & 3 \end{bmatrix}, \begin{bmatrix} 5 & 0 \\ 0 & 5 \end{bmatrix} \right\}$ for M_{22}. Compute
    $P_{C \leftarrow B}$.

13. Use the bases $\mathcal{B} = \{ 5x^2 - 3x + 5, -12x^2 - 44x, 7x^2 + 31x - 1 \}$ and
    $\mathcal{C} = \{ 3x^2 + 7x + 1, -2x^2 + 2x - 2, -8x + 2 \}$ for P_2. Compute
    $P_{C \leftarrow B}$.

14. Use the bases $\mathcal{B} = \{ 14x^2 - 9x + 1, 19x^2 - 3x - 4, 9x^2 + x - 10 \}$
    and $\mathcal{C} = \{ x^2 - x + 4, 5x^2 - 3, -2x^2 + x + 3 \}$ for P_2. Compute
    $P_{C \leftarrow B}$.

15. Let $\mathcal{B} = \left\{ \begin{bmatrix} 3 \\ 7 \\ -1 \end{bmatrix}, \begin{bmatrix} 1 \\ 0 \\ -3 \end{bmatrix}, \begin{bmatrix} 5 \\ 9 \\ 1 \end{bmatrix} \right\}$ be a basis for $\mathbb{R}^3$ and $\mathcal{C}$ be the
    standard basis. Compute $P = P_{C \leftarrow B}$.

16. Let $\mathcal{B} = \left\{ \begin{bmatrix} 10 \\ -6 \\ 0 \end{bmatrix}, \begin{bmatrix} -3 \\ 4 \\ 7 \end{bmatrix}, \begin{bmatrix} 9 \\ 2 \\ -5 \end{bmatrix} \right\}$ be a basis for $\mathbb{R}^3$ and $\mathcal{C}$ be the
    standard basis. Compute $P = P_{C \leftarrow B}$.

17. Let $\mathcal{B} = \left\{ \begin{bmatrix} 8 \\ 0 \\ 9 \\ 1 \end{bmatrix}, \begin{bmatrix} -11 \\ 6 \\ 3 \\ -3 \end{bmatrix}, \begin{bmatrix} 4 \\ -5 \\ -1 \\ 2 \end{bmatrix}, \begin{bmatrix} -2 \\ 3 \\ 4 \\ 11 \end{bmatrix} \right\}$ be a basis for $\mathbb{R}^4$ and $\mathcal{C}$ be

the standard basis. Compute $P = P_{\mathcal{C} \leftarrow \mathcal{B}}$.

18. Let $\mathcal{B} = \left\{ \begin{bmatrix} 4 \\ -1 \\ 5 \\ 0 \end{bmatrix}, \begin{bmatrix} -2 \\ 7 \\ -1 \\ 3 \end{bmatrix}, \begin{bmatrix} 0 \\ 3 \\ -6 \\ 5 \end{bmatrix}, \begin{bmatrix} 9 \\ 0 \\ 2 \\ 1 \end{bmatrix} \right\}$ be a basis for $\mathbb{R}^4$ and $\mathcal{C}$ be

the standard basis. Compute $P = P_{\mathcal{C} \leftarrow \mathcal{B}}$.

19. Let $\mathcal{B} = \left\{ \begin{bmatrix} 2 \\ -4 \\ 2 \end{bmatrix}, \begin{bmatrix} -3 \\ 1 \\ 0 \end{bmatrix}, \begin{bmatrix} 0 \\ -1 \\ 1 \end{bmatrix} \right\}$ be a basis for $\mathbb{R}^3$ and $\mathcal{C}$ be the

standard basis. Compute $P^{-1} = P_{\mathcal{B} \leftarrow \mathcal{C}}$.

20. Let $\mathcal{B} = \left\{ \begin{bmatrix} -2 \\ 0 \\ 1 \end{bmatrix}, \begin{bmatrix} 6 \\ 4 \\ 2 \end{bmatrix}, \begin{bmatrix} 2 \\ 0 \\ 1 \end{bmatrix} \right\}$ be a basis for $\mathbb{R}^3$ and $\mathcal{C}$ be the

standard basis. Compute $P^{-1} = P_{\mathcal{B} \leftarrow \mathcal{C}}$.

21. Let $\mathcal{B} = \left\{ \begin{bmatrix} 3 \\ 0 \\ -2 \\ 1 \end{bmatrix}, \begin{bmatrix} 2 \\ 1 \\ 0 \\ 0 \end{bmatrix}, \begin{bmatrix} -1 \\ 1 \\ 1 \\ 0 \end{bmatrix}, \begin{bmatrix} 0 \\ 0 \\ 1 \\ 1 \end{bmatrix} \right\}$ be a basis for $\mathbb{R}^4$ and $\mathcal{C}$ be

the standard basis. Compute $P^{-1} = P_{\mathcal{B} \leftarrow \mathcal{C}}$.

22. Let $\mathcal{B} = \left\{ \begin{bmatrix} 1 \\ 0 \\ 1 \\ 0 \end{bmatrix}, \begin{bmatrix} -2 \\ 1 \\ 0 \\ 1 \end{bmatrix}, \begin{bmatrix} 0 \\ -1 \\ 2 \\ 0 \end{bmatrix}, \begin{bmatrix} 1 \\ 2 \\ 0 \\ 1 \end{bmatrix} \right\}$ be a basis for $\mathbb{R}^4$ and $\mathcal{C}$ be

the standard basis. Compute $P^{-1} = P_{\mathcal{B} \leftarrow \mathcal{C}}$.

23. Let $\mathcal{B} = \left\{ \begin{bmatrix} 4 \\ 1 \\ 0 \end{bmatrix}, \begin{bmatrix} -1 \\ 0 \\ 6 \end{bmatrix}, \begin{bmatrix} 3 \\ -4 \\ 2 \end{bmatrix} \right\}$ be a basis for $\mathbb{R}^3$ made up of

eigenvectors of A and $\mathcal{C}$ be the standard basis. If the eigenvalues of
the basis vectors of $\mathcal{B}$ are (in order) $\lambda_1 = -7$, $\lambda_2 = 2$, and $\lambda_3 = -1$,
compute the matrices P and D used in our formula $A = PDP^{-1}$.

24. Let $\mathcal{B} = \left\{ \begin{bmatrix} 5 \\ 1 \\ 0 \end{bmatrix}, \begin{bmatrix} -2 \\ 6 \\ 1 \end{bmatrix}, \begin{bmatrix} 0 \\ -4 \\ 1 \end{bmatrix} \right\}$ be a basis for $\mathbb{R}^3$ made up of

eigenvectors of A and $\mathcal{C}$ be the standard basis. If the eigenvalues of
the basis vectors of $\mathcal{B}$ are (in order) $\lambda_1 = -4$, $\lambda_2 = 5$, and $\lambda_3 = 0$,
compute the matrices P and D used in our formula $A = PDP^{-1}$.

25. Let $\mathcal{B} = \left\{ \begin{bmatrix} -13 \\ 10 \\ 0 \\ 1 \end{bmatrix}, \begin{bmatrix} 4 \\ -1 \\ 0 \\ 2 \end{bmatrix}, \begin{bmatrix} 0 \\ 3 \\ 2 \\ 0 \end{bmatrix}, \begin{bmatrix} -2 \\ 1 \\ 5 \\ -3 \end{bmatrix} \right\}$ be a basis for $\mathbb{R}^4$ made up

of eigenvectors of A and C be the standard basis. If the eigenvalues of the basis vectors of $\mathcal{B}$ are (in order) $\lambda_1 = -1$, $\lambda_2 = -1$, $\lambda_3 = \frac{1}{2}$, and $\lambda_4 = 6$, compute the matrices P and D used in our formula $A = PDP^{-1}$.

26. Let $\mathcal{B} = \left\{ \begin{bmatrix} -8 \\ 1 \\ 0 \\ 0 \end{bmatrix}, \begin{bmatrix} 3 \\ 0 \\ 1 \\ 0 \end{bmatrix}, \begin{bmatrix} -1 \\ 0 \\ -4 \\ 1 \end{bmatrix}, \begin{bmatrix} 5 \\ 0 \\ -2 \\ 6 \end{bmatrix} \right\}$ be a basis for $\mathbb{R}^4$ made up of eigenvectors of A and C be the standard basis. If the eigenvalues of the basis vectors of $\mathcal{B}$ are (in order) $\lambda_1 = 3$, $\lambda_2 = -6$, $\lambda_3 = 0$, and $\lambda_4 = 5$, compute the matrices P and D used in our formula $A = PDP^{-1}$.

27. Describe a situation where you might want to diagonalize a matrix.

5

Computational Vector Geometry

5.1 Length

In this chapter, we'll return to $\mathbb{R}^n$ and start developing a set of tools that will allow us to easily compute geometric quantities without needing to visualize them first. This is most obviously useful in dealing with $\mathbb{R}^n$ for $n > 3$, but can also be easier than drawing a picture even in complicated situations in $\mathbb{R}^2$ or $\mathbb{R}^3$. (It is also appreciated by less than stellar artists like me!) The two basic geometric quantities we'll work with are length and angle. These are both scalar quantities, so we'll need a way to create scalars from vectors. Our basic tool is the following.

Definition. Let $\vec{x} = \begin{bmatrix} x_1 \\ x_2 \\ \vdots \\ x_n \end{bmatrix}$ and $\vec{y} = \begin{bmatrix} y_1 \\ y_2 \\ \vdots \\ y_n \end{bmatrix}$ be n vectors. Their *dot product* is

$$\vec{v} \cdot \vec{w} = x_1 y_1 + x_2 y_2 + \cdots + x_n y_n.$$

Note that we cannot take the dot product of two vectors unless they are both the same size.

Example 1. Compute $\begin{bmatrix} -2 \\ 1 \\ 6 \end{bmatrix} \cdot \begin{bmatrix} 3 \\ 7 \\ -1 \end{bmatrix}$.

Since both of our vectors are from $\mathbb{R}^3$, this dot product makes sense. To compute it, we multiply corresponding entries of the two vectors and add up those products to get

$$\begin{bmatrix} -2 \\ 1 \\ 6 \end{bmatrix} \cdot \begin{bmatrix} 3 \\ 7 \\ -1 \end{bmatrix} = -2(3) + 1(7) + 6(-1) = -5.$$

As with any other new vector operation, we want to explore its properties, including how it interacts with addition and scalar multiplication.

The first nice property to notice about the dot product is that unlike matrix multiplication it is commutative, i.e., $\vec{x} \cdot \vec{y} = \vec{y} \cdot \vec{x}$. This is because

$$\vec{x} \cdot \vec{y} = x_1 y_1 + x_2 y_2 + \cdots + x_n y_n$$

which equals

$$y_1 x_1 + y_2 x_2 + \cdots + y_n x_n = \vec{y} \cdot \vec{x}$$

since multiplication of real numbers commutes.

Next, let's see how the dot product interacts with vector addition. Suppose $\vec{x}$, $\vec{v}$, and $\vec{w}$ are in $\mathbb{R}^n$. If we take the dot product of one vector with the sum of the other two vectors we get

$$\vec{x} \cdot (\vec{v} + \vec{w}) = \begin{bmatrix} x_1 \\ x_2 \\ \vdots \\ x_n \end{bmatrix} \cdot \left(\begin{bmatrix} v_1 \\ v_2 \\ \vdots \\ v_n \end{bmatrix} + \begin{bmatrix} w_1 \\ w_2 \\ \vdots \\ w_n \end{bmatrix} \right) = \begin{bmatrix} x_1 \\ x_2 \\ \vdots \\ x_n \end{bmatrix} \cdot \begin{bmatrix} v_1 + w_1 \\ v_2 + w_2 \\ \vdots \\ v_n + w_n \end{bmatrix}$$

$$= x_1(v_1 + w_1) + x_2(v_2 + w_2) + \cdots + x_n(v_n + w_n)$$
$$= x_1 v_1 + x_1 w_1 + x_2 v_2 + x_2 w_2 + \cdots + x_n v_n + x_n w_n$$
$$= (x_1 v_1 + x_2 v_2 + \cdots + x_n v_n) + (x_1 w_1 + x_2 w_2 + \cdots + x_n w_n)$$
$$= \vec{x} \cdot \vec{v} + \vec{x} \cdot \vec{w}$$

so $\vec{x} \cdot (\vec{v} + \vec{w}) = \vec{x} \cdot \vec{v} + \vec{x} \cdot \vec{w}$. Thus the dot product distributes over vector addition.

Example 2. Check that $\vec{x} \cdot (\vec{v} + \vec{w}) = \vec{x} \cdot \vec{v} + \vec{x} \cdot \vec{w}$ where $\vec{x} = \begin{bmatrix} 3 \\ -2 \end{bmatrix}$, $\vec{v} = \begin{bmatrix} 8 \\ 5 \end{bmatrix}$, and $\vec{w} = \begin{bmatrix} -6 \\ 1 \end{bmatrix}$.

The left-hand side of this equation is $\vec{x} \cdot (\vec{v} + \vec{w})$ which is

$$\begin{bmatrix} 3 \\ -2 \end{bmatrix} \cdot \left(\begin{bmatrix} 8 \\ 5 \end{bmatrix} + \begin{bmatrix} -6 \\ 1 \end{bmatrix} \right) = \begin{bmatrix} 3 \\ -2 \end{bmatrix} \cdot \begin{bmatrix} 2 \\ 6 \end{bmatrix} = 3(2) + (-2)(6) = -6.$$

The right-hand side is $\vec{x} \cdot \vec{v} + \vec{x} \cdot \vec{w}$ which is

$$\begin{bmatrix} 3 \\ -2 \end{bmatrix} \cdot \begin{bmatrix} 8 \\ 5 \end{bmatrix} + \begin{bmatrix} 3 \\ -2 \end{bmatrix} \cdot \begin{bmatrix} -6 \\ 1 \end{bmatrix} = (3(8) + (-2)(5)) + (3(-6) + (-2)(1)) = -6.$$

Both sides match, so we've verified that the equation holds.

Finally, let's explore how dot products interact with scalar multiplication. Suppose $\vec{x}$ and $\vec{y}$ are in $\mathbb{R}^n$ and r is a scalar. Multiplying the dot product of

$\vec{x}$ and $\vec{y}$ by r gives us

$$r(\vec{x} \cdot \vec{y}) = r \left(\begin{bmatrix} x_1 \\ x_2 \\ \vdots \\ x_n \end{bmatrix} \cdot \begin{bmatrix} y_1 \\ y_2 \\ \vdots \\ y_n \end{bmatrix} \right) = r(x_1 y_1 + x_2 y_2 + \cdots + x_n y_n)$$

$$= r x_1 y_1 + r x_2 y_2 + \cdots + r x_n y_n.$$

This is interesting, because we can split this up one of two ways: as

$$r x_1 y_1 + r x_2 y_2 + \cdots + r x_n y_n = (r x_1) y_1 + (r x_2) y_2 + \cdots + (r x_n) y_n$$

$$= \begin{bmatrix} r x_1 \\ r x_2 \\ \vdots \\ r x_n \end{bmatrix} \cdot \begin{bmatrix} y_1 \\ y_2 \\ \vdots \\ y_n \end{bmatrix} = (r\vec{x}) \cdot \vec{y}$$

or as

$$x_1 r y_1 + x_2 r y_2 + \cdots + x_n r y_n = x_1 (r y_1) + x_2 (r y_2) + \cdots + x_n (r y_n)$$

$$= \begin{bmatrix} x_1 \\ x_2 \\ \vdots \\ x_n \end{bmatrix} \cdot \begin{bmatrix} r y_1 \\ r y_2 \\ \vdots \\ r y_n \end{bmatrix} = \vec{x} \cdot (r\vec{y}).$$

This means that scalar multiplication can be thought of as halfway distributing over the dot product, in that multiplying a dot product by a scalar is the same as multiplying one of the vectors in the dot product by that scalar, i.e.,

$$r(\vec{x} \cdot \vec{y}) = (r\vec{x}) \cdot \vec{y} = \vec{x} \cdot (r\vec{y}).$$

Example 3. Check that $r(\vec{x} \cdot \vec{y}) = (r\vec{x}) \cdot \vec{y} = \vec{x} \cdot (r\vec{y})$ holds for $\vec{x} = \begin{bmatrix} 2 \\ -3 \\ 4 \end{bmatrix}$,
$\vec{y} = \begin{bmatrix} 5 \\ 1 \\ 2 \end{bmatrix}$, and $r = 10$.

As in the previous example, we'll compute each part of the equation to check that they match. The left piece is $r(\vec{x} \cdot \vec{y})$, which in our case is

$$10 \left(\begin{bmatrix} 2 \\ -3 \\ 4 \end{bmatrix} \cdot \begin{bmatrix} 5 \\ 1 \\ 2 \end{bmatrix} \right) = 10(2(5) + (-3)(1) + 4(2)) = 150.$$

The middle piece is $(r\vec{x}) \cdot \vec{y}$, which is

$$\left(10 \begin{bmatrix} 2 \\ -3 \\ 4 \end{bmatrix} \right) \cdot \begin{bmatrix} 5 \\ 1 \\ 2 \end{bmatrix} = \begin{bmatrix} 20 \\ -30 \\ 40 \end{bmatrix} \cdot \begin{bmatrix} 5 \\ 1 \\ 2 \end{bmatrix} = 20(5) + (-30)(1) + 40(2) = 150.$$

The right piece is $\vec{x} \cdot (r\vec{y})$, which here is

$$\begin{bmatrix} 2 \\ -3 \\ 4 \end{bmatrix} \cdot \left(10 \begin{bmatrix} 5 \\ 1 \\ 2 \end{bmatrix} \right) = \begin{bmatrix} 2 \\ -3 \\ 4 \end{bmatrix} \cdot \begin{bmatrix} 50 \\ 10 \\ 20 \end{bmatrix} = 2(50) + (-3)(10) + 4(20) = 150.$$

All three parts are equal, so we are done.

Now that we understand dot products, we can start using them to explore a computational version of geometry in $\mathbb{R}^n$.

In $\mathbb{R}^2$, we have a nice formula for the length of a vector because we can view our vector as the hypotenuse of a right triangle.

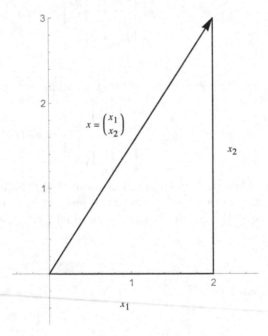

Figure 5.1: Vector length in $\mathbb{R}^2$

Thus the length of $\vec{x} = \begin{bmatrix} x_1 \\ x_2 \end{bmatrix}$ is $\sqrt{(x_1)^2 + (x_2)^2}$. We have a similar formula in $\mathbb{R}^3$ which says that the length of $\vec{x} = \begin{bmatrix} x_1 \\ x_2 \\ x_3 \end{bmatrix}$ is $\sqrt{(x_1)^2 + (x_2)^2 + (x_3)^2}$.

If we think about dot products, the quantities inside these square roots should look suggestive since they are sums of products of vector entries. In fact, they are the dot products of the vectors with themselves. This allows us to make the following definition.

Definition. Let $\vec{x}$ be in $\mathbb{R}^n$. The *norm* of $\vec{x}$ is $||\vec{x}|| = \sqrt{\vec{x} \cdot \vec{x}}$.

This could also be called the length of $\vec{x}$, but we'll follow the usual conventions and use norm. Notice that this definition does not require us to have any picture of $\vec{x}$, but is purely computational. As mentioned at the start of this section, this is extremely exciting if we want to talk about the length of vectors in $\mathbb{R}^n$ for $n > 3$.

Example 4. Compute $\left|\left| \begin{bmatrix} -2 \\ 1 \\ 0 \\ 4 \end{bmatrix} \right|\right|$.

From the definition above, we know

$$\left|\left| \begin{bmatrix} -2 \\ 1 \\ 0 \\ 4 \end{bmatrix} \right|\right| = \sqrt{\begin{bmatrix} -2 \\ 1 \\ 0 \\ 4 \end{bmatrix} \cdot \begin{bmatrix} -2 \\ 1 \\ 0 \\ 4 \end{bmatrix}}.$$

Computing the dot product gives us

$$\sqrt{(-2)^2 + 1^2 + 0^2 + 4^2} = \sqrt{21}.$$

Therefore

$$\left|\left| \begin{bmatrix} -2 \\ 1 \\ 0 \\ 4 \end{bmatrix} \right|\right| = \sqrt{21}.$$

Another formula with a similar flavor to the length formula from $\mathbb{R}^2$ is the formula for the distance between two points. From calculus, we recall that the distance between (x_1, x_2) and (y_1, y_2) is given by the formula $\sqrt{(x_1 - y_1)^2 + (x_2 - y_2)^2}$. If we think of these two points as vectors, the distance formula starts to look like a norm - specifically the norm of $\vec{x} - \vec{y}$ where $\vec{x} = \begin{bmatrix} x_1 \\ x_2 \end{bmatrix}$ and $\vec{y} = \begin{bmatrix} y_1 \\ y_2 \end{bmatrix}$. To see this more precisely, we can rewrite $||\vec{x} - \vec{y}||$ as

$$||\vec{x} - \vec{y}|| = \left|\left| \begin{bmatrix} x_1 - y_1 \\ x_2 - y_2 \end{bmatrix} \right|\right| = \sqrt{(x_1 - y_1)^2 + (x_2 - y_2)^2}$$

as claimed.

Geometrically, we can look at this in Figure 5.2. Notice that the distance between $\vec{x}$ and $\vec{y}$, labeled d, is exactly the same length as the vector $\vec{x} - \vec{y}$ geometrically constructed via our parallelogram rule as the sum of $\vec{x}$ and $-\vec{y}$.

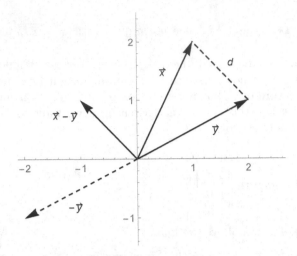

Figure 5.2: The distance between vectors

There is nothing special here about $\mathbb{R}^2$. In fact we get the following general way to compute distances between vectors in any $\mathbb{R}^n$.

Theorem 1. The distance between $\vec{x}$ and $\vec{y}$ in $\mathbb{R}^n$ is $\|\vec{x} - \vec{y}\|$.

Example 5. Find the distance between $\vec{x} = \begin{bmatrix} -2 \\ 0 \\ 6 \\ 1 \end{bmatrix}$ and $\vec{y} = \begin{bmatrix} -3 \\ 1 \\ 4 \\ 0 \end{bmatrix}$.

Theorem 1's formula tells us the distance between $\vec{x}$ and $\vec{y}$ is

$$\|\vec{x} - \vec{y}\| = \left\| \begin{bmatrix} -2 \\ 0 \\ 6 \\ 1 \end{bmatrix} - \begin{bmatrix} -3 \\ 1 \\ 4 \\ 0 \end{bmatrix} \right\| = \left\| \begin{bmatrix} 1 \\ -1 \\ 2 \\ 1 \end{bmatrix} \right\|$$

$$= \sqrt{1^2 + (-1)^2 + 2^2 + 1^2} = \sqrt{7} \approx 2.65.$$

As we did with the dot product, we next want to explore the basic properties of the norm.

One of the first major things to notice about the norm is that it is always the square root of a sum of squares. This means that no matter what the entries of $\vec{x}$ are, $\|\vec{x}\|$ can never be negative. Additionally, since the sum of squares doesn't allow for any cancelation within the square root, the only way we can have $\|\vec{x}\| = 0$ is to have all entries of $\vec{x}$ be zero. In other words:

Theorem 2. $||\vec{x}|| \geq 0$ for any $\vec{x}$, and $||\vec{x}|| = 0$ if and only if $\vec{x} = \vec{0}$.

This result also makes sense geometrically, since the length of a vector can't be negative and the only way to get a vector $\vec{x}$ to have length 0 is to make $\vec{x}$ the point at the origin, i.e., $\vec{0}$.

To see how the norm interacts with scalar multiplication, let's consider $\vec{x}$ in $\mathbb{R}^n$ and a scalar r. Then

$$||r\vec{x}|| = \left|\left| r \begin{bmatrix} x_1 \\ x_2 \\ \vdots \\ x_n \end{bmatrix} \right|\right| = \left|\left| \begin{bmatrix} rx_1 \\ rx_2 \\ \vdots \\ rx_n \end{bmatrix} \right|\right| = \sqrt{(rx_1)^2 + (rx_2)^2 + \cdots (rx_n)^2}$$

$$= \sqrt{r^2(x_1)^2 + r^2(x_2)^2 + \cdots r^2(x_n)^2} = \sqrt{r^2((x_1)^2 + (x_2)^2 + \cdots (x_n)^2)}$$

$$= \sqrt{r^2}\sqrt{(x_1)^2 + (x_2)^2 + \cdots (x_n)^2} = |r|\sqrt{(x_1)^2 + (x_2)^2 + \cdots (x_n)^2}$$

$$= |r|(||\vec{x}||).$$

Thus multiplication by a positive scalar can be done before or after the norm without changing the answer. Multiplication by a negative scalar multiplies the norm by the absolute value of the scalar. Geometrically this makes sense, because scalar multiplication multiplies the length of the vector by r and reverses its direction if r is negative. Since reversing the direction of a vector does nothing to its length, only the magnitude of r matters when figuring out the effect on the length of the vector.

Example 6. Verify $||r\vec{x}|| = |r|(||\vec{x}||)$ when $\vec{x} = \begin{bmatrix} 3 \\ 4 \end{bmatrix}$ and $r = -2$.

The left-hand side is $||r\vec{x}||$, which in our case is

$$\left|\left| -2 \begin{bmatrix} 3 \\ 4 \end{bmatrix} \right|\right| = \left|\left| \begin{bmatrix} -6 \\ -8 \end{bmatrix} \right|\right| = \sqrt{(-6)^2 + (-8)^2} = \sqrt{100} = 10.$$

The right-hand side is $|r|(||\vec{x}||)$, which is

$$|-2| \left(\left|\left| \begin{bmatrix} 3 \\ 4 \end{bmatrix} \right|\right| \right) = 2\sqrt{3^2 + 4^2} = 2\sqrt{25} = 2(5) = 10.$$

Since both sides are equal, we are done.

The interaction between vector addition and the norm is more complicated. This shouldn't surprise us, since we know from 2D geometry that the length of the sum of two vectors is unlikely to be the sum of their lengths. However, we are all familiar with the idea that for two vectors $\vec{x}$ and $\vec{y}$ in $\mathbb{R}^2$ at right angles we can use the Pythagorean theorem to say that

$$||\vec{x}||^2 + ||\vec{y}||^2 = ||\vec{x} + \vec{y}||^2$$

as shown in Figure 5.3.

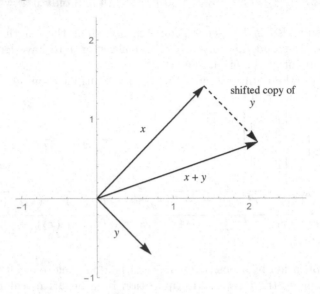

Figure 5.3: Pythagorean theorem with vectors

In the next section we'll generalize our idea of right angle as we have our idea of length, after which we can tackle a generalization of the Pythagorean Theorem. For now, let's finish up this section by introducing and exploring the idea of normalizing a vector.

The main idea here is that many computations are easier if the length of your vector is 1. This is why if you ask a mathematician to embed a square in $\mathbb{R}^2$, they'll usually choose to make each of its sides have length 1. Vectors of length 1 are usually called *unit vectors*. If we aren't lucky enough to start out with $||\vec{x}|| = 1$, we can fix that by multiplying $\vec{x}$ by the scalar $\frac{1}{||\vec{x}||}$. This is called *normalizing* $\vec{x}$. Geometrically, normalizing a vector changes its length but not its direction, which we can see because the normalized vector lies along the same line as $\vec{x}$ since it is in $\vec{x}$'s span. This is illustrated in Figure 5.4.

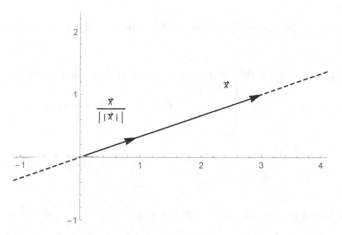

Figure 5.4: Normalizing a vector

However, this is generally done computationally even when we can draw a picture, because it is both quicker and more accurate.

Example 7. Find the unit vector in the same direction as $\vec{x} = \begin{bmatrix} -4 \\ 0 \\ 3 \end{bmatrix}$.

If we think about it, the unit vector in the same direction as a given vector is just the normalization of that vector, so this question is really asking us to normalize $\vec{x}$.

To do this, we first need to find $\|\vec{x}\|$ which is

$$\sqrt{(-4)^2 + 0^2 + 3^2} = \sqrt{25} = 5.$$

Next we need to divide $\vec{x}$ by 5, i.e., multiply by $\frac{1}{5}$. This gives us

$$\frac{1}{5} \begin{bmatrix} -4 \\ 0 \\ 3 \end{bmatrix} = \begin{bmatrix} \frac{-4}{5} \\ 0 \\ \frac{3}{5} \end{bmatrix}.$$

Therefore the unit vector in the same direction as $\begin{bmatrix} -4 \\ 0 \\ 3 \end{bmatrix}$ is $\begin{bmatrix} \frac{-4}{5} \\ 0 \\ \frac{3}{5} \end{bmatrix}$.

Recall from Chapter 4 that the stable stage distribution of a population is the eigenvector of length 1 with the largest eigenvector, aka population growth rate. At that point, we couldn't compute that, because we didn't have a way to find a vector of length 1 in the span of a given vector. Now we can do this by normalizing the basis vector of that eigenspace.

Example 8. Find the stable stage distribution of the population from 4.3's Example 7.

We computed in 4.3 that our population's stable stage distribution was in the eigenspace $E_1 = \text{Span}\left\{ \begin{bmatrix} 0.5 \\ 1 \\ 1 \end{bmatrix} \right\}$. This vector has length

$$\left\| \begin{bmatrix} 0.5 \\ 1 \\ 1 \end{bmatrix} \right\| = \sqrt{(0.5)^2 + (1)^2 + (1)^2} = \sqrt{2.25} = 1.5.$$

Normalizing our spanning vector gives us

$$\frac{1}{1.5} \begin{bmatrix} 0.5 \\ 1 \\ 1 \end{bmatrix} = \begin{bmatrix} \frac{1}{3} \\ \frac{2}{3} \\ \frac{2}{3} \end{bmatrix}.$$

This means our stable stage distribution is $\begin{bmatrix} \frac{1}{3} \\ \frac{2}{3} \\ \frac{2}{3} \end{bmatrix}$.

If we are dealing with more than one vector, we can't normalize each of them without changing the relationship between their lengths. However, we can preserve the basics of the situation while normalizing one of the vectors, say $\vec{x}$, by multiplying all our vectors by the scalar needed to normalize $\vec{x}$.

Example 9. Suppose we are looking at the triangle formed by the two vectors $\vec{v}_1 = \begin{bmatrix} 4 \\ 5 \end{bmatrix}$ and $\vec{v}_2 = \begin{bmatrix} 2 \\ 1 \end{bmatrix}$, but we would really prefer to look at the similar triangle whose longest side has length 1. Find the two vectors which define that similar triangle.

To give ourselves a better idea what's going on here, let's look at our original triangle.

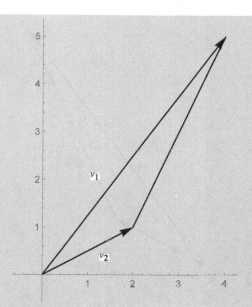

The vector $\vec{v}_1$ is clearly the longest side, so we can solve this problem by finding the scalar needed to normalize $\vec{v}_1$ and then multiplying both vectors by it. Since

$$\|\vec{v}_1\| = \sqrt{4^2 + 5^2} = \sqrt{41}$$

we'll need to multiply both of our vectors by $\frac{1}{\sqrt{41}}$. This gives us

$$\frac{1}{\sqrt{41}} \begin{bmatrix} 4 \\ 5 \end{bmatrix} = \begin{bmatrix} \frac{4}{\sqrt{41}} \\ \frac{5}{\sqrt{41}} \end{bmatrix} \approx \begin{bmatrix} .63 \\ .78 \end{bmatrix}$$

and

$$\frac{1}{\sqrt{41}} \begin{bmatrix} 2 \\ 1 \end{bmatrix} = \begin{bmatrix} \frac{2}{\sqrt{41}} \\ \frac{1}{\sqrt{41}} \end{bmatrix} \approx \begin{bmatrix} .31 \\ .16 \end{bmatrix}.$$

Thus our new similar triangle with longest side of length 1 is given (approximately) by the vectors $\begin{bmatrix} .63 \\ .78 \end{bmatrix}$ and $\begin{bmatrix} .31 \\ .16 \end{bmatrix}$. We can check this geometrically by plotting both triangles together as seen below.

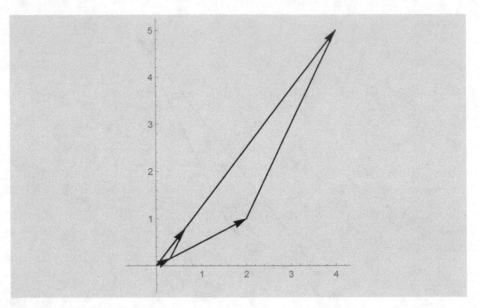

Exercises 5.1.

1. Compute $\begin{bmatrix} -3 \\ 1 \\ 5 \end{bmatrix} \cdot \begin{bmatrix} 2 \\ 7 \\ -2 \end{bmatrix}$ or say it is impossible.

2. Compute $\begin{bmatrix} -1 \\ 1 \\ 13 \end{bmatrix} \cdot \begin{bmatrix} 4 \\ 6 \end{bmatrix}$ or say it is impossible.

3. Compute $\begin{bmatrix} 2 \\ -1 \\ 8 \end{bmatrix} \cdot \begin{bmatrix} 1 \\ 2 \\ 0 \end{bmatrix}$ or say it is impossible.

4. Compute $\begin{bmatrix} 5 \\ 7 \end{bmatrix} \cdot \begin{bmatrix} -2 \\ 1 \end{bmatrix}$ or say it is impossible.

5. Suppose $\vec{v} \cdot \vec{u} = 8$. Find $\vec{v} \cdot (4\vec{u})$.

6. Suppose $\vec{v} \cdot \vec{u} = 4$, $\vec{v} \cdot \vec{w} = -9$, and $\vec{u} \cdot \vec{w} = 12$. Find $\vec{v} \cdot (\vec{u} + \vec{w})$.

7. Compute $\left\| \begin{bmatrix} 4 \\ -1 \\ 2 \end{bmatrix} \right\|$.

8. Compute $\left\| \begin{bmatrix} -1 \\ 1 \\ 1 \\ -1 \end{bmatrix} \right\|$.

9. Compute $\left\| \begin{bmatrix} 0 \\ 5 \\ 7 \end{bmatrix} \right\|$.

10. Compute $\left\| \begin{bmatrix} -2 \\ 6 \end{bmatrix} \right\|$.

11. Find the distance between $\begin{bmatrix} 7 \\ -2 \end{bmatrix}$ and $\begin{bmatrix} 3 \\ -1 \end{bmatrix}$.

12. Find the distance between $\begin{bmatrix} 2 \\ 8 \\ -3 \end{bmatrix}$ and $\begin{bmatrix} 3 \\ 9 \\ 1 \end{bmatrix}$.

13. Find the distance between $\begin{bmatrix} 4 \\ -5 \\ 0 \end{bmatrix}$ and $\begin{bmatrix} 7 \\ -2 \\ 4 \end{bmatrix}$.

14. Find the distance between $\begin{bmatrix} 5 \\ -2 \\ 0 \\ 6 \end{bmatrix}$ and $\begin{bmatrix} 4 \\ 0 \\ -1 \\ 8 \end{bmatrix}$.

15. Suppose $\|\vec{v}\| = 18$. Find $\left\| -\frac{1}{2}\vec{v} \right\|$.

16. Normalize $\begin{bmatrix} 5 \\ 12 \end{bmatrix}$.

17. Normalize $\begin{bmatrix} 6 \\ 0 \\ -6 \end{bmatrix}$.

18. Normalize $\begin{bmatrix} 1 \\ 0 \\ 1 \\ -1 \end{bmatrix}$.

19. Normalize $\begin{bmatrix} -2 \\ 1 \\ 0 \end{bmatrix}$.

20. Give an example of a situation when it is convenient to be able to compute length without a picture?

5.2   Orthogonality

In the last section we explored a way to compute the length of a vector in a non-geometric way. Toward the end of the section we wanted to generalize the Pythagorean theorem, but didn't have a good notion of how to tell vectors are at right angles without using a picture. In this section we'll develop a computational test for that and use it to explore several related ideas.

We've already developed a quick algebraic way to compute the dot product of two vectors, but it turns out there is also a geometric formula for $\vec{x} \cdot \vec{y}$ which computes the dot product in terms of the lengths of the vectors and the angle between them.

Theorem 1. Let $\vec{x}$ and $\vec{y}$ be in $\mathbb{R}^n$. Then $\vec{x} \cdot \vec{y} = ||\vec{x}||\,||\vec{y}|| \cos(\theta)$ where θ is the angle between $\vec{x}$ and $\vec{y}$.

In other words, the dot product of two vectors can also be found by multiplying together their lengths and the cosine of the angle between them.

While this formula holds for any n, I'll provide an explanation here in $\mathbb{R}^2$ so we can draw pictures. The first step in our explanation is to rewrite our vectors $\vec{x}$ and $\vec{y}$ in polar coordinates. If we let α_x be the angle between $\vec{x}$ and the x-axis as shown in Figure 5.5, then we have $\vec{x} = \begin{bmatrix} ||\vec{x}|| \cos(\alpha_x) \\ ||\vec{x}|| \sin(\alpha_x) \end{bmatrix}$.

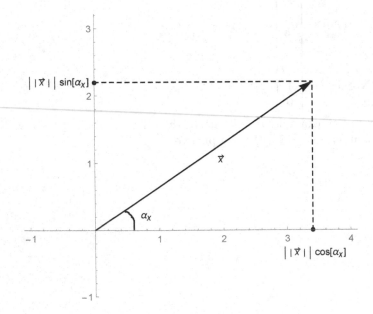

Figure 5.5: $\vec{x}$ in polar coordinates

Similarly we can rewrite $\vec{y}$ as $\begin{bmatrix} ||\vec{y}|| \cos(\alpha_y) \\ ||\vec{y}|| \sin(\alpha_y) \end{bmatrix}$. Taking the dot product of $\vec{x}$ and $\vec{y}$ using our algebraic formula from 5.1 now gives us

$$\vec{x} \cdot \vec{y} = \begin{bmatrix} ||\vec{x}|| \cos(\alpha_x) \\ ||\vec{x}|| \sin(\alpha_x) \end{bmatrix} \cdot \begin{bmatrix} ||\vec{y}|| \cos(\alpha_y) \\ ||\vec{y}|| \sin(\alpha_y) \end{bmatrix}$$

$$= ||\vec{x}|| \cos(\alpha_x) ||\vec{y}|| \cos(\alpha_y) + ||\vec{x}|| \sin(\alpha_x) ||\vec{y}|| \sin(\alpha_y)$$

$$= ||\vec{x}|| ||\vec{y}|| \cos(\alpha_x) \cos(\alpha_y) + ||\vec{x}|| ||\vec{y}|| \sin(\alpha_x) \sin(\alpha_y)$$

$$= ||\vec{x}|| ||\vec{y}|| \Big(\cos(\alpha_x) \cos(\alpha_y) + \sin(\alpha_x) \sin(\alpha_y) \Big).$$

Using the trigonometric identity $\cos(\beta) \cos(\gamma) + \sin(\beta) \sin(\gamma) = \cos(\gamma - \beta)$, this gives us

$$\vec{x} \cdot \vec{y} = ||\vec{x}|| ||\vec{y}|| \cos(\alpha_y - \alpha_x).$$

Now we need to relate the difference between the angles associated with our two vectors to the angle between them.

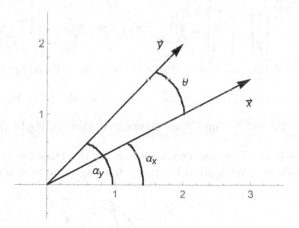

Figure 5.6: The angle between vectors

If we look at Figure 5.6, we can see that the angle between $\vec{x}$ and $\vec{y}$ is $\theta = \alpha_y - \alpha_x$. Therefore $\vec{x} \cdot \vec{y} = ||\vec{x}|| ||\vec{y}|| \cos(\theta)$ as claimed.

This relates back to our original goal of checking when vectors are at right angles because when $\theta = 90° = \pi/2$ we have $\cos(\theta) = 0$. This means we can use the dot product to define perpendicular vectors as follows.

Definition. Two vectors $\vec{v}$ and $\vec{w}$ in $\mathbb{R}^n$ are *orthogonal* if $\vec{v} \cdot \vec{w} = 0$.

As with saying norm instead of length, we'll typically use orthogonal instead of perpendicular or at right angles.

Computational Vector Geometry

Example 1. Show $\begin{bmatrix} 1 \\ 2 \\ 3 \end{bmatrix}$ and $\begin{bmatrix} 5 \\ -1 \\ -1 \end{bmatrix}$ are orthogonal.

The dot product of these two vectors is

$$\begin{bmatrix} 1 \\ 2 \\ 3 \end{bmatrix} \cdot \begin{bmatrix} 5 \\ -1 \\ -1 \end{bmatrix} = 1(5) + 2(-1) + 3(-1) = 0.$$

Since their dot product is zero, they are orthogonal.

Example 2. Show $\begin{bmatrix} 1 \\ 2 \\ 3 \end{bmatrix}$ and $\begin{bmatrix} 1 \\ 1 \\ -5 \end{bmatrix}$ are not orthogonal.

As in the previous example, we take the dot product of these vectors and get

$$\begin{bmatrix} 1 \\ 2 \\ 3 \end{bmatrix} \cdot \begin{bmatrix} 1 \\ 1 \\ -5 \end{bmatrix} = 1(1) + 2(1) + 3(-5) = -12.$$

Since the dot product is nonzero, these vectors are not orthogonal.

With this analog for right angles, we can state the following theorem.

Theorem 2. If $\vec{v}$ and $\vec{w}$ from $\mathbb{R}^n$ are orthogonal, then $||\vec{v}||^2 + ||\vec{w}||^2 = ||\vec{v} + \vec{w}||^2$.

In $\mathbb{R}^2$, this is the Pythagorean Theorem for the triangle with sides a, $\vec{v}$, and $\vec{v} + \vec{w}$ as shown by Figure 5.7, since a has the same length as $\vec{w}$.

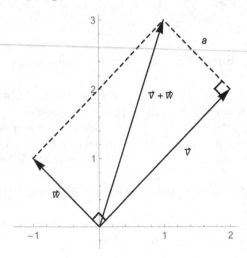

Figure 5.7: Vector Pythagorean theorem

Example 3. Check that Theorem 2 holds for $\begin{bmatrix} 1 \\ 2 \\ 3 \end{bmatrix}$ and $\begin{bmatrix} 5 \\ -1 \\ -1 \end{bmatrix}$.

These are our vectors from Example 1, so we already know they are orthogonal. To check that they satisfy Theorem 2, let's compute each side of its equation to make sure they match. For our vectors,

$$||\vec{v}||^2 + ||\vec{w}||^2 = \left|\left| \begin{bmatrix} 1 \\ 2 \\ 3 \end{bmatrix} \right|\right|^2 + \left|\left| \begin{bmatrix} 5 \\ -1 \\ -1 \end{bmatrix} \right|\right|^2$$

$$= \left(\sqrt{1^2 + 2^2 + 3^2} \right)^2 + \left(\sqrt{5^2 + (-1)^2 + (-1)^2} \right)^2$$

$$= 14 + 27 = 41.$$

The other side of the equation is

$$||\vec{v} + \vec{w}||^2 = \left|\left| \begin{bmatrix} 1 \\ 2 \\ 3 \end{bmatrix} + \begin{bmatrix} 5 \\ -1 \\ -1 \end{bmatrix} \right|\right|^2 = \left|\left| \begin{bmatrix} 6 \\ 1 \\ 2 \end{bmatrix} \right|\right|^2$$

$$= \left(\sqrt{6^2 + 1^2 + 2^2} \right)^2 = 41$$

Since the two sides of the equation are equal, the theorem is satisfied.

Example 4. Check that Theorem 2 does not hold for $\begin{bmatrix} 1 \\ 2 \\ 3 \end{bmatrix}$ and $\begin{bmatrix} 1 \\ 1 \\ -5 \end{bmatrix}$.

These are our vectors from Example 2, so we know they aren't orthogonal. As in the previous example, we can tackle this by computing each side of the theorem's equation. For our vectors,

$$||\vec{v}||^2 + ||\vec{w}||^2 = \left|\left| \begin{bmatrix} 1 \\ 2 \\ 3 \end{bmatrix} \right|\right|^2 + \left|\left| \begin{bmatrix} 1 \\ 1 \\ -5 \end{bmatrix} \right|\right|^2$$

$$= \left(\sqrt{1^2 + 2^2 + 3^2} \right)^2 + \left(\sqrt{1^2 + 1^2 + 5^2} \right)^2$$

$$= 14 + 27 = 41.$$

The other side of the equation is

$$\|\vec{v} + \vec{w}\|^2 = \left\| \begin{bmatrix} 1 \\ 2 \\ 3 \end{bmatrix} + \begin{bmatrix} 1 \\ 1 \\ -5 \end{bmatrix} \right\|^2 = \left\| \begin{bmatrix} 2 \\ 3 \\ -2 \end{bmatrix} \right\|^2$$

$$= \left(\sqrt{2^2 + 3^2 + (-2)^2} \right)^2 = 17.$$

Since the two sides of the equation are different, the theorem doesn't hold.

Suppose $\vec{v}$ is orthogonal to $\vec{w}$, i.e., $\vec{v} \cdot \vec{w} = 0$. This also means any multiple of $\vec{v}$ is orthogonal to $\vec{w}$ since $(r\vec{v}) \cdot \vec{w} = r(\vec{v} \cdot \vec{w}) = 0$. Therefore it makes sense to talk about a set of vectors which are orthogonal to $\vec{w}$. Additionally, we know $\vec{v}$ is orthogonal to any multiple of $\vec{w}$ since $\vec{v} \cdot (r\vec{w}) = r(\vec{v} \cdot \vec{w}) = 0$. Geometrically, this is the same as saying that if $\vec{v}$ is orthogonal to $\vec{w}$, then any vector on the line spanned by $\vec{v}$ is orthogonal to any vector on the line spanned by $\vec{w}$ as shown in Figure 5.8.

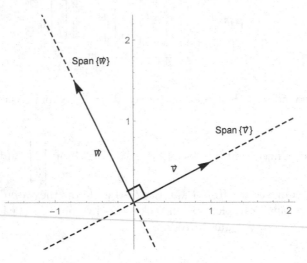

Figure 5.8: Orthogonal spans

Now that we understand what it means for two vectors to be orthogonal to each other, let's extend that notion to talk about when a vector is orthogonal to a whole set of vectors.

Definition. Let W be a subset of $\mathbb{R}^n$. The *orthogonal complement* of W is $W^{\perp} = \{\vec{v} \text{ in } \mathbb{R}^n | \vec{v} \cdot \vec{w} = 0 \text{ for every } \vec{w} \text{ in } W\}$.

If this notation is confusing, we can restate this in words by saying that the orthogonal complement of W is the set of all vectors which are orthogonal to everything in W.

Example 5. Show $\vec{v} = \begin{bmatrix} -6 \\ 3 \end{bmatrix}$ is orthogonal to W where W is the line $y = 2x$ in $\mathbb{R}^2$.

Before we can check that $\vec{v}$ is in $W^\perp$, we need to write W as a set of vectors in $\mathbb{R}^2$. Since we often write 2-vectors as $\begin{bmatrix} x \\ y \end{bmatrix}$, adding in the condition that $y = 2x$ means $W = \left\{ \begin{bmatrix} x \\ 2x \end{bmatrix} \right\}$.

Now that we have W written as a set of 2-vectors, we can check that $\vec{v}$ is orthogonal to anything in W by taking the dot product of $\vec{v}$ with a generic vector from W. In our case, this is

$$\begin{bmatrix} -6 \\ 3 \end{bmatrix} \cdot \begin{bmatrix} x \\ 2x \end{bmatrix} = -6(x) + 2(3x) = 0.$$

Since this dot product is always zero, $\vec{v}$ is orthogonal to everything in W. Therefore $\vec{v}$ is in $W^\perp$.

Example 6. Show $\vec{v} = \begin{bmatrix} 1 \\ 2 \\ 3 \end{bmatrix}$ is not in $W^\perp$ for $W = \left\{ \begin{bmatrix} 5x_1 + x_2 \\ -x_1 + 2x_2 \\ -x_1 - x_2 \end{bmatrix} \right\}$

A vector is in $W^\perp$ only if it is orthogonal to everything in W. Therefore to show $\vec{v}$ is not in $W^\perp$, we simply need to find some $\vec{w}$ in W which is not orthogonal to $\vec{v}$.

If we let $x_1 = 0$ and $x_2 = 1$, we get $\vec{w} = \begin{bmatrix} 1 \\ 2 \\ -1 \end{bmatrix}$. Taking the dot product with $\vec{v}$ gives us

$$\begin{bmatrix} 1 \\ 2 \\ 3 \end{bmatrix} \cdot \begin{bmatrix} 1 \\ 2 \\ -1 \end{bmatrix} = 1(1) + 2(2) + 3(-1) = 2.$$

This means $\vec{v}$ is not orthogonal to $\vec{w}$, and hence $\vec{v}$ is not orthogonal to everything in W. Therefore $\vec{v}$ is not in $W^\perp$.

Notice that setting $x_1 = 1$ and $x_2 = 0$ actually gives us $\vec{w} = \begin{bmatrix} 5 \\ -1 \\ -1 \end{bmatrix}$ which we showed in Example 1 was orthogonal to $\vec{v}$. However, this is irrelevant, since $W^\perp$ is the set of vectors orthogonal to all of W.

In the previous examples we were only testing whether or not a single vector was in $W^\perp$, and our set W was fairly simple. In general this is a complicated problem, however, if W is a span then we have the following easier criteria for $W^\perp$.

Theorem 3. If $W = \text{Span}\{\vec{w}_1, \ldots, \vec{w}_k\}$, then $\vec{v}$ is in $W^\perp$ if $\vec{v} \cdot \vec{w}_i = 0$ for $\vec{w}_1, \ldots, \vec{w}_k$.

In other words, if W is a span then it is enough to check whether $\vec{v}$ is orthogonal to each spanning vector. If it is, $\vec{v}$ is automatically orthogonal to everything else in W. You will check that this is true in Exercise 22 of this section.

Example 7. Show that $\vec{v} = \begin{bmatrix} 1 \\ 2 \\ 3 \end{bmatrix}$ is in $W^\perp$ for $W = \text{Span}\left\{ \begin{bmatrix} -3 \\ 0 \\ 1 \end{bmatrix}, \begin{bmatrix} 7 \\ -2 \\ -1 \end{bmatrix} \right\}$.

Rather than checking that $\vec{v}$ is orthogonal to every $\vec{w}$ in W, we can instead just show $\vec{v}$ is orthogonal to the two spanning vectors of W. This is

$$\begin{bmatrix} 1 \\ 2 \\ 3 \end{bmatrix} \cdot \begin{bmatrix} -3 \\ 0 \\ 1 \end{bmatrix} = 1(-3) + 2(0) + 3(1) = 0$$

and

$$\begin{bmatrix} 1 \\ 2 \\ 3 \end{bmatrix} \cdot \begin{bmatrix} 7 \\ -2 \\ -1 \end{bmatrix} = 1(7) + 2(-2) + 3(-1) = 0.$$

Since both of these dot products are zero, $\vec{v}$ is orthogonal to all spanning vectors of W and hence by Theorem 2 we know $\vec{v}$ is in $W^\perp$.

Example 8. Find $W^\perp$ where $W = \text{Span}\left\{ \begin{bmatrix} 4 \\ -2 \\ 1 \\ 0 \end{bmatrix}, \begin{bmatrix} -1 \\ 5 \\ 2 \\ 9 \end{bmatrix} \right\}$.

To be in $W^\perp$, a vector $\vec{x} = \begin{bmatrix} x_1 \\ x_2 \\ x_3 \\ x_4 \end{bmatrix}$ needs to satisfy the equations

$$\vec{w}_1 \cdot \vec{x} = 4x_1 - 2x_2 + x_3 = 0$$
$$\vec{w}_2 \cdot \vec{x} = -x_1 + 5x_2 + 2x_3 + 9x_4 = 0$$

This means we are really solving the linear system whose augmented coefficient matrix is

$$\begin{bmatrix} 4 & -2 & 1 & 0 & | & 0 \\ -1 & 5 & 2 & 9 & | & 0 \end{bmatrix}.$$

This matrix has reduced echelon form

$$\left[\begin{array}{cccc|c} 1 & 0 & \frac{1}{2} & 1 & 0 \\ 0 & 1 & \frac{1}{2} & 2 & 0 \end{array}\right]$$

so we have $x_1 = -\frac{1}{2}x_3 - x_4$ and $x_2 = -\frac{1}{2}x_3 - 2x_4$ where x_3 and x_4 are free variables. Thus

$$W^{\perp} = \left\{ \left[\begin{array}{c} -\frac{1}{2}x_3 - x_4 \\ -\frac{1}{2}x_3 - 2x_4 \\ x_3 \\ x_4 \end{array}\right] \right\}.$$

In the case where W is a subspace of $\mathbb{R}^n$ (which is automatically true if W is a span), we also get the following fact.

Theorem 4. If W is a subspace of $\mathbb{R}^n$, then $W^{\perp}$ is also a subspace of $\mathbb{R}^n$.

To see that this is true, we can use the subspace test. Since $\vec{0}$ is orthogonal to everything ($\vec{v} \cdot \vec{0} = 0$ for every $\vec{v}$ in $\mathbb{R}^n$), we clearly have $\vec{0}$ in $W^{\perp}$. If $\vec{x}$ and $\vec{y}$ are in $W^{\perp}$, then we have $\vec{x} \cdot \vec{w} = 0$ and $\vec{y} \cdot \vec{w} = 0$ for all $\vec{w}$ in W. Therefore

$$(\vec{x} + \vec{y}) \cdot \vec{w} = (\vec{x} \cdot \vec{w}) + (\vec{y} \cdot \vec{w}) = 0 + 0 = 0$$

so $\vec{x} + \vec{y}$ is in $W^{\perp}$. Finally, suppose $\vec{x}$ is in $W^{\perp}$, so $\vec{x} \cdot \vec{w} = 0$ for every $\vec{w}$ in W, and r is any scalar. Then

$$(r\vec{x}) \cdot \vec{w} = r(\vec{x} \cdot \vec{w}) = r(0) = 0$$

so $r\vec{x}$ is in $W^{\perp}$. Since $W^{\perp}$ satisfies all three conditions of our subspace check, it is a subspace of $\mathbb{R}^n$.

Exercises 5.2.

1. Are $\begin{bmatrix} 2 \\ -4 \\ 1 \end{bmatrix}$ and $\begin{bmatrix} 1 \\ 1 \\ 2 \end{bmatrix}$ orthogonal?

2. Are $\begin{bmatrix} -3 \\ 0 \\ 6 \end{bmatrix}$ and $\begin{bmatrix} 1 \\ 5 \\ -1 \end{bmatrix}$ orthogonal?

3. Are $\begin{bmatrix} 8 \\ -1 \\ 3 \\ 2 \end{bmatrix}$ and $\begin{bmatrix} 1 \\ 1 \\ -2 \\ 1 \end{bmatrix}$ orthogonal?

4. Are $\begin{bmatrix} 4 \\ -4 \\ 1 \\ 0 \end{bmatrix}$ and $\begin{bmatrix} 1 \\ 1 \\ 1 \\ 1 \end{bmatrix}$ orthogonal?

5. Does Theorem 2 hold for $\vec{v} = \begin{bmatrix} 1 \\ 0 \\ -1 \end{bmatrix}$ and $\vec{w} = \begin{bmatrix} 7 \\ 12 \\ 7 \end{bmatrix}$?

6. Does Theorem 2 hold for $\vec{v} = \begin{bmatrix} 3 \\ 4 \end{bmatrix}$ and $\vec{w} = \begin{bmatrix} -6 \\ 3 \end{bmatrix}$?

7. Find a nonzero vector which is orthogonal to $\begin{bmatrix} -5 \\ 5 \\ 1 \end{bmatrix}$.

8. Find a nonzero vector which is orthogonal to $\begin{bmatrix} 1 \\ 2 \\ -9 \end{bmatrix}$.

9. Is $\begin{bmatrix} -2 \\ 1 \end{bmatrix}$ in $W^{\perp}$ for $W = \left\{ \begin{bmatrix} 2x_1 - 4x_2 \\ 4x_1 + x_2 \end{bmatrix} \right\}$?

10. Is $\begin{bmatrix} 3 \\ -1 \end{bmatrix}$ in $W^{\perp}$ for $W = \left\{ \begin{bmatrix} 2x_1 - 4x_2 \\ -6x_1 + 12x_2 \end{bmatrix} \right\}$?

11. Is $\begin{bmatrix} 1 \\ 1 \\ -1 \end{bmatrix}$ in $W^{\perp}$ for $W = \left\{ \begin{bmatrix} x_1 + x_2 \\ x_1 - 6x_2 \\ 2x_1 - x_2 \end{bmatrix} \right\}$?

12. Is $\begin{bmatrix} -3 \\ 4 \\ 1 \end{bmatrix}$ in $W^{\perp}$ for $W = \left\{ \begin{bmatrix} 5x_1 + x_2 \\ 2x_1 - x_2 \\ 7x_1 + 7x_2 \end{bmatrix} \right\}$?

13. Is $\begin{bmatrix} -1 \\ -1 \\ 2 \end{bmatrix}$ in $W^{\perp}$ for $W = \text{Span} \left\{ \begin{bmatrix} 3 \\ 5 \\ 4 \end{bmatrix}, \begin{bmatrix} 2 \\ 0 \\ 1 \end{bmatrix} \right\}$?

14. Is $\begin{bmatrix} 0 \\ 1 \\ 1 \end{bmatrix}$ in $W^{\perp}$ for $W = \text{Span} \left\{ \begin{bmatrix} -3 \\ 3 \\ 4 \end{bmatrix}, \begin{bmatrix} 4 \\ -3 \\ 3 \end{bmatrix} \right\}$?

15. Is $\begin{bmatrix} 4 \\ -2 \\ -1 \end{bmatrix}$ in $W^{\perp}$ for $W = \text{Span} \left\{ \begin{bmatrix} 0 \\ 1 \\ 2 \end{bmatrix}, \begin{bmatrix} 2 \\ 3 \\ 5 \end{bmatrix} \right\}$?

16. Is $\begin{bmatrix} 1 \\ 3 \\ 1 \end{bmatrix}$ in $W^{\perp}$ for $W = \text{Span} \left\{ \begin{bmatrix} -1 \\ 2 \\ -5 \end{bmatrix}, \begin{bmatrix} 2 \\ -1 \\ 1 \end{bmatrix} \right\}$?

17. Compute $W^{\perp}$ for $W = \text{Span} \left\{ \begin{bmatrix} -1 \\ 2 \\ 1 \end{bmatrix}, \begin{bmatrix} 3 \\ -1 \\ 0 \end{bmatrix} \right\}$?

18. Compute $W^{\perp}$ for $W = \text{Span} \left\{ \begin{bmatrix} 4 \\ 2 \\ -1 \end{bmatrix}, \begin{bmatrix} -3 \\ 1 \\ 2 \end{bmatrix} \right\}$?

19. Compute $W^\perp$ for $W = \text{Span}\left\{ \begin{bmatrix} 6 \\ -2 \\ -1 \\ 1 \end{bmatrix}, \begin{bmatrix} -3 \\ 2 \\ 0 \\ 1 \end{bmatrix} \right\}$?

20. Compute $W^\perp$ for $W = \text{Span}\left\{ \begin{bmatrix} 1 \\ 0 \\ -2 \\ 5 \end{bmatrix}, \begin{bmatrix} 0 \\ 1 \\ 5 \\ -6 \end{bmatrix} \right\}$?

21. Compute $W^\perp$ for $W = \text{Span}\left\{ \begin{bmatrix} 2 \\ -2 \\ 0 \\ 6 \end{bmatrix}, \begin{bmatrix} -1 \\ 2 \\ 4 \\ 1 \end{bmatrix}, \begin{bmatrix} 4 \\ -4 \\ -5 \\ 2 \end{bmatrix} \right\}$?

22. (a) Let $W = \text{Span}\{\vec{w}_1, \vec{w}_2\}$ be a subspace of $\mathbb{R}^n$, and $\vec{v}$ be any n-vector. Show that if $\vec{v} \cdot \vec{w}_1 = 0$ and $\vec{v} \cdot \vec{w}_2 = 0$ then $\vec{v}$ is in $W^\perp$.

    (b) Let $W = \text{Span}\{\vec{w}_1, \ldots, \vec{w}_k\}$ be a subspace of $\mathbb{R}^n$, and let $\vec{v}$ be an n-vector. Show that if $\vec{v} \cdot \vec{w}_i = 0$ for $\vec{w}_1, \ldots, \vec{w}_k$ then $\vec{v}$ is in $W^\perp$.

5.3   Orthogonal Projection

In this section we'll use our concept of orthogonality to develop a technique which allows us to decompose a vector into two parts: one in the span of another vector or set of vectors and the other orthogonal to it. This is particularly applicable to physics, where we often want to decompose a force vector into its parts along a certain line or plane of motion and orthogonal to that motion. (There the part orthogonal to the line or plane of motion is commonly called the *normal vector*.)

Let's start in the simplest possible case, where our span W has only one spanning vector $\vec{y}$. Here we want to take our starting vector $\vec{x}$ and write it as a sum of two pieces: $\vec{x} = \vec{w} + \vec{v}$. The first piece $\vec{w}$ should be in the span of $\vec{y}$, which we can think of either computationally as all multiples of $\vec{y}$ or geometrically as the line defined by $\vec{y}$. The second piece $\vec{v}$ should be orthogonal to W, i.e., in $W^{\perp}$. Since W is the span of $\vec{y}$, Theorem 2 from the last section tells us we simply need $\vec{v}$ orthogonal to $\vec{y}$. Geometrically, this means we want to find $\vec{w}$ along $\vec{y}$'s line and $\vec{v}$ orthogonal to $\vec{y}$'s line so that $\vec{x}$ is the diagonal of the parallelogram formed by $\vec{w}$ and $\vec{v}$ as in Figure 5.9.

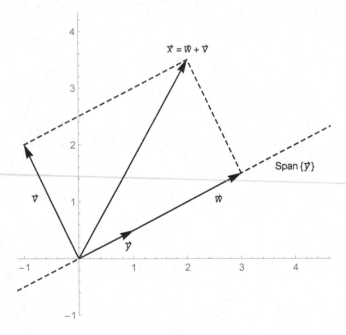

Figure 5.9: Decomposing $\vec{x}$

Putting this more computationally, we want to find two vectors $\vec{w}$ and $\vec{v}$ so that

$$\vec{x} = \vec{w} + \vec{v}, \ \vec{w} = r\vec{y} \text{ for some scalar } r, \text{ and } \vec{v} \cdot \vec{y} = 0.$$

We know the values of $\vec{x}$ and $\vec{y}$, and want to solve for r (and hence $\vec{w}$) and $\vec{v}$. We can use the first two equations to solve for $\vec{v}$ by plugging $\vec{w} = r\vec{y}$ into $\vec{x} = \vec{w} + \vec{v}$ to get

$$\vec{x} = r\vec{y} + \vec{v}.$$

Solving this for $\vec{v}$ gives us

$$\vec{v} = \vec{x} - r\vec{y}.$$

It may feel as if we are stuck, but remember we have one more equation we haven't used yet: $\vec{v} \cdot \vec{y} = 0$. Plugging $\vec{v} = \vec{x} - r\vec{y}$ into this equation gives us

$$(\vec{x} - r\vec{y}) \cdot \vec{y} = 0.$$

Using the properties of the dot product explored in 5.1 to expand the left-hand side of this equation gives us

$$(\vec{x} - r\vec{y}) \cdot \vec{y} = \vec{x} \cdot \vec{y} - (r\vec{y}) \cdot \vec{y} = \vec{x} \cdot \vec{y} - r(\vec{y} \cdot \vec{y}).$$

This means we have

$$\vec{x} \cdot \vec{y} - r(\vec{y} \cdot \vec{y}) = 0.$$

Now that r is the only unknown quantity in this equation, we can solve for it to get

$$r(\vec{y} \cdot \vec{y}) = \vec{x} \cdot \vec{y}$$

or

$$r = \frac{\vec{x} \cdot \vec{y}}{\vec{y} \cdot \vec{y}}.$$

(Note that since dot products are real numbers, this value of r is a scalar.) Now that we know the value of r, we can plug it back into the equations for $\vec{w}$ and $\vec{v}$ to get

$$\vec{w} = \left(\frac{\vec{x} \cdot \vec{y}}{\vec{y} \cdot \vec{y}} \right) \vec{y}$$

and

$$\vec{v} = \vec{x} - \left(\frac{\vec{x} \cdot \vec{y}}{\vec{y} \cdot \vec{y}} \right) \vec{y}.$$

Both of these vectors have special names which are defined below.

Definition.   The *orthogonal projection of $\vec{x}$ onto $\vec{y}$* is $\left(\frac{\vec{x} \cdot \vec{y}}{\vec{y} \cdot \vec{y}} \right) \vec{y}$ and the *component of $\vec{x}$ orthogonal to $\vec{y}$* is $\vec{x} - \left(\frac{\vec{x} \cdot \vec{y}}{\vec{y} \cdot \vec{y}} \right) \vec{y}$.

These names should intuitively make sense, because the orthogonal projection of $\vec{x}$ onto $\vec{y}$ is the piece of $\vec{x}$ that lies along the line defined by $\vec{y}$ and the component of $\vec{x}$ orthogonal to $\vec{y}$ is the piece of $\vec{x}$ that lies in the orthogonal complement of $\vec{y}$'s span.

Example 1. Find the orthogonal projection of $\begin{bmatrix} 2 \\ 9 \end{bmatrix}$ onto $\begin{bmatrix} 3 \\ 1 \end{bmatrix}$ and the component of $\begin{bmatrix} 2 \\ 9 \end{bmatrix}$ orthogonal to $\begin{bmatrix} 3 \\ 1 \end{bmatrix}$.

Since these formulas are not symmetric in $\vec{x}$ and $\vec{y}$, it is important to start by identifying which vector to plug in for $\vec{x}$ and which to plug in for $\vec{y}$. The vector that we're computing the projection of is always $\vec{x}$ and the vector we're projecting onto is always $\vec{y}$, so in this example

$$\vec{x} = \begin{bmatrix} 2 \\ 9 \end{bmatrix} \text{ and } \vec{y} = \begin{bmatrix} 3 \\ 1 \end{bmatrix}.$$

Next we need to compute $\vec{x} \cdot \vec{y}$ and $\vec{y} \cdot \vec{y}$. Plugging in our $\vec{x}$ and $\vec{y}$ gives us

$$\vec{x} \cdot \vec{y} = \begin{bmatrix} 2 \\ 9 \end{bmatrix} \cdot \begin{bmatrix} 3 \\ 1 \end{bmatrix} = 2(3) + 9(1) = 15$$

and

$$\vec{y} \cdot \vec{y} = \begin{bmatrix} 3 \\ 1 \end{bmatrix} \cdot \begin{bmatrix} 3 \\ 1 \end{bmatrix} = 3^2 + 1^2 = 10.$$

Plugging these dot products into our formula for the orthogonal projection gives us

$$\left(\frac{15}{10} \right) \begin{bmatrix} 3 \\ 1 \end{bmatrix} = (1.5) \begin{bmatrix} 3 \\ 1 \end{bmatrix} = \begin{bmatrix} 4.5 \\ 1.5 \end{bmatrix}.$$

To find the component of $\begin{bmatrix} 2 \\ 9 \end{bmatrix}$ orthogonal to $\begin{bmatrix} 3 \\ 1 \end{bmatrix}$, we just need to subtract the orthogonal projection from $\begin{bmatrix} 2 \\ 9 \end{bmatrix}$. This gives us

$$\begin{bmatrix} 2 \\ 9 \end{bmatrix} - \begin{bmatrix} 4.5 \\ 1.5 \end{bmatrix} = \begin{bmatrix} -2.5 \\ 7.5 \end{bmatrix}.$$

Therefore, the orthogonal projection of $\begin{bmatrix} 2 \\ 9 \end{bmatrix}$ onto $\begin{bmatrix} 3 \\ 1 \end{bmatrix}$ is $\begin{bmatrix} 4.5 \\ 1.5 \end{bmatrix}$, and the component of $\begin{bmatrix} 2 \\ 9 \end{bmatrix}$ orthogonal to $\begin{bmatrix} 3 \\ 1 \end{bmatrix}$ is $\begin{bmatrix} -2.5 \\ 7.5 \end{bmatrix}$.

Geometrically, you can think of finding the orthogonal projection of $\vec{x}$ onto $\vec{y}$ as shining a light perpendicularly down onto the line spanned by $\vec{y}$ and looking at the shadow cast by $\vec{x}$. This is shown in the picture below.

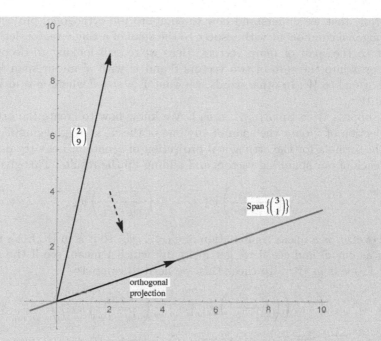

The component of $\vec{x}$ orthogonal to $\vec{y}$ is then the vector perpendicular to y's line that is the right length so its sum with the orthogonal projection is $\vec{x}$, which is shown in the following picture.

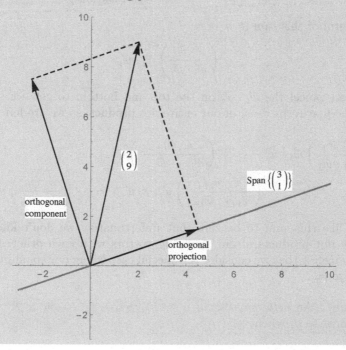

Now that we understand how to compute the orthogonal projection and orthogonal component with respect to the span of a single vector, let's extend that to the span of more vectors. Here we're still looking to decompose a vector $\vec{x}$ into the sum of two vectors $\vec{v}$ and $\vec{w}$ with $\vec{w}$ in our span W and $\vec{v}$ orthogonal to W. In other words, we want $\vec{x} = \vec{w} + \vec{v}$ where $\vec{w}$ is in W and $\vec{v}$ is in $W^{\perp}$.

Suppose $W = \text{Span}\{\vec{w}_1, \ldots, \vec{w}_k\}$. We know how to create the orthogonal projection of $\vec{x}$ onto the span of any one of the $\vec{w}_i$'s, by plugging $\vec{w}_i$ in for $\vec{y}$ in the formula for the orthogonal projection of $\vec{x}$ onto $\vec{y}$. Let's try doing that for each of our spanning vectors and adding up the results. This gives us

$$\left(\frac{\vec{x} \cdot \vec{w}_1}{\vec{w}_1 \cdot \vec{w}_1} \right) \vec{w}_1 + \cdots + \left(\frac{\vec{x} \cdot \vec{w}_k}{\vec{w}_k \cdot \vec{w}_k} \right) \vec{w}_k.$$

This vector is a linear combination of $\vec{w}_1, \ldots, \vec{w}_k$, so it is in W. Let's try using this as our $\vec{w}$ and see if we get lucky, by which I mean, see if the resulting $\vec{v} = \vec{x} - \vec{w}$ is in $W^{\perp}$. To check this, we need to compute

$$\left(\vec{x} - \left(\left(\frac{\vec{x} \cdot \vec{w}_1}{\vec{w}_1 \cdot \vec{w}_1} \right) \vec{w}_1 + \cdots + \left(\frac{\vec{x} \cdot \vec{w}_k}{\vec{w}_k \cdot \vec{w}_k} \right) \vec{w}_k \right) \right) \cdot \vec{w}_i$$

for $\vec{w}_1, \ldots, \vec{w}_k$. This dot product can be expanded to give

$$\vec{x} \cdot \vec{w}_i - \left(\frac{\vec{x} \cdot \vec{w}_1}{\vec{w}_1 \cdot \vec{w}_1} \right) \vec{w}_1 \cdot \vec{w}_i - \cdots - \left(\frac{\vec{x} \cdot \vec{w}_k}{\vec{w}_k \cdot \vec{w}_k} \right) \vec{w}_k \cdot \vec{w}_i.$$

The ith term of this sum is

$$-\left(\frac{\vec{x} \cdot \vec{w}_i}{\vec{w}_i \cdot \vec{w}_i} \right) \vec{w}_i \cdot \vec{w}_i$$

and we can cancel the $\vec{w}_i \cdot \vec{w}_i$ on the top and bottom to get $-\vec{x} \cdot \vec{w}_i$. This cancels the first term $\vec{x} \cdot \vec{w}_i$ of our entire dot product, so we are left with

$$-\left(\frac{\vec{x} \cdot \vec{w}_1}{\vec{w}_1 \cdot \vec{w}_1} \right) \vec{w}_1 \cdot \vec{w}_i - \cdots - \left(\frac{\vec{x} \cdot \vec{w}_{i-1}}{\vec{w}_{i-1} \cdot \vec{w}_{i-1}} \right) \vec{w}_{i-1} \cdot \vec{w}_i$$

$$-\left(\frac{\vec{x} \cdot \vec{w}_{i+1}}{\vec{w}_{i+1} \cdot \vec{w}_{i+1}} \right) \vec{w}_{i+1} \cdot \vec{w}_i - \cdots - \left(\frac{\vec{x} \cdot \vec{w}_k}{\vec{w}_k \cdot \vec{w}_k} \right) \vec{w}_k \cdot \vec{w}_i.$$

We'd like this sum to be zero, but unfortunately we don't know much about the dot products of our spanning vectors with each other. However, this does suggest that we consider the special case where our spanning vectors are orthogonal.

Definition. An *orthogonal set* is a set of vectors $\vec{v}_1, \ldots, \vec{v}_k$ in $\mathbb{R}^n$ where $\vec{v}_i$ is orthogonal to $\vec{v}_j$ whenever $i \neq j$.

Geometrically, we can think of an orthogonal set as one where each vector is at right angles to all the other vectors, which means each points in a completely different direction from the others. (This is harder to imagine once we move beyond 3 dimensions.)

Example 2. Show that $\begin{bmatrix} 0 \\ -2 \\ 4 \end{bmatrix}$, $\begin{bmatrix} -3 \\ 2 \\ 1 \end{bmatrix}$, and $\begin{bmatrix} 5 \\ 6 \\ 3 \end{bmatrix}$ form an orthogonal set.

To show this we need to verify that all three possible dot products are zero. These three dot products are

$$\begin{bmatrix} 0 \\ -2 \\ 4 \end{bmatrix} \cdot \begin{bmatrix} -3 \\ 2 \\ 1 \end{bmatrix} = 0(-3) + (-2)(2) + 4(1) = 0,$$

$$\begin{bmatrix} -3 \\ 2 \\ 1 \end{bmatrix} \cdot \begin{bmatrix} 5 \\ 6 \\ 3 \end{bmatrix} = -3(5) + 2(6) + 1(3) = 0,$$

and

$$\begin{bmatrix} 0 \\ -2 \\ 4 \end{bmatrix} \cdot \begin{bmatrix} 5 \\ 6 \\ 3 \end{bmatrix} = 0(5) + (-2)(6) + 4(3) = 0.$$

Since all three dot products are zero, this is an orthogonal set.

Example 3. Show $\begin{bmatrix} 0 \\ -2 \\ 4 \end{bmatrix}$, $\begin{bmatrix} -3 \\ 2 \\ 1 \end{bmatrix}$, and $\begin{bmatrix} 1 \\ 4 \\ 2 \end{bmatrix}$ are not an orthogonal set.

Unlike in the previous example where we needed to show that all possible dot products equaled zero, here we just need to find one dot product which isn't zero. The first two vectors are the same as in Example 2, so their dot product is zero. That doesn't help, so we'll need to compute dot products involving the third vector.
These are

$$\begin{bmatrix} 0 \\ -2 \\ 4 \end{bmatrix} \cdot \begin{bmatrix} 1 \\ 4 \\ 2 \end{bmatrix} = 0(1) + (-2)(4) + 4(2) = 0$$

and

$$\begin{bmatrix} -3 \\ 2 \\ 1 \end{bmatrix} \cdot \begin{bmatrix} 1 \\ 4 \\ 2 \end{bmatrix} (-3)(1) + 2(4) + 1(2) = 7.$$

Since the dot product of the second and third vectors is nonzero, this is not an orthogonal set.

Now that we understand orthogonal sets, let's return to our attempt to decompose $\vec{x}$ with respect to $W = \text{Span}\{\vec{w}_1, \ldots, \vec{w}_k\}$ but add the condition that the spanning vectors are an orthogonal set. In this case, $\vec{w}_i \cdot \vec{w}_j = 0$ whenever $i \neq j$, which we can use to simplify our previous computation. There we were trying to use

$$\vec{w} = \left(\frac{\vec{x} \cdot \vec{w}_1}{\vec{w}_1 \cdot \vec{w}_1}\right) \vec{w}_1 + \cdots + \left(\frac{\vec{x} \cdot \vec{w}_k}{\vec{w}_k \cdot \vec{w}_k}\right) \vec{w}_k$$

as our orthogonal projection of $\vec{x}$ onto W. We were checking whether $\vec{v} = \vec{x} - \vec{w}$ was in $W^{\perp}$ by computing the dot product $\vec{v} \cdot \vec{w}_i$, and had reduced this down to

$$-\left(\frac{\vec{x} \cdot \vec{w}_1}{\vec{w}_1 \cdot \vec{w}_1}\right) \vec{w}_1 \cdot \vec{w}_i - \cdots - \left(\frac{\vec{x} \cdot \vec{w}_{i-1}}{\vec{w}_{i-1} \cdot \vec{w}_{i-1}}\right) \vec{w}_{i-1} \cdot \vec{w}_i$$

$$-\left(\frac{\vec{x} \cdot \vec{w}_{i+1}}{\vec{w}_{i+1} \cdot \vec{w}_{i+1}}\right) \vec{w}_{i+1} \cdot \vec{w}_i - \cdots - \left(\frac{\vec{x} \cdot \vec{w}_k}{\vec{w}_k \cdot \vec{w}_k}\right) \vec{w}_k \cdot \vec{w}_i$$

before we got stuck. Every term of this sum has as its rightmost factor a dot product of the form $\vec{w}_j \cdot \vec{w}_i$ with $j \neq i$, and now all such dot products are zero because the $\vec{w}_i$s are an orthogonal set. Therefore we get $\vec{v} \cdot \vec{w}_i = 0$, so $\vec{v} = \vec{x} - \vec{w}$ is in $W^{\perp}$. This allows us to make the following definition.

Definition. Let $W = \text{Span}\{\vec{w}_1, \ldots, \vec{w}_k\}$ where $\vec{w}_1, \ldots, \vec{w}_k$ are an orthogonal set. The *orthogonal projection of $\vec{x}$ onto W* is $(\frac{\vec{x} \cdot \vec{w}_1}{\vec{w}_1 \cdot \vec{w}_1})\vec{w}_1 + \cdots + (\frac{\vec{x} \cdot \vec{w}_k}{\vec{w}_k \cdot \vec{w}_k})\vec{w}_k$ and the *component of $\vec{x}$ orthogonal to W* is $\vec{x} - \left((\frac{\vec{x} \cdot \vec{w}_1}{\vec{w}_1 \cdot \vec{w}_1})\vec{w}_1 + \cdots + (\frac{\vec{x} \cdot \vec{w}_k}{\vec{w}_k \cdot \vec{w}_k})\vec{w}_k\right)$.

Notice that the orthogonal projection's coefficient on each $\vec{w}_i$ is the same as the one for the orthogonal projection onto a single vector $\vec{y}$.

Example 4. Let $W = \text{Span}\left\{\begin{bmatrix} -3 \\ 2 \\ 1 \end{bmatrix}, \begin{bmatrix} 5 \\ 6 \\ 3 \end{bmatrix}\right\}$. Find the orthogonal projection of $\begin{bmatrix} 1 \\ 1 \\ 8 \end{bmatrix}$ onto W and the component of $\begin{bmatrix} 1 \\ 1 \\ 8 \end{bmatrix}$ orthogonal to W.

In the language of our formulas for the orthogonal projection onto W and component orthogonal to W, we have

$$\vec{w}_1 = \begin{bmatrix} -3 \\ 2 \\ 1 \end{bmatrix}, \vec{w}_2 = \begin{bmatrix} 5 \\ 6 \\ 3 \end{bmatrix}, \text{ and } \vec{x} = \begin{bmatrix} 1 \\ 1 \\ 8 \end{bmatrix}.$$

In order to use the formulas in the previous definition, we first need to check that the spanning vectors of W are an orthogonal set. Taking their dot product gives us

$$\begin{bmatrix} -3 \\ 2 \\ 1 \end{bmatrix} \cdot \begin{bmatrix} 5 \\ 6 \\ 3 \end{bmatrix} = -3(5) + 2(6) + 1(3) = 0$$

so $\vec{w}_1$ and $\vec{w}_2$ are orthogonal and we can proceed as planned.

The formulas for the orthogonal projection of $\vec{x}$ onto W and component of $\vec{x}$ orthogonal to W involve the dot products $\vec{w}_1 \cdot \vec{x}$, $\vec{w}_2 \cdot \vec{x}$, $\vec{w}_1 \cdot \vec{w}_1$, and $\vec{w}_2 \cdot \vec{w}_2$. For our vectors, we get

$$\vec{w}_1 \cdot \vec{x} = \begin{bmatrix} -3 \\ 2 \\ 1 \end{bmatrix} \cdot \begin{bmatrix} 1 \\ 1 \\ 8 \end{bmatrix} = (-3)(1) + 2(1) + 1(8) = 7,$$

$$\vec{w}_2 \cdot \vec{x} = \begin{bmatrix} 5 \\ 6 \\ 3 \end{bmatrix} \cdot \begin{bmatrix} 1 \\ 1 \\ 8 \end{bmatrix} = 5(1) + 6(1) + 3(8) = 35,$$

$$\vec{w}_1 \cdot \vec{w}_1 = \begin{bmatrix} -3 \\ 2 \\ 1 \end{bmatrix} \cdot \begin{bmatrix} -3 \\ 2 \\ 1 \end{bmatrix} = (-3)^2 + 2^2 + 1^2 = 14,$$

and

$$\vec{w}_2 \cdot \vec{w}_2 = \begin{bmatrix} 5 \\ 6 \\ 3 \end{bmatrix} \cdot \begin{bmatrix} 5 \\ 6 \\ 3 \end{bmatrix} = 5^2 + 6^2 + 3^2 = 70.$$

Plugging these dot products into the formulas from our definition, we get that the orthogonal projection of $\vec{x}$ onto W is

$$\left(\frac{\vec{x} \cdot \vec{w}_1}{\vec{w}_1 \cdot \vec{w}_1} \right) \vec{w}_1 + \left(\frac{\vec{x} \cdot \vec{w}_2}{\vec{w}_2 \cdot \vec{w}_2} \right) \vec{w}_2 = \left(\frac{7}{14} \right) \begin{bmatrix} -3 \\ 2 \\ 1 \end{bmatrix} + \left(\frac{35}{70} \right) \begin{bmatrix} 5 \\ 6 \\ 3 \end{bmatrix} = \begin{bmatrix} 1 \\ 4 \\ 2 \end{bmatrix}$$

and the component of $\vec{x}$ orthogonal to W is

$$\vec{x} - \left(\left(\frac{\vec{x} \cdot \vec{w}_1}{\vec{w}_1 \cdot \vec{w}_1} \right) \vec{w}_1 + \left(\frac{\vec{x} \cdot \vec{w}_2}{\vec{w}_2 \cdot \vec{w}_2} \right) \vec{w}_2 \right) = \begin{bmatrix} 1 \\ 1 \\ 8 \end{bmatrix} - \begin{bmatrix} 1 \\ 4 \\ 2 \end{bmatrix} = \begin{bmatrix} 0 \\ -3 \\ 6 \end{bmatrix}.$$

If we want to check that these are correct, we can check that the component orthogonal to W is in $W^\perp$. To do this we need to check its dot products with $\vec{w}_1$ and $\vec{w}_2$. These are

$$\begin{bmatrix} 0 \\ -3 \\ 6 \end{bmatrix} \cdot \begin{bmatrix} -3 \\ 2 \\ 1 \end{bmatrix} = 0(-3) + (-3)(2) + 6(1) = 0$$

and

$$\begin{bmatrix} 0 \\ -3 \\ 6 \end{bmatrix} \cdot \begin{bmatrix} 5 \\ 6 \\ 3 \end{bmatrix} = 0(5) + (-3)(6) + 3(6) = 0.$$

Since they are both zero, $\begin{bmatrix} 0 \\ -3 \\ 6 \end{bmatrix}$ is in $W^{\perp}$ as required.

Let's wrap up this section by doing a physics problem of the type discussed at the beginning of the section.

Example 5. A 1.5 kg metal sled is sliding down a hill with a flat slope of 45° downward. The only force acting on the sled is gravity, which pulls it straight downward with a force of −14.7 N. Find the component of the gravitational force which is pushing the sled along the slope of the hill.

Often in examples like this it can be helpful to sketch the situation. Ours looks like this:

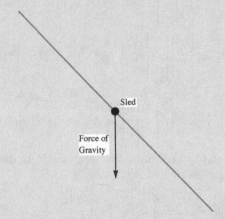

We can view this as an orthogonal projection problem by placing the sled at the origin and viewing the hill as W. In this context, W is the line through the origin with a negative 45° slope, i.e., the line $y = -x$. This means we can view W as the span of $\begin{bmatrix} 1 \\ -1 \end{bmatrix}$. (Note that there are literally infinitely many other choices for our spanning vector.) We can view the force of gravity acting on the sled as the vector $\begin{bmatrix} 0 \\ -14.7 \end{bmatrix}$ since it pulls straight downward with a total force of −14.7.

Within this framework, the component of gravity pulling the sled along the hill is just the orthogonal projection of gravity's vector $\vec{x} = \begin{bmatrix} 0 \\ -14.7 \end{bmatrix}$ onto the hill's spanning vector $\vec{y} = \begin{bmatrix} 1 \\ -1 \end{bmatrix}$. As in Example 1, we can start by computing

$$\vec{x} \cdot \vec{y} = \begin{bmatrix} 0 \\ -14.7 \end{bmatrix} \cdot \begin{bmatrix} 1 \\ -1 \end{bmatrix} = 0(1) + (-14.7)(-1) = 14.7$$

and

$$\vec{y} \cdot \vec{y} = \begin{bmatrix} 1 \\ -1 \end{bmatrix} \cdot \begin{bmatrix} 1 \\ -1 \end{bmatrix} = 1^2 + (-1)^2 = 2.$$

Plugging these into our orthogonal projection formula we get

$$\left(\frac{\vec{x} \cdot \vec{y}}{\vec{y} \cdot \vec{y}} \right) \vec{y} = \left(\frac{14.7}{2} \right) \begin{bmatrix} 1 \\ -1 \end{bmatrix} = \begin{bmatrix} 7.35 \\ -7.35 \end{bmatrix}.$$

Recalling the application which motivated this orthogonal projection computation, we can now say that the component of the gravitational force pulling the sled along the slope of the hill is $\begin{bmatrix} 7.35 \\ -7.35 \end{bmatrix}$. We can even go a step further and compute the norm of this vector which is

$$\sqrt{(7.35)^2 + (-7.35)^2} \approx 10.39$$

to conclude that a little more than 2/3 of the magnitude of our gravitational force is moving the sled along the hill.

Exercises 5.3.

1. If we want to compute the orthogonal projection of $\begin{bmatrix} -3 \\ 4 \\ -4 \end{bmatrix}$ onto $\begin{bmatrix} 0 \\ 2 \\ -2 \end{bmatrix}$, which vector is $\vec{x}$ in our formula and which is $\vec{y}$?

2. If we want to compute the orthogonal projection of $\begin{bmatrix} 15 \\ -2 \end{bmatrix}$ onto $\begin{bmatrix} 4 \\ 9 \end{bmatrix}$, which vector is $\vec{x}$ in our formula and which is $\vec{y}$?

3. Compute the orthogonal projection of $\begin{bmatrix} 8 \\ -1 \end{bmatrix}$ onto $\begin{bmatrix} 2 \\ 1 \end{bmatrix}$ and the component of $\begin{bmatrix} 8 \\ -1 \end{bmatrix}$ orthogonal to $\begin{bmatrix} 2 \\ 1 \end{bmatrix}$.

4. Compute the orthogonal projection of $\begin{bmatrix} -1 \\ 7 \end{bmatrix}$ onto $\begin{bmatrix} 1 \\ -1 \end{bmatrix}$ and the component of $\begin{bmatrix} -1 \\ 7 \end{bmatrix}$ orthogonal to $\begin{bmatrix} 1 \\ -1 \end{bmatrix}$.

5. Compute the orthogonal projection of $\begin{bmatrix} 3 \\ -1 \\ 2 \end{bmatrix}$ onto $\begin{bmatrix} 1 \\ -1 \\ 1 \end{bmatrix}$ and the component of $\begin{bmatrix} 3 \\ -1 \\ 2 \end{bmatrix}$ orthogonal to $\begin{bmatrix} 1 \\ -1 \\ 1 \end{bmatrix}$.

6. Compute the orthogonal projection of $\begin{bmatrix} 6 \\ -5 \\ 8 \end{bmatrix}$ onto $\begin{bmatrix} 1 \\ 0 \\ -1 \end{bmatrix}$ and the

   component of $\begin{bmatrix} 6 \\ -5 \\ 8 \end{bmatrix}$ orthogonal to $\begin{bmatrix} 1 \\ 0 \\ -1 \end{bmatrix}$.

7. Is $\left\{ \begin{bmatrix} 8 \\ -4 \\ 0 \end{bmatrix}, \begin{bmatrix} 0 \\ 0 \\ 1 \end{bmatrix}, \begin{bmatrix} 1 \\ 2 \\ 0 \end{bmatrix} \right\}$ an orthogonal set?

8. Is $\left\{ \begin{bmatrix} -4 \\ 6 \\ 2 \end{bmatrix}, \begin{bmatrix} -1 \\ 3 \\ -11 \end{bmatrix}, \begin{bmatrix} 5 \\ 2 \\ 4 \end{bmatrix} \right\}$ an orthogonal set?

9. Is $\left\{ \begin{bmatrix} 3 \\ 0 \\ -5 \\ -1 \end{bmatrix}, \begin{bmatrix} 2 \\ 4 \\ 3 \\ -9 \end{bmatrix}, \begin{bmatrix} 4 \\ 13 \\ 1 \\ 7 \end{bmatrix} \right\}$ an orthogonal set?

10. Is $\left\{ \begin{bmatrix} 6 \\ 1 \\ 2 \\ 3 \end{bmatrix}, \begin{bmatrix} -2 \\ 6 \\ -3 \\ 4 \end{bmatrix}, \begin{bmatrix} 4 \\ 3 \\ -7 \\ -5 \end{bmatrix} \right\}$ an orthogonal set?

11. Let $W = \text{Span} \left\{ \begin{bmatrix} 1 \\ 0 \\ -1 \end{bmatrix}, \begin{bmatrix} 0 \\ -2 \\ 0 \end{bmatrix} \right\}$ and $\vec{x} = \begin{bmatrix} 3 \\ 8 \\ 5 \end{bmatrix}$. Compute the
    orthogonal projection of $\vec{x}$ onto W and the component of $\vec{x}$
    orthogonal to W.

12. Let $W = \text{Span} \left\{ \begin{bmatrix} 1 \\ -2 \\ 1 \end{bmatrix}, \begin{bmatrix} 1 \\ 1 \\ 1 \end{bmatrix} \right\}$ and $\vec{x} = \begin{bmatrix} 3 \\ -4 \\ -5 \end{bmatrix}$. Compute the
    orthogonal projection of $\vec{x}$ onto W and the component of $\vec{x}$
    orthogonal to W.

13. Let $W = \text{Span} \left\{ \begin{bmatrix} 1 \\ -1 \\ 2 \\ 0 \end{bmatrix}, \begin{bmatrix} -2 \\ 0 \\ 1 \\ -1 \end{bmatrix} \right\}$ and $\vec{x} = \begin{bmatrix} 10 \\ 3 \\ 4 \\ -6 \end{bmatrix}$. Compute the
    orthogonal projection of $\vec{x}$ onto W and the component of $\vec{x}$
    orthogonal to W.

14. Let $W = \text{Span} \left\{ \begin{bmatrix} -1 \\ -1 \\ 1 \\ 0 \end{bmatrix}, \begin{bmatrix} 0 \\ 1 \\ -1 \\ 2 \end{bmatrix} \right\}$ and $\vec{x} = \begin{bmatrix} 2 \\ -2 \\ -3 \\ 1 \end{bmatrix}$. Compute the
    orthogonal projection of $\vec{x}$ onto W and the component of $\vec{x}$
    orthogonal to W.

15. What is the shortest distance between $\begin{bmatrix} 2 \\ 9 \end{bmatrix}$ and the line spanned by $\begin{bmatrix} 1 \\ 2 \end{bmatrix}$?

16. What is the shortest distance between $\begin{bmatrix} -2 \\ 3 \\ -1 \end{bmatrix}$ and the plane spanned by $\begin{bmatrix} -1 \\ 1 \\ 2 \end{bmatrix}$ and $\begin{bmatrix} 2 \\ 0 \\ 1 \end{bmatrix}$?

17. A man is dragging a sled up a hill by pulling on a rope attached to the sled. The line formed by the rope has a slope of 3, and the hill has a slope of 1. The man exerts 225 N of force in the direction of the rope.

    (a) What is the vector of the man's force?
    (b) What set of vectors represents the hill?
    (c) How much of that force is exerted along the hill in the plane of motion of the sled?

5.4   Orthogonal Basis

In the last section, we happened onto the idea of an orthogonal set as part of our exploration of orthogonal projections. In this section, we'll merge the ideas of an orthogonal set and a basis to create orthogonal bases for $\mathbb{R}^n$, and see why that might make our lives easier. Our observation from the previous section that orthogonal vectors point in completely different directions should remind you of our original geometric idea of linear independence as having no vector in the span of any of the others. The following theorem formalizes that connection.

Theorem 1. If $\vec{v}_1, \ldots, \vec{v}_k$ are nonzero vectors in $\mathbb{R}^n$ which form an orthogonal set, then they are linearly independent.

To see why this is true, suppose we have nonzero $\vec{v}_1, \ldots, \vec{v}_k$ in $\mathbb{R}^n$ which are an orthogonal set. We want to show that they are also linearly independent, i.e., none of these vectors is in the span of the others. We'll do this by assuming one of the $\vec{v}$s is in the span of the others, and showing that assumption produces an impossible consequence. For notational ease I'll assume $\vec{v}_1$ is in the span of $\vec{v}_2, \ldots, \vec{v}_k$. This means

$$\vec{v}_1 = a_2\vec{v}_2 + \cdots + a_k\vec{v}_k$$

for some scalars $a_2, \ldots, a_k$. Because $\vec{v}_1$ is orthogonal to $\vec{v}_2$, we have

$$0 = \vec{v}_1 \cdot \vec{v}_2 = (a_2\vec{v}_2 + \cdots + a_k\vec{v}_k) \cdot \vec{v}_2 = (a_2\vec{v}_2) \cdot \vec{v}_2 + \cdots + (a_k\vec{v}_k) \cdot \vec{v}_2$$
$$= a_2(\vec{v}_2 \cdot \vec{v}_2) + \cdots + a_k(\vec{v}_k \cdot \vec{v}_2).$$

The fact that the $\vec{v}$s are an orthogonal set means $\vec{v}_i \cdot \vec{v}_2 = 0$ for $i \neq 2$, so our equation reduces to

$$0 = a_2(\vec{v}_2 \cdot \vec{v}_2).$$

Since $\vec{v}_2 \neq \vec{0}$, we know $\vec{v}_2 \cdot \vec{v}_2 \neq 0$. Therefore we must have $a_2 = 0$. A similar argument shows that each of our scalars $a_2, \ldots, a_k$ must be zero. But this means $\vec{v}_1 = \vec{0}$, which is impossible since we required our orthogonal set to be made up of nonzero vectors! Therefore we cannot have any of the $\vec{v}$s in the span of the others, and our orthogonal set must also be linearly independent.

While this theorem is great, we don't want to get carried away and try to reverse it. There are plenty of linearly independent sets which aren't orthogonal. Geometrically, we can see this by thinking of any two vectors in $\mathbb{R}^2$ which aren't on the same line but don't lie at right angles to each other.

Since a basis is defined as a linearly independent spanning set, Theorem 1 tells us that having an orthogonal set gets us halfway to having a basis. This leads to the following special case of a basis.

Definition. $\vec{b}_1, \ldots, \vec{b}_n$ are an *orthogonal basis* for $\mathbb{R}^n$ if they are a basis for $\mathbb{R}^n$ and an orthogonal set.

Example 1. Our standard basis for $\mathbb{R}^n$ is an orthogonal basis.

The standard basis for $\mathbb{R}^n$ is $\begin{bmatrix} 1 \\ 0 \\ \vdots \\ 0 \end{bmatrix}, \begin{bmatrix} 0 \\ 1 \\ \vdots \\ 0 \end{bmatrix}, \cdots, \begin{bmatrix} 0 \\ \vdots \\ 0 \\ 1 \end{bmatrix}$, which we already know

form a basis. The 1s do not occur in the same entry of any pair of these basis vectors, so the dot product of any two of them is zero. This means they form an orthogonal set and hence an orthogonal basis.

We could check that $\vec{b}_1, \ldots, \vec{b}_n$ form an orthogonal basis by checking that they are linearly independent and span $\mathbb{R}^n$ (and hence are a basis) and then checking that they are an orthogonal set. However, Theorem 1 provides a shortcut since orthogonality automatically implies linear independence. Therefore we can check that $\vec{b}_1, \ldots, \vec{b}_n$ form an orthogonal basis by checking that they span $\mathbb{R}^n$ and are an orthogonal set. If we like, we can do even less work by using Theorem 3 from 3.1. That theorem said that any linearly independent set of n vectors was a basis for $\mathbb{R}^n$. Combining this with Theorem 1 from this section we get the following:

Theorem 2. Any orthogonal set in $\mathbb{R}^n$ which contains n vectors is an orthogonal basis for $\mathbb{R}^n$.

Example 2. Show that $\begin{bmatrix} 0 \\ -2 \\ 4 \end{bmatrix}, \begin{bmatrix} -3 \\ 2 \\ 1 \end{bmatrix}$, and $\begin{bmatrix} 5 \\ 6 \\ 3 \end{bmatrix}$ form an orthogonal basis for $\mathbb{R}^3$.

These are the three vectors from 5.3's Example 2, so we've seen that they form an orthogonal set. Since there are three of them and we are in $\mathbb{R}^3$, Theorem 2 tells us that our three vectors are an orthogonal basis for $\mathbb{R}^3$.

Example 3. Show that $\begin{bmatrix} 1 \\ 1 \end{bmatrix}$ and $\begin{bmatrix} 1 \\ -2 \end{bmatrix}$ do not form an orthogonal basis for $\mathbb{R}^2$.

A set of vectors can fail to be an orthogonal basis either by failing to be orthogonal or failing to be a basis. It's easier to check whether or not these two vectors are orthogonal, so let's start there. Since

$$\begin{bmatrix} 1 \\ 1 \end{bmatrix} \cdot \begin{bmatrix} 1 \\ -2 \end{bmatrix} = 1(1) + 1(-2) = -1 \neq 0$$

these vectors are not orthogonal which means they do not form an orthogonal basis for $\mathbb{R}^2$.

(You can check if you like that they do form a basis for $\mathbb{R}^2$, perhaps the easiest way to see this is to show that the matrix with those two columns has nonzero determinant.)

One reason to use an orthogonal basis is that it makes finding coordinate vectors much faster. If you recall from 3.1, we can find the coordinate vector of $\vec{v}$ with respect to a basis $\mathcal{B} = \{\vec{b}_1, \ldots, \vec{b}_n\}$ by solving the vector equation

$$\vec{v} = x_1 \vec{b}_1 + \cdots + x_n \vec{b}_n$$

and setting

$$[\vec{v}]_{\mathcal{B}} = \begin{bmatrix} x_1 \\ \vdots \\ x_n \end{bmatrix}.$$

With row reduction, this is fairly straightforward, if sometimes a bit tedious. It also requires us to compute every entry of the coordinate vector, i.e., we cannot just compute x_2 and ignore the rest of the problem. If our basis $\mathcal{B}$ is orthogonal, then we have another way to compute coordinate vectors which is both quicker and allows us to compute any entry of the coordinate vector without computing the other entries.

To do this, let's forget about the fact that $\mathcal{B}$ is a basis and simply think of it as an orthogonal spanning set for $\mathbb{R}^n$. We can use the techniques developed in the last section to compute the orthogonal projection of any vector $\vec{v}$ onto the space spanned by $\vec{b}_1, \ldots, \vec{b}_n$. Since $W = \text{Span}\{\vec{b}_1, \ldots, \vec{b}_n\} = \mathbb{R}^n$, we know $\vec{v}$ is in W. This means when we decompose $\vec{v}$ as the sum of its orthogonal projection onto W and its component orthogonal to W, the component orthogonal to W will be $\vec{0}$ since no part of $\vec{v}$ lies outside of $W = \mathbb{R}^n$. This means $\vec{v}$'s $\mathcal{B}$-coordinate vector is simply its orthogonal projection onto $\mathbb{R}^n$ using the orthogonal spanning set $\vec{b}_1, \ldots, \vec{b}_n$. As we saw in 5.3, this orthogonal projection is

$$\left(\frac{\vec{v} \cdot \vec{b}_1}{\vec{b}_1 \cdot \vec{b}_1} \right) \vec{b}_1 + \cdots + \left(\frac{\vec{v} \cdot \vec{b}_n}{\vec{b}_n \cdot \vec{b}_n} \right) \vec{b}_n$$

which gives the following formula for $\vec{v}$'s $\mathcal{B}$-coordinate vector.

Theorem 3. If $\mathcal{B}$ is an orthogonal basis for $\mathbb{R}^n$, then $[\vec{v}]_{\mathcal{B}} = \begin{bmatrix} \frac{\vec{v} \cdot \vec{b}_1}{\vec{b}_1 \cdot \vec{b}_1} \\ \vdots \\ \frac{\vec{v} \cdot \vec{b}_n}{\vec{b}_n \cdot \vec{b}_n} \end{bmatrix}.$

Not only is this computation faster, but each entry's computation is independent of the computation of the other entries, so we can choose to compute only those components of $[\vec{v}]_{\mathcal{B}}$ that we want.

Example 4. Let $\mathcal{B} = \left\{ \begin{bmatrix} 0 \\ -2 \\ 4 \end{bmatrix}, \begin{bmatrix} -3 \\ 2 \\ 1 \end{bmatrix}, \begin{bmatrix} 5 \\ 6 \\ 3 \end{bmatrix} \right\}$ be an orthogonal basis for $\mathbb{R}^3$.

Compute the $\mathcal{B}$-coordinate vector of $\vec{v} = \begin{bmatrix} -1 \\ 1 \\ 2 \end{bmatrix}$.

Here we have $\vec{b}_1 = \begin{bmatrix} 0 \\ -2 \\ 4 \end{bmatrix}$, $\vec{b}_2 = \begin{bmatrix} -3 \\ 2 \\ 1 \end{bmatrix}$, and $\vec{b}_3 = \begin{bmatrix} 5 \\ 6 \\ 3 \end{bmatrix}$. To use our formula

for $[\vec{v}]_{\mathcal{B}}$, we need to compute the six dot products $\vec{v} \cdot \vec{b}_1$, $\vec{v} \cdot \vec{b}_2$, $\vec{v} \cdot \vec{b}_3$, $\vec{b}_1 \cdot \vec{b}_1$, $\vec{b}_2 \cdot \vec{b}_2$, and $\vec{b}_3 \cdot \vec{b}_3$. For our vectors, these are

$$\vec{v} \cdot \vec{b}_1 = \begin{bmatrix} -1 \\ 1 \\ 2 \end{bmatrix} \cdot \begin{bmatrix} 0 \\ -2 \\ 4 \end{bmatrix} = (-1)(0) + 1(-2) + 2(4) = 6,$$

$$\vec{v} \cdot \vec{b}_2 = \begin{bmatrix} -1 \\ 1 \\ 2 \end{bmatrix} \cdot \begin{bmatrix} -3 \\ 2 \\ 1 \end{bmatrix} = (-1)(-3) + 1(2) + 2(1) = 7,$$

$$\vec{v} \cdot \vec{b}_3 = \begin{bmatrix} -1 \\ 1 \\ 2 \end{bmatrix} \cdot \begin{bmatrix} 5 \\ 6 \\ 3 \end{bmatrix} = (-1)(5) + 1(6) + 2(3) = 7,$$

$$\vec{b}_1 \cdot \vec{b}_1 = \begin{bmatrix} 0 \\ -2 \\ 4 \end{bmatrix} \cdot \begin{bmatrix} 0 \\ -2 \\ 4 \end{bmatrix} = 0^2 + (-2)^2 + 4^2 = 20,$$

$$\vec{b}_2 \cdot \vec{b}_2 = \begin{bmatrix} -3 \\ 2 \\ 1 \end{bmatrix} \cdot \begin{bmatrix} -3 \\ 2 \\ 1 \end{bmatrix} = (-3)^2 + 2^2 + 1^2 = 14,$$

and

$$\vec{b}_3 \cdot \vec{b}_3 = \begin{bmatrix} 5 \\ 6 \\ 3 \end{bmatrix} \cdot \begin{bmatrix} 5 \\ 6 \\ 3 \end{bmatrix} = 5^2 + 6^2 + 3^2 = 70.$$

Plugging these into our coordinate vector formula gives us

$$[\vec{v}]_{\mathcal{B}} = \begin{bmatrix} \frac{\vec{v} \cdot \vec{b}_1}{\vec{b}_1 \cdot \vec{b}_1} \\ \frac{\vec{v} \cdot \vec{b}_2}{\vec{b}_2 \cdot \vec{b}_2} \\ \frac{\vec{v} \cdot \vec{b}_3}{\vec{b}_3 \cdot \vec{b}_3} \end{bmatrix} = \begin{bmatrix} \frac{6}{20} \\ \frac{7}{14} \\ \frac{7}{70} \end{bmatrix} = \begin{bmatrix} \frac{3}{10} \\ \frac{1}{2} \\ \frac{1}{10} \end{bmatrix}.$$

Example 5. Let $\mathcal{B} = \left\{ \begin{bmatrix} 0 \\ -2 \\ 4 \end{bmatrix}, \begin{bmatrix} -3 \\ 2 \\ 1 \end{bmatrix}, \begin{bmatrix} 5 \\ 6 \\ 3 \end{bmatrix} \right\}$. Compute the third entry of

the $\mathcal{B}$-coordinate vector of $\vec{v} = \begin{bmatrix} 9 \\ 8 \\ 4 \end{bmatrix}$.

The third entry of $[\vec{v}]_{\mathcal{B}}$ is $\dfrac{\vec{v} \cdot \vec{b}_3}{\vec{b}_3 \cdot \vec{b}_3}$, so we need to compute $\vec{v} \cdot \vec{b}_3$ and $\vec{b}_3 \cdot \vec{b}_3$.
For our $\vec{v}$, we have

$$\vec{v} \cdot \vec{b}_3 = \begin{bmatrix} 9 \\ 8 \\ 4 \end{bmatrix} \cdot \begin{bmatrix} 5 \\ 6 \\ 3 \end{bmatrix} = 9(5) + 8(6) + 4(3) = 105$$

and from Example 4 we know $\vec{b}_3 \cdot \vec{b}_3 = 70$. Therefore the third entry of $[\vec{v}]_{\mathcal{B}}$ is

$$\frac{\vec{v} \cdot \vec{b}_3}{\vec{b}_3 \cdot \vec{b}_3} = \frac{105}{70} = 1.5.$$

Now that we've computed a few coordinate vectors with respect to an orthogonal basis, you may have noticed that each entry of those coordinate vectors is a fraction. The ith denominator from that fraction is the dot product of the basis vector $\vec{b}_i$ with itself. Since

$$\vec{b}_i \cdot \vec{b}_i = \left(\sqrt{\vec{b}_i \cdot \vec{b}_i} \right)^2 = ||\vec{b}_i||^2$$

the denominators in our coordinate vectors are the squares of the norms of our basis vectors. This suggests an easy way to simplify this computation: normalize each basis vector so that its length is 1. Then we'd have

$$\vec{b}_i \cdot \vec{b}_i = ||\vec{b}_i||^2 = 1$$

for every i, which would make

$$[\vec{v}]_{\mathcal{B}} = \begin{bmatrix} \vec{v} \cdot \vec{b}_1 \\ \vdots \\ \vec{v} \cdot \vec{b}_n \end{bmatrix}.$$

An orthogonal basis which has all basis vectors of length 1 is often called an *orthonormal basis.*

It turns out that we can take any basis $\vec{v}_1, \ldots, \vec{v}_n$ for $\mathbb{R}^n$ and transform it into an orthonormal basis using an algorithm called the Gram-Schmidt process. Like row reduction, it has two stages. The first stage transforms our original basis into an orthogonal basis $\vec{b}_1, \ldots, \vec{b}_n$, and the second normalizes each of the orthogonal basis vectors produced in the first stage to produce the orthonormal basis $\vec{u}_1, \ldots, \vec{u}_n$. Formally, this can be stated as follows:

Gram-Schmidt Process:

Part 1:

- Let $\vec{b}_1 = \vec{v}_1$.

- Starting with $i = 2$ and repeating with each successive i until you reach $i = n$, let

$$\vec{b}_i = \vec{v}_i - \left(\left(\frac{\vec{v}_i \cdot \vec{b}_1}{\vec{b}_1 \cdot \vec{b}_1} \right) \vec{b}_1 + \cdots + \left(\frac{\vec{v}_i \cdot \vec{b}_{i-1}}{\vec{b}_{i-1} \cdot \vec{b}_{i-1}} \right) \vec{b}_{i-1} \right).$$

(This means $\vec{b}_i$ is the component of $\vec{v}_i$ orthogonal to the span of $\vec{b}_1, \ldots, \vec{b}_{i-1}$.)

Part 2:

- Let $\vec{u}_i = \dfrac{1}{\|\vec{b}_i\|} \vec{b}_i$.

Example 6. Use the Gram-Schmidt process to create an orthonormal basis from the basis $\begin{bmatrix} 1 \\ 0 \\ -1 \end{bmatrix}, \begin{bmatrix} -3 \\ 4 \\ 1 \end{bmatrix}, \begin{bmatrix} -1 \\ 7 \\ -7 \end{bmatrix}$ for $\mathbb{R}^3$.

We'll start by applying the first part of the process to create an orthogonal basis $\vec{b}_1, \vec{b}_2, \vec{b}_3$ for $\mathbb{R}^3$. Since $\vec{b}_1 = \vec{v}_1$, here we have $\vec{b}_1 = \begin{bmatrix} 1 \\ 0 \\ -1 \end{bmatrix}$.

Next we need to compute

$$\vec{b}_2 = \vec{v}_2 - \left(\frac{\vec{v}_2 \cdot \vec{b}_1}{\vec{b}_1 \cdot \vec{b}_1} \right) \vec{b}_1.$$

To do this, we need the dot products $\vec{v}_2 \cdot \vec{b}_1$ and $\vec{b}_1 \cdot \vec{b}_1$. Plugging in our $\vec{v}_2$ and $\vec{b}_1$ gives us

$$\vec{v}_2 \cdot \vec{b}_1 = \begin{bmatrix} -3 \\ 4 \\ 1 \end{bmatrix} \cdot \begin{bmatrix} 1 \\ 0 \\ -1 \end{bmatrix} = (-3)(1) + 4(0) + 1(-1) = -4$$

and

$$\vec{b}_1 \cdot \vec{b}_1 = \begin{bmatrix} 1 \\ 0 \\ -1 \end{bmatrix} \cdot \begin{bmatrix} 1 \\ 0 \\ -1 \end{bmatrix} = 1^2 + 0^2 + (-1)^2 = 2.$$

Thus

$$\vec{b}_2 = \vec{v}_2 - \left(\frac{\vec{v}_2 \cdot \vec{b}_1}{\vec{b}_1 \cdot \vec{b}_1}\right)\vec{b}_1 = \begin{bmatrix} -3 \\ 4 \\ 1 \end{bmatrix} - \left(\frac{-4}{2}\right)\begin{bmatrix} 1 \\ 0 \\ -1 \end{bmatrix} = \begin{bmatrix} -1 \\ 4 \\ -1 \end{bmatrix}.$$

(We can check our work by computing $\vec{b}_1 \cdot \vec{b}_2 = 0$ to see that $\vec{b}_1$ and $\vec{b}_2$ are orthogonal.)

To complete part 1, we need to compute

$$\vec{b}_3 = \vec{v}_3 - \left(\left(\frac{\vec{v}_3 \cdot \vec{b}_1}{\vec{b}_1 \cdot \vec{b}_1}\right)\vec{b}_1 + \left(\frac{\vec{v}_3 \cdot \vec{b}_2}{\vec{b}_2 \cdot \vec{b}_2}\right)\vec{b}_2\right).$$

Here we need the dot products $\vec{v}_3 \cdot \vec{b}_1$, $\vec{b}_1 \cdot \vec{b}_1$, $\vec{v}_3 \cdot \vec{b}_2$, and $\vec{b}_2 \cdot \vec{b}_2$. From the previous step we know $\vec{b}_1 \cdot \vec{b}_1 = 2$, and we can compute

$$\vec{v}_3 \cdot \vec{b}_1 = \begin{bmatrix} -1 \\ 7 \\ -7 \end{bmatrix} \cdot \begin{bmatrix} 1 \\ 0 \\ -1 \end{bmatrix} = (-1)(1) + 7(0) + (-7)(-1) = 6,$$

$$\vec{v}_3 \cdot \vec{b}_2 = \begin{bmatrix} -1 \\ 7 \\ -7 \end{bmatrix} \cdot \begin{bmatrix} -1 \\ 4 \\ -1 \end{bmatrix} = (-1)(-1) + 7(4) + (-7)(-1) = 36,$$

and

$$\vec{b}_2 \cdot \vec{b}_2 = \begin{bmatrix} -1 \\ 4 \\ -1 \end{bmatrix} \cdot \begin{bmatrix} -1 \\ 4 \\ -1 \end{bmatrix} = (-1)^2 + 4^2 + (-1)^2 = 18.$$

Plugging these into our formula for $\vec{b}_3$ gives us

$$\vec{b}_3 = \vec{v}_3 - \left(\left(\frac{\vec{v}_3 \cdot \vec{b}_1}{\vec{b}_1 \cdot \vec{b}_1}\right)\vec{b}_1 + \left(\frac{\vec{v}_3 \cdot \vec{b}_2}{\vec{b}_2 \cdot \vec{b}_2}\right)\vec{b}_2\right)$$

$$= \begin{bmatrix} -1 \\ 7 \\ -7 \end{bmatrix} - \left(\left(\frac{6}{2}\right)\begin{bmatrix} 1 \\ 0 \\ -1 \end{bmatrix} + \left(\frac{36}{18}\right)\begin{bmatrix} -1 \\ 4 \\ -1 \end{bmatrix}\right)$$

$$= \begin{bmatrix} -1 \\ 7 \\ -7 \end{bmatrix} - \begin{bmatrix} 1 \\ 8 \\ -5 \end{bmatrix} = \begin{bmatrix} -2 \\ -1 \\ -2 \end{bmatrix}.$$

(Again, we can check our work by computing $\vec{b}_1 \cdot \vec{b}_3 = 0$ and $\vec{b}_2 \cdot \vec{b}_3 = 0$ to see that $\vec{b}_1, \vec{b}_2, \vec{b}_3$ are an orthogonal set.)

We are now done with Part 1 of the Gram-Schmidt process and have created the orthogonal basis

$$\vec{b}_1 = \begin{bmatrix} 1 \\ 0 \\ -1 \end{bmatrix}, \ \vec{b}_2 = \begin{bmatrix} -1 \\ 4 \\ -1 \end{bmatrix}, \ \vec{b}_3 = \begin{bmatrix} -2 \\ -1 \\ -2 \end{bmatrix}$$

for $\mathbb{R}^3$.

Moving on to part 2 of the Gram-Schmidt process, we want to normalize each of our orthogonal basis vectors using the formula $\vec{u}_i = \dfrac{1}{||\vec{b}_i||}\vec{b}_i$ to create our orthonormal basis $\vec{u}_1, \vec{u}_2, \vec{u}_3$ for $\mathbb{R}^3$. To do this, we need to compute the norm of each $\vec{b}_i$. These are

$$||\vec{b}_1|| = \left\|\begin{bmatrix} 1 \\ 0 \\ -1 \end{bmatrix}\right\| = \sqrt{1^2 + 0^2 + (-1)^2} = \sqrt{2},$$

$$||\vec{b}_2|| = \left\|\begin{bmatrix} -1 \\ 4 \\ -1 \end{bmatrix}\right\| = \sqrt{(-1)^2 + 4^2 + (-1)^2} = \sqrt{18} = 3\sqrt{2},$$

and

$$||\vec{b}_3|| = \left\|\begin{bmatrix} -2 \\ 1 \\ -2 \end{bmatrix}\right\| = \sqrt{(-2)^2 + (-1)^2 + (-2)^2} = \sqrt{9} = 3.$$

Plugging these norms into our formula for $\vec{u}_i$ gives us

$$\vec{u}_1 = \frac{1}{||\vec{b}_1||}\vec{b}_1 = \frac{1}{\sqrt{2}}\begin{bmatrix} 1 \\ 0 \\ -1 \end{bmatrix} = \begin{bmatrix} \frac{1}{\sqrt{2}} \\ 0 \\ -\frac{1}{\sqrt{2}} \end{bmatrix},$$

$$\vec{u}_2 = \frac{1}{||\vec{b}_2||}\vec{b}_2 = \frac{1}{3\sqrt{2}}\begin{bmatrix} -1 \\ 4 \\ -1 \end{bmatrix} = \begin{bmatrix} -\frac{1}{3\sqrt{2}} \\ \frac{4}{3\sqrt{2}} \\ -\frac{1}{3\sqrt{2}} \end{bmatrix},$$

and

$$\vec{u}_3 = \frac{1}{||\vec{b}_3||}\vec{b}_3 = \frac{1}{3}\begin{bmatrix} -2 \\ -1 \\ -2 \end{bmatrix} = \begin{bmatrix} -\frac{2}{3} \\ -\frac{1}{3} \\ -\frac{2}{3} \end{bmatrix}.$$

(You can check on your own that each of these vectors has norm 1.)

This means our orthonormal basis for $\mathbb{R}^3$ is

$$\vec{u}_1 = \begin{bmatrix} \frac{1}{\sqrt{2}} \\ 0 \\ -\frac{1}{\sqrt{2}} \end{bmatrix}, \vec{u}_2 = \begin{bmatrix} -\frac{1}{3\sqrt{2}} \\ \frac{4}{3\sqrt{2}} \\ -\frac{1}{3\sqrt{2}} \end{bmatrix}, \vec{u}_3 = \begin{bmatrix} -\frac{2}{3} \\ -\frac{1}{3} \\ -\frac{2}{3} \end{bmatrix}.$$

Our newfound ability to create orthogonal bases also allows us to show the following interesting fact.

Theorem 4. Let W be a subset of $\mathbb{R}^n$. Then $\dim(W) + \dim(W^\perp) = n$.

Suppose $\dim(W) = k$. This means any basis for W contains k vectors, so pick a basis $\vec{w}_1, \ldots, \vec{w}_k$ for W. Similarly, suppose $\dim(W^\perp) = \ell$ and $\vec{v}_1, \ldots, \vec{v}_\ell$ is a basis for $W^\perp$.

We can use the Gram-Schmidt process on $\vec{w}_1, \ldots, \vec{w}_k$ to create an orthonormal basis $\vec{u}_1, \cdots, \vec{u}_k$ for W. Similarly, we can use the Gram-Schmidt process on $\vec{v}_1, \ldots, \vec{v}_\ell$ to create an orthonormal basis $\vec{y}_1, \ldots, \vec{y}_\ell$ for $W^\perp$. The set $\vec{u}_1, \cdots, \vec{u}_k, \vec{y}_1, \ldots, \vec{y}_\ell$ is an orthogonal set, since the sets $\vec{u}_1, \cdots, \vec{u}_k$ and $\vec{y}_1, \ldots, \vec{y}_\ell$ are orthogonal by Gram-Schmidt and $\vec{u}_i \cdot \vec{y}_j = 0$ for all i and j because $\vec{u}_i$ is in W and $\vec{y}_j$ is in $W^\perp$. Therefore by Theorem 1, we know $\vec{u}_1, \cdots, \vec{u}_k, \vec{y}_1, \ldots, \vec{y}_\ell$ is a linearly independent set.

If we take any $\vec{x}$ in $\mathbb{R}^n$, we can use our orthogonal projection process from 5.3 to write $\vec{x} = \vec{w} + \vec{v}$ where $\vec{w}$ is in W and $\vec{v}$ is in $W^\perp$. This means we can write $\vec{w}$ as a linear combination of $\vec{u}_1, \cdots, \vec{u}_k$ and $\vec{v}$ as a linear combination of $\vec{y}_1, \ldots, \vec{y}_\ell$, so $\vec{x}$ is a linear combination of $\vec{u}_1, \cdots, \vec{u}_k, \vec{y}_1, \ldots, \vec{y}_\ell$. Thus $\text{Span}\{\vec{u}_1, \cdots, \vec{u}_k, \vec{y}_1, \ldots, \vec{y}_\ell\} = \mathbb{R}^n$.

Putting this together, we see that $\vec{u}_1, \cdots, \vec{u}_k, \vec{y}_1, \ldots, \vec{y}_\ell$ is a basis for $\mathbb{R}^n$ and therefore must contain n vectors. This means $k + \ell = n$, so we are done.

Example 7. Show that Theorem 4 holds for $W = \text{Span}\left\{ \begin{bmatrix} 4 \\ -2 \\ 1 \\ 0 \end{bmatrix}, \begin{bmatrix} -1 \\ 5 \\ 2 \\ 9 \end{bmatrix} \right\}$.

Since W is a subset of $\mathbb{R}^4$, we need to show that $\dim(W) + \dim(W^\perp) = 4$. Our W is the set from 5.2's Example 8, where we showed that

$$W^\perp = \left\{ \begin{bmatrix} -\frac{1}{2}x_3 - x_4 \\ -\frac{1}{2}x_3 - 2x_4 \\ x_3 \\ x_4 \end{bmatrix} \right\}.$$

Therefore our next step is to compute $\dim(W)$ and $\dim(W^\perp)$.

One way to compute the dimension of a span like W is to row reduce the matrix whose columns are the spanning vectors. The dimension is then the number of leading 1s in that reduced echelon form. (For more details, see 2.8's discussion of rank as it relates to linear independence.) In our case this matrix is

$$\begin{bmatrix} 4 & -1 \\ -2 & 5 \\ 1 & 2 \\ 0 & 9 \end{bmatrix}$$

which row reduces to

$$\begin{bmatrix} 1 & 0 \\ 0 & 1 \\ 0 & 0 \\ 0 & 0 \end{bmatrix}.$$

Since our reduced echelon form has two leading 1s, we know $\dim(W) = 2$.

To compute the dimension of $W^\perp$, notice that it is written as the solution set of a matrix equation of the form $A\vec{x} = \vec{0}$. This means we can find a basis for $W^\perp$ which has one basis vector corresponding to each of the free variables x_3 and x_4 which appear in the entries of the vectors in $W^\perp$. (See Example 2 of 3.1 for more discussion of this idea.) Since we have two free variables, we have $\dim(W^\perp) = 2$.

Thus
$$\dim(W) + \dim(W^\perp) - 2 + 2 = 4$$

as claimed.

Exercises 5.4.

1. Is $\left\{ \begin{bmatrix} 3 \\ -1 \\ 4 \end{bmatrix}, \begin{bmatrix} 2 \\ 10 \\ 1 \end{bmatrix} \right\}$ an orthogonal basis for $\mathbb{R}^3$?

2. Is $\left\{ \begin{bmatrix} 2 \\ 10 \\ 1 \end{bmatrix}, \begin{bmatrix} 41 \\ -5 \\ 32 \end{bmatrix}, \begin{bmatrix} 3 \\ -1 \\ 4 \end{bmatrix} \right\}$ an orthogonal basis for $\mathbb{R}^3$?

3. Is $\left\{ \begin{bmatrix} 0 \\ 6 \\ -2 \\ 1 \end{bmatrix}, \begin{bmatrix} -1 \\ 1 \\ 3 \\ 0 \end{bmatrix}, \begin{bmatrix} 9 \\ 0 \\ 3 \\ 6 \end{bmatrix}, \begin{bmatrix} 41 \\ 14 \\ 9 \\ -66 \end{bmatrix} \right\}$ an orthogonal basis for $\mathbb{R}^4$?

4. Is $\left\{ \begin{bmatrix} -4 \\ 0 \\ 1 \\ 1 \end{bmatrix}, \begin{bmatrix} 1 \\ 4 \\ -1 \\ -1 \end{bmatrix}, \begin{bmatrix} 3 \\ -2 \\ 6 \\ 6 \end{bmatrix}, \begin{bmatrix} 0 \\ 5 \\ -1 \\ 1 \end{bmatrix} \right\}$ an orthogonal basis for $\mathbb{R}^4$?

5. Let $\mathcal{B} = \left\{ \begin{bmatrix} -3 \\ 1 \end{bmatrix}, \begin{bmatrix} 1 \\ 3 \end{bmatrix} \right\}$ be an orthogonal basis for $\mathbb{R}^2$. Compute the first component of $[\vec{v}]_\mathcal{B}$ for $\vec{v} = \begin{bmatrix} 4 \\ 6 \end{bmatrix}$.

6. Let $\mathcal{B} = \left\{ \begin{bmatrix} 1 \\ 1 \end{bmatrix}, \begin{bmatrix} 1 \\ -1 \end{bmatrix} \right\}$ be an orthogonal basis for $\mathbb{R}^2$. Compute the second component of $[\vec{v}]_\mathcal{B}$ for $\vec{v} = \begin{bmatrix} 8 \\ -2 \end{bmatrix}$.

7. Let $\mathcal{B} = \left\{ \begin{bmatrix} -3 \\ 4 \\ 1 \end{bmatrix}, \begin{bmatrix} 1 \\ 0 \\ 3 \end{bmatrix}, \begin{bmatrix} -6 \\ -5 \\ 2 \end{bmatrix} \right\}$ be an orthogonal basis for $\mathbb{R}^3$.

Compute the second component of $[\vec{v}]_\mathcal{B}$ for $\vec{v} = \begin{bmatrix} 4 \\ 7 \\ 6 \end{bmatrix}$.

8. Let $\mathcal{B} = \left\{ \begin{bmatrix} 1 \\ 1 \\ -2 \end{bmatrix}, \begin{bmatrix} 4 \\ -2 \\ 1 \end{bmatrix}, \begin{bmatrix} 1 \\ 3 \\ 2 \end{bmatrix} \right\}$ be an orthogonal basis for $\mathbb{R}^3$.

Compute the first component of $[\vec{v}]_{\mathcal{B}}$ for $\vec{v} = \begin{bmatrix} 7 \\ -2 \\ 2 \end{bmatrix}$.

9. Let $\mathcal{B} = \left\{ \begin{bmatrix} 0 \\ 1 \\ -1 \end{bmatrix}, \begin{bmatrix} -1 \\ 1 \\ 1 \end{bmatrix}, \begin{bmatrix} 2 \\ 1 \\ 1 \end{bmatrix} \right\}$ be an orthogonal basis for $\mathbb{R}^3$.

Compute $[\vec{v}]_{\mathcal{B}}$ for $\vec{v} = \begin{bmatrix} 4 \\ -2 \\ 3 \end{bmatrix}$.

10. Let $\mathcal{B} = \left\{ \begin{bmatrix} 1 \\ 1 \\ 1 \end{bmatrix}, \begin{bmatrix} -2 \\ -1 \\ 3 \end{bmatrix}, \begin{bmatrix} -4 \\ 5 \\ -1 \end{bmatrix} \right\}$ be an orthogonal basis for $\mathbb{R}^3$.

Compute $[\vec{v}]_{\mathcal{B}}$ for $\vec{v} = \begin{bmatrix} 2 \\ 1 \\ 4 \end{bmatrix}$.

11. Let $\mathcal{B} = \left\{ \begin{bmatrix} 1 \\ 1 \\ -2 \end{bmatrix}, \begin{bmatrix} 4 \\ -2 \\ 1 \end{bmatrix}, \begin{bmatrix} 1 \\ 3 \\ 2 \end{bmatrix} \right\}$ be an orthogonal basis for $\mathbb{R}^3$.

Compute $[\vec{v}]_{\mathcal{B}}$ for $\vec{v} = \begin{bmatrix} 2 \\ 1 \\ 1 \end{bmatrix}$.

12. Let $\mathcal{B} = \left\{ \begin{bmatrix} -1 \\ 0 \\ 1 \end{bmatrix}, \begin{bmatrix} 1 \\ 2 \\ 1 \end{bmatrix}, \begin{bmatrix} -1 \\ 1 \\ -1 \end{bmatrix} \right\}$ be an orthogonal basis for $\mathbb{R}^3$.

Compute $[\vec{v}]_{\mathcal{B}}$ for $\vec{v} = \begin{bmatrix} 3 \\ 3 \\ 1 \end{bmatrix}$.

13. Use Part 1 of the Gram-Schmidt process on $\mathcal{B} = \left\{ \begin{bmatrix} 1 \\ 3 \end{bmatrix}, \begin{bmatrix} 9 \\ 7 \end{bmatrix} \right\}$ to create an orthogonal basis for $\mathbb{R}^2$.

14. Use Part 1 of the Gram-Schmidt process on $\mathcal{B} = \left\{ \begin{bmatrix} 4 \\ 2 \\ 0 \end{bmatrix}, \begin{bmatrix} 10 \\ 10 \\ -1 \end{bmatrix}, \begin{bmatrix} -8 \\ 6 \\ -2 \end{bmatrix} \right\}$ to create an orthogonal basis for $\mathbb{R}^3$.

15. Use the Gram-Schmidt process on $\mathcal{B} = \left\{ \begin{bmatrix} -1 \\ 2 \end{bmatrix}, \begin{bmatrix} -4 \\ 3 \end{bmatrix} \right\}$ to create an orthonormal basis for $\mathbb{R}^2$.

16. Use the Gram-Schmidt process on $\mathcal{B} = \left\{ \begin{bmatrix} -1 \\ 0 \\ 1 \end{bmatrix}, \begin{bmatrix} 1 \\ -1 \\ 3 \end{bmatrix}, \begin{bmatrix} 4 \\ 3 \\ 2 \end{bmatrix} \right\}$ to create an orthonormal basis for $\mathbb{R}^3$.

17. Check that Theorem 2 holds for $W = \text{Span} \left\{ \begin{bmatrix} 3 \\ -1 \\ 5 \end{bmatrix}, \begin{bmatrix} 2 \\ 0 \\ 4 \end{bmatrix} \right\}$.

18. Check that Theorem 2 holds for $W = \text{Span} \left\{ \begin{bmatrix} 2 \\ -1 \\ 0 \\ 1 \end{bmatrix}, \begin{bmatrix} -8 \\ 3 \\ 2 \\ 5 \end{bmatrix} \right\}$.

19. Suppose W is a subset of $\mathbb{R}^4$ with $\dim(W) = 1$. What is $\dim(W^\perp)$?

20. Suppose W is a subset of $\mathbb{R}^3$ with $\dim(W) = 2$. What is $\dim(W^\perp)$?

21. Suppose W is a subset of $\mathbb{R}^8$ with $\dim(W) = 2$. What is $\dim(W^\perp)$?

22. Suppose W is a subset of $\mathbb{R}^5$ with $\dim(W) = 3$. What is $\dim(W^\perp)$?

23. Show that the standard basis for $\mathbb{R}^n$ is an orthogonal basis.

24. Can you find an orthogonal set of 3-vectors which is not a basis for $\mathbb{R}^3$?

A

Appendices

A.1 Complex Numbers

We briefly introduced the complex numbers, $\mathbb{C}$, in 3.3 as another example of a vector space. In this appendix, we'll explore them as an interesting space in their own right. We'll show that we can redo much of what we did in this book with scalars from $\mathbb{C}$ instead of $\mathbb{R}$. At the end, we'll provide a justification for the practical need to study $\mathbb{C}$, and I'll make an argument that this is really the correct set of scalars for a vector space. (If you want to skip straight to that, see Theorem 1 below.)

Let's start by restating our definition of $\mathbb{C}$ as $\mathbb{R}$ with the addition of the number i from 3.3.

Definition. The *complex numbers* are $\mathbb{C} = \{a + bi \mid a, b \text{ in } \mathbb{R}\}$. Here a is called the *real part* and bi is called the *imaginary part* of the complex number $a + bi$.

Example 1. Find the real and imaginary parts of $14 - 7i$.

The real part is the piece without a factor of i, which is 14. The imaginary part is the piece with a factor of i, which is $-7i$.

Since we saw in 3.3 that the complex numbers are a 2-dimensional vector space over $\mathbb{R}$, it shouldn't be surprising to learn that $\mathbb{C}$ is often identified with the plane $\mathbb{R}^2$. This is usually done by identifying a complex number $a + bi$ with its coordinate vector $\begin{bmatrix} a \\ b \end{bmatrix}$ with respect to the standard basis $\{1, i\}$ for $\mathbb{C}$. In other words, we use the real part a as the x-coordinate and coefficient b on i in the imaginary part bi as the y-coordinate. This sometimes leads people to refer to the x-axis as the *real axis* and the y-axis as the *imaginary axis*. This is illustrated by the following figure.

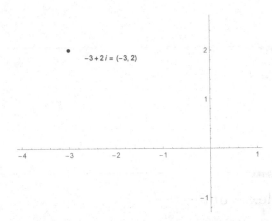

Definition. Two complex numbers are *equal* if their real parts are equal and their imaginary parts are equal.

This makes sense geometrically as well, since two complex numbers with equal real and imaginary parts occupy the same point in the complex plane.

In 3.3, we mostly treated i as a placeholder variable. However, it is really the key to $\mathbb{C}$'s good algebraic properties, but to explore $\mathbb{C}$ more deeply we need to connect it to $\mathbb{R}$ as follows.

Definition. The complex number i is defined as $i = \sqrt{-1}$.

We can state this in words by saying that i is the positive square root of -1. (The negative square root of -1 is therefore $-i$.)

We'll use the same definition for addition of complex numbers as we did in 3.3.

Definition. If $a + bi$ and $x + yi$ are in $\mathbb{C}$, then *complex addition* is defined to be $(a + bi) + (x + yi) = (a + x) + (b + y)i$.

Note that this says we add component-wise, i.e., the real part of the sum is the sum of the real parts and similarly for the imaginary parts.

Example 2. Find the sum of $3 + 5i$ and $2 - 9i$.

This sum is $(3 + 5i) + (2 - 9i) = (3 + 2) + (5 - 9)i = 5 - 4i$.

Since we're identifying $a + bi$ with the 2-vector $\begin{bmatrix} a \\ b \end{bmatrix}$, complex addition can be visualized in the plane exactly the same way as addition in $\mathbb{R}^2$.

Complex addition behaves very similarly to addition of real numbers. The most important properties are commutativity, associativity, additive identity, and additive inverses, which are explained below. (We also showed these properties in 3.3 as part of our check that $\mathbb{C}$ is a vector space.)

The order in which we add complex numbers doesn't matter since

$$(a+bi)+(x+yi) = (a+x)+(b+y)i = (x+a)+(y+b)i = (x+yi)+(a+bi)$$

so complex addition is commutative.

If we have three complex numbers $a + bi$, $x + yi$, and $w + vi$, which pair we add first doesn't matter since

$$
\begin{aligned}
[(a + bi) + (x + yi)] + (w + vi) &= [(a + x) + (b + y)i] + (w + vi) \\
&= (a + x + w) + (b + y + v)i \\
&= (a + bi) + [(x + w) + (y + v)i] \\
&= (a + bi) + [(x + yi) + (w + vi)]
\end{aligned}
$$

so complex addition is associative.

The complex number 0, which can be thought of as $0+0i$, has the property that for any complex number $a + bi$ we have

$$(a + bi) + 0 = (a + bi) + (0 + 0i) = (a + 0) + (b + 0)i = a + bi.$$

Thus 0 is our additive identity for complex addition.

Since complex numbers add component-wise and the additive identity is 0, the additive inverse of a complex number will be one whose values for a and b are the opposites of our original complex number. This means the additive inverse of $a+bi$ is $-a-bi$. Thus every complex number has an additive inverse.

Unlike in 3.3, here we want to define multiplication between two complex numbers. Let's explore this first by thinking again of i as acting like a variable. This means $a + bi$ times $x + yi$ can be expanded to get

$$(a + bi)(x + yi) = ax + ayi + bix + biyi.$$

Since a, b, x, y are real numbers, this can be rearranged as

$$ax + ayi + bix + biyi = ax + ayi + bxi + byi^2.$$

But remember that $i^2 = -1$, so we have

$$ax + ayi + bxi + byi^2 = ax + ayi + bxi - by.$$

Now we can write this in our standard complex number format by grouping the terms without a factor of i to form the real part and the terms with a factor of i to form the imaginary part. This gives us

$$ax + ayi + bxi - by = (ax - by) + (ay + bx)i$$

so we can create the following definition.

Definition. Let $a + bi$ and $x + yi$ be in $\mathbb{C}$, then *complex multiplication* is defined to be $(a + bi)(x + yi) = (ax - by) + (ay + bx)i$.

Example 3. Compute $(1 - 3i)(4 + 2i)$.

We can either find the values of a, b, x, y and plug them into the formula above, or we can expand out as we would for $(1 - 3x)(4 + 2x)$ and remember that $i^2 = -1$. I prefer the latter option so I don't have to memorize a formula, but feel free to do this from the formula if you prefer.

Expanding $(1 - 3i)(4 + 2i)$ gives us

$$4 + 2i - 12i - 6i^2 = 4 + 2i - 12i + 6.$$

Combining like terms now gives

$$(1 - 3i)(4 + 2i) = 10 - 10i.$$

As with complex addition, complex multiplication also behaves much like multiplication in $\mathbb{R}$. Some of the most important properties are commutativity, associativity, a multiplicative identity, and multiplicative inverses for nonzero complex numbers. These properties are explained below.

To see that the order in which we multiply complex numbers doesn't matter, let's compute $(a + bi)(x + yi)$ and $(x + yi)(a + bi)$. This gives us

$$(a + bi)(x + yi) = (ax - by) + (ay + bx)i$$

and

$$(x + yi)(a + bi) = (xa - yb) + (xb + ya)i.$$

Since multiplication of real numbers commutes we know $ax - by = xa - yb$ and $ay + bx = ya + bx$, so the real and imaginary parts of $(a + bi)(x + yi)$ and $(x + yi)(a + bi)$ are equal. Therefore

$$(a + bi)(x + yi) = (x + yi)(a + bi)$$

so complex multiplication is commutative.

To check that it doesn't matter which pair of a trio we multiply first, consider the complex numbers $a + bi$, $x + yi$, and $w + vi$. We can see that

$$
\begin{aligned}
[(a + bi)(x + yi)](w + vi) &= [(ax - by) + (ay + bx)i](w + vi) \\
&= ((ax - by)w - (ay + bx)v) + ((ax - by)v + (ay + bx)w)i \\
&= (axw - byw - ayv - bxv) + (axv - byv + ayw + bxw)i \\
&= (axw - ayv - bxv - byw) + (axv + ayw + bxw - byv)i \\
&= (a(xw - yv) - b(xv + yw)) + (a(xv + yw) + b(xw - yv))i \\
&= (a + bi)[(xw - yv) + (xv + yw)i] \\
&= (a + bi)[(x + yi)(w + vi)]
\end{aligned}
$$

so complex multiplication is associative.

As in $\mathbb{R}$, we have $1(a + bi) = a + bi$ so 1, or $1 + 0i$, is the identity element for complex multiplication.

If $a + bi \neq 0$, then

$$(a + bi) \left(\frac{a}{a^2 + b^2} - \frac{b}{a^2 + b^2} i \right)$$

$$= a \left(\frac{a}{a^2 + b^2} \right) + a \left(-\frac{b}{a^2 + b^2} i \right) + bi \left(\frac{a}{a^2 + b^2} \right) + bi \left(-\frac{b}{a^2 + b^2} i \right)$$

$$= \frac{a^2}{a^2 + b^2} - \frac{ab}{a^2 + b^2} i + \frac{ba}{a^2 + b^2} i - \frac{b^2}{a^2 + b^2} i^2$$

$$= \frac{a^2}{a^2 + b^2} - \frac{ab}{a^2 + b^2} i + \frac{ba}{a^2 + b^2} i + \frac{b^2}{a^2 + b^2}$$

$$= \left(\frac{a^2}{a^2 + b^2} + \frac{b^2}{a^2 + b^2} \right) + \left(-\frac{ab}{a^2 + b^2} + \frac{ba}{a^2 + b^2} \right) i$$

$$= \frac{a^2 + b^2}{a^2 + b^2} + \frac{-ab + ba}{a^2 + b^2} i = 1 + 0i = 1.$$

This means that every nonzero complex number $a + bi$ has the multiplicative inverse

$$\frac{a}{a^2 + b^2} - \frac{b}{a^2 + b^2} i.$$

As long as $a + bi$ is nonzero, this inverse is well defined, since $a + bi \neq 0$ means $a \neq 0$ or $b \neq 0$ and hence $a^2 + b^2 \neq 0$.

Finally on our list of parallels between these complex operations and the operations in $\mathbb{R}$, we get the expected relationship between multiplication and addition. To see this, we can compute

$$(a + bi)[(x + yi) + (v + wi)] = (a + bi)[(x + v) + (y + w)i]$$
$$= (a(x + v) - b(y + w)) + (a(y + w) + b(x + v))i$$
$$= (ax + av - by - bw) + (ay + aw + bx + bv)i$$
$$= (ax - by + av - bw) + (ay + bx + aw + bv)i$$
$$= [(ax - by) + (ay + bx)i] + [(av - bw) + (aw + bv)i]$$
$$= (a + bi)(x + yi) + (a + bi)(v + wi).$$

This means complex multiplication distributes over complex addition.

Visualizing complex multiplication using coordinates with respect to the standard basis $1, i$ is very difficult. However, our identification of $\mathbb{C}$ with the plane also gives rise to an alternate way to write complex numbers via a radius and angle. We can connect this to our $a + bi$ notation by identifying $a + bi$ with $\begin{bmatrix} a \\ b \end{bmatrix}$, then with the point in the plane (a, b) and finally with that point's polar coordinates r and θ. In this r and θ notation, multiplying complex numbers is much easier. If we want to multiply two complex numbers z_1 with polar coordinates r_1 and θ_1 and z_2 with polar coordinates r_2 and θ_2, then their product $z_1 z_2$ has polar coordinates $r_1 r_2$ and $\theta_1 + \theta_2$. In other words, we multiply radii and add angles.

Generally speaking, complex multiplication problems are easier in this polar notation while complex addition problems are easier in terms of the $a + bi$ notation. The major downside of using the polar notation for complex numbers is that it isn't unique, since angles of the form θ and $\theta + 2\pi k$ give the same position.

Perhaps the biggest complication of working with $\mathbb{C}$ rather than $\mathbb{R}$ is that geometric methods break down in most cases. For example, we can't draw the graph of a function $f : \mathbb{C} \to \mathbb{C}$. This is because the dimension of a graph of $f : V \to V$ is twice the dimension of V, since we need enough axes to describe the domain and then enough more to describe the codomain. (This is why the graph of $f : \mathbb{R} \to \mathbb{R}$ requires the 2D plane to draw even though $\dim(\mathbb{R}) = 1$.) Since $\dim(\mathbb{C}) = 2$, this means the graph of a complex-valued function will be 4-dimensional. Similarly, this means as soon as we have a complex vector space with dimension over $\mathbb{C}$ at least 2, we'll be unable to graph it. Fortunately, we've spent some time in this book working in $\mathbb{R}^n$ for $n > 3$, so have developed computational tools which can replace geometric arguments in these situations.

And now, finally, here is the real reason that I think linear algebra is a better place if you're willing to use complex numbers as your scalars: the fantastic Fundamental Theorem of Algebra. This is basically a polynomial analog of the Unique Factorization Theorem (also known as the Fundamental Theorem of Arithmetic) which states that every integer can be completely factored into a unique set of prime numbers. Here, instead of prime numbers, we factor into linear polynomials, i.e., $x - a$ for some scalar a.

Theorem 1. Let $f(x)$ be a polynomial with coefficients in $\mathbb{C}$. If f has degree n, then we can always factor f as $f(x) = b(x - a_1)(x - a_2) \cdots (x - a_n)$ for some complex numbers $b, a_1, \ldots, a_n$.

This means that unlike polynomials with real coefficients, our complex polynomials factor completely into linear factors. In fact, since $\mathbb{R}$ is a subset of $\mathbb{C}$, if we view a real polynomial like $x^2 + 1$ which doesn't factor completely over $\mathbb{R}$ as a complex polynomial, it will factor completely over $\mathbb{C}$. (In this case some of the roots will be complex numbers.)

Example 4. Factor $x^2 + 1$ into linear factors.

We can use the factorization formula

$$x^2 - a^2 = (x + a)(x - a)$$

where $a^2 = -1$ since then

$$x^2 - (a)^2 = x^2 - (-1) = x^2 + 1.$$

This shows that it is impossible to factor x^2+1 over $\mathbb{R}$, because no real number squares to -1. However, in $\mathbb{C}$, we have $i^2 = -1$ so we get

$$x^2 + 1 = (x + i)(x - i).$$

This comes up in linear algebra when we are solving $\det(A - \lambda I_n) = 0$ to find the eigenvalues of A. We know $\det(A - \lambda I_n)$ is a polynomial, so solving $\det(A - \lambda I_n) = 0$ is equivalent to factoring. If we allow ourselves to think over $\mathbb{C}$ instead of $\mathbb{R}$, we are guaranteed to be able to solve for all the eigenvalues! Even if we restrict our matrix entries and therefore our polynomial's coefficients to $\mathbb{R}$, there will still be times when our eigenvalues will be in $\mathbb{C}$ because our polynomial didn't factor completely over $\mathbb{R}$. In fact this is one of my main counterarguments to the notion that studying $\mathbb{C}$ isn't relevant to a practical mathematics course. Even if all your input data is from the observable world, i.e., $\mathbb{R}$, there will still be times when you need to consider complex numbers to solve your problem.

Now that you're hopefully convinced that it is worth it to let your scalars be complex numbers instead of real numbers, let's spend a moment thinking about that situation. We can still define vectors and matrices, except that their entries are now complex numbers instead of being limited to $\mathbb{R}$. The set of n-vectors with complex entries is called $\mathbb{C}^n$, which generalizes our notation for $\mathbb{R}^n$ as the set of n-vectors with real number entries. More general vector spaces with $\mathbb{C}$ as their set of scalars, for example, the set of $m \times n$ complex matrices, are usually called complex vector spaces, while the vector spaces we discuss in the body of this book are called real vector spaces.

The good news is that other than smoother sailing when we find eigenvalues, not much is different structurally. The complex numbers share all the good properties of $\mathbb{R}$ that enabled us to construct vector spaces: $\mathbb{C}$ has addition, subtraction, multiplication, and division with all the good properties of $\mathbb{R}$'s similar operations. In fact, $\mathbb{R}$ and $\mathbb{C}$ are both examples of an algebraic object called a field, and any field can be used as the scalars for a family of vector spaces which behave in the same way real vector spaces do.

A.2 Mathematica

The vast majority of calculations from this book have been implemented in Mathematica. Of course, this is also true of many graphing calculators and other mathematical software packages. I've chosen to discuss Mathematica over other software packages because it is the software package available at the school where I teach, and I've chosen not to discuss graphing calculators because there are so many different makes and models that having a unified discussion is too difficult. If you prefer to use another software package or a graphing calculator, my advice is to google the linear algebra computation and the name of your technology of choice. In general, the resulting resources will be fairly good and accurate.

I'll start our tour of Mathematica by saying that its help section is usually excellent. In my version, this is accessed through the Help menu under Wolfram Documentation where you can then search by topic. Usually examples are provided to walk you through a sample problem. However, a little bit of knowledge starting out can help you avoid some of my past frustrations with this program.

Before we get into linear algebra topics, here are a few basic tips for using Mathematica.

- Mathematica documents are organized into cells which are each evaluated as one unit. You can tell where each cell starts and ends by looking at the right edge of the document where a bracket will indicate what is in that cell. To leave a cell and start a new cell, simply click below that cell. Just start typing to default to a Wolfram Language Input which is the basic mode for executing mathematical commands, or click on the tab with the + at the left of the page to choose which type of new cell you want. I often use Plain Text cells for times when I'm trying to write up my work nicely and want to access the formatting options we commonly associate with a program like Microsoft Word.

- All commands start with a capital letter. (Not knowing this made me cry as a student, but hopefully I've spared you!)

- All inputs to commands are put in square brackets rather than parentheses, i.e., Factor$[x^2 - 1]$ instead of Factor$(x^2 - 1)$.

- To execute the contents of a cell, press Shift and Enter at the same time. The output(s) will appear in another cell below the one you executed. Just pressing Enter will simply take you down to the next line without executing anything. If you don't want your commands to execute all at once, put each of them in a separate cell. If you want to execute a command without seeing its output, put a semicolon (;) after that command.

- You can copy and paste commands either within the same cell or into a new cell. This is very useful when you're using the same type of command several times. You can also copy and paste an entire cell, which is especially useful if you've customized that cell's format. To select a cell, click on its bracket on the right-hand side of the document.

- While you can type in most things manually, many of the most common commands can be entered using buttons on the Classroom Assistant which is accessed through the Palettes menu. Of particular use are the Calculator and Basic Commands sections.

  Now that you've had a whirlwind tour of the basics of Mathematica, let's discuss some topics of particular relevance to our study of linear algebra.

- One of the first things you'll want to do if you're using Mathematica for linear algebra computations is to enter vectors and matrices.

  Mathematica doesn't care whether you write your vector as a column of numbers (which we've been doing) or as a row of numbers. Either way, Mathematica treats a vector as an ordered list of numbers. This means the easiest way to enter a vector is as a list. Mathematica denotes lists using curly braces to contain the list and commas to separate the entries as in {1,2,3}.

  The default format used by Mathematica is to write a matrix as a "list of lists." Mathematica treats each row of the matrix as a list and then lists those row lists in order as in {{1,2,3},{3,2,1}}. Each row is listed from left to right and the rows are listed from top to bottom, basically mirroring the way English is read. This means that the 3 × 3 identity matrix would be written {{1,0,0},{0,1,0},{0,0,1}}. Personally I find this presentation of a matrix less digestible than our "grid of numbers" approach, but it is important to understand what it means if Mathematica gives you something in this format.

  If you prefer to enter your matrices as grids of numbers, you can use the piece of the Classroom Assistant devoted to matrices. This is under the Basic Commands section and is accessed by clicking on the button which is highlighted blue in the picture below.

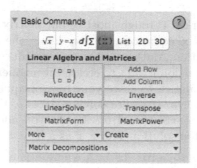

To create a matrix, click on the button that looks like a 2×2 matrix of squares directly under the word Algebra in the previous picture. The default size is 2×2, but you can add rows and columns using the buttons to the right of the matrix button.

Now that you've created the template for your matrix, you can fill in its entries by clicking on one of the boxes in the matrix and entering the appropriate number. You can get from one box to the next by hitting Tab, which will take you through the matrix by going across each row from left to right and then down to the left of the row below (again as you would read English).

To make Mathematica give its outputs in our usual format instead of a list of lists, use the command `MatrixForm`. Simply add it as the outermost command when you do a matrix computation as in `MatrixForm[other commands]`.

- Now that you can enter vectors and matrices, let's turn to getting Mathematica to do our matrix operations. (Remember that you may want to layer on the MatrixForm command.) Matrix addition is pretty straightforward, simply use $+$ between your matrices to add them. To multiply a matrix A by a scalar r, you can type rA or $r * A$.

Multiplication of matrices and vectors or matrices and other matrices is slightly trickier because the natural impulse is to use $*$, but DO NOT do that! Using $*$ causes Mathematica to multiply corresponding entries as if we'd used the format for matrix addition in our multiplication. This is not proper matrix multiplication, so avoid using $*$ at all costs! Instead, use $\cdot$ as your multiplication symbol. This can be used to multiply a matrix by a vector or a matrix by another matrix. When multiplying a matrix by a vector, Mathematica allows you to enter that vector as a row using the list format instead of a column and still gives the correct answer.

You can ask Mathematica to find the inverse of a matrix A using the command `Inverse[A]`. Again you may want to add on the MatrixForm command.

Mathematica will also find the determinant of a matrix A via the command `Det[A]`.

- Now that we've worked through most of our matrix operations, let's move on to solving matrix equations. If you know the augmented coefficient matrix of your equation, Mathematica can row reduce it using the command `RowReduce[ ]`. If you don't want to create the augmented coefficient matrix, you can ask Mathematica to solve your matrix equation directly using the command `LinearSolve`. If you're solving $A\vec{x} = \vec{b}$, then your Mathematica input would be `LinearSolve[A, b⃗]`. Unless you add `MatrixForm`, Mathematica will return the solution vector $\vec{x}$ as a list.

Sometimes we want to solve $A\vec{x} = \vec{b}$ where $\vec{b}$ is a vector of variables, for example when finding the range of a linear function. If A is $m \times n$ with $m \leq n$, then `RowReduce` handles this just fine. However, if $m > n$, then `RowReduce` will give numeric results that aren't useful. In this case, you can use the more complicated command `Reduce`. To solve $A\vec{x} = \vec{b}$ where the entries of $\vec{x}$ are $x_1, \ldots, x_n$ and the entries of $\vec{b}$ are $b_1, \ldots, b_m$, your Mathematica input would be

`Reduce[A.{`$x_1, \ldots, x_n$`}=={`$b_1, \ldots, b_m$`},{`$x_1, \ldots, x_n$`}]`

The output will list any conditions on the entries of $\vec{b}$ needed to ensure that there is a solution, and then the values of $x_1, \ldots, x_n$ in terms of $b_1, \ldots, b_m$.

In[5]:= `Reduce`$\left[\begin{pmatrix} 1 & 2 \\ -1 & 5 \\ 0 & -1 \end{pmatrix} . \{x_1, x_2\} == \{b_1, b_2, b_3\}, \{x_1, x_2\} \right]$

Out[5]= $b_1 == -b_2 - 7\,b_3$ `&&` $x_1 == -b_2 - 5\,b_3$ `&&` $x_2 == -b_3$

In the example above, we need $b_1 = -b_2 - 7b_3$ in order to have the solution $x_1 = -b_2 - 5b_3$ and $x_2 = -b_3$.

If no condition on the entries of $\vec{b}$ is needed to ensure a solution, then Mathematica will simply return the solution itself.

In[6]:= `Reduce`$\left[\begin{pmatrix} 1 & 2 & 3 \\ 0 & 1 & 0 \\ 1 & 0 & 1 \end{pmatrix} . \{x_1, x_2, x_3\} == \{b_1, b_2, b_3\}, \{x_1, x_2, x_3\} \right]$

Out[6]= $x_1 == -\dfrac{b_1}{2} + b_2 + \dfrac{3\,b_3}{2}$ `&&` $x_2 == b_2$ `&&` $x_3 == \dfrac{b_1}{2} - b_2 - \dfrac{b_3}{2}$

In the example above, there is no condition involving only $b_1, \ldots, b_m$, so we have the solution $x_1 = -\frac{b_1}{2} + b_2 + 3\frac{b_3}{2}$, $x_2 = b_2$, and $x_3 = \frac{b_1}{2} - b_2 - \frac{b_3}{2}$ for any value of $\vec{b}$.

- Mathematica has several commands to help you find eigenvalues and eigenvectors depending on which information you want and how you prefer to view it.

The command `Eigenvalues[A]` gives you all the eigenvalues of your matrix A as a list. Note that these values are the roots of a polynomial, and therefore might be complicated in their exact formats. For this reason, it is sometimes easier to use the command `Eigenvalues[N[A]]` which gives you a list of decimal representations of the eigenvalues.

The command `Eigenvectors[A]` gives you a list of linearly independent eigenvectors of A as a list of lists. This corresponds to our compilation of basis vectors from each of our eigenspaces. For an $n \times n$ matrix A,

Mathematica always returns n vectors in this list. As we've seen, sometimes an $n \times n$ matrix A will not have n linearly independent eigenvectors. In this case, Mathematica will simply add enough zero vectors to the end of the list to make the list contain n vectors.

If you want to find both the eigenvalues and eigenvectors at once, then you can use the command **Eigensystem[A]**. The formatting of the output can initially be a little confusing, so I will discuss it in slightly more detail. The overall format of Mathematica's output is a list with two entries. The first entry is a list of all the eigenvalues of A and the second entry is a list of eigenvectors. The eigenvalues and eigenvectors are listed in corresponding order, i.e., the first eigenvector listed has the first eigenvalue listed and so on.

In[1]:= **Eigensystem**$\left[\left(\begin{smallmatrix} 1 & 2 \\ 4 & 3 \end{smallmatrix}\right)\right]$

Out[1]= {{5, -1}, {{1, 2}, {-1, 1}}}

In the example above, the matrix has eigenvalues 5 and -1 and eigenspaces $E_5 = \text{Span}\left\{ \begin{bmatrix} 1 \\ 2 \end{bmatrix} \right\}$ and $E_{-1} = \text{Span}\left\{ \begin{bmatrix} -1 \\ 1 \end{bmatrix} \right\}$.

An eigenvalue which has multiplicity greater than one, i.e., is a multiple root of $\det(A - \lambda I_n)$, will be listed more than once. In fact, it will be listed as many times as it is a root of $\det(A - \lambda I_n)$. If the eigenspace for the multiple eigenvalue has dimension large enough to match the multiplicity of the eigenvalue, then Mathematica will list the basis eigenvectors of that eigenspace.

In[3]:= **Eigensystem**$\left[\left(\begin{smallmatrix} 5 & 0 \\ 0 & 5 \end{smallmatrix}\right)\right]$

Out[3]= {{5, 5}, {{0, 1}, {1, 0}}}

In the previous example, the matrix has the eigenvalue 5, which is a double root of $\det(A - \lambda I_n)$, and eigenspace $E_5 = \text{Span}\left\{ \begin{bmatrix} 0 \\ 1 \end{bmatrix}, \begin{bmatrix} 1 \\ 0 \end{bmatrix} \right\}$.

If the eigenspace for the multiple eigenvalue doesn't have enough basis vectors, Mathematica will put in zero vectors as placeholders.

In[4]:= **Eigensystem**$\left[\left(\begin{smallmatrix} 5 & 1 \\ 0 & 5 \end{smallmatrix}\right)\right]$

Out[4]= {{5, 5}, {{1, 0}, {0, 0}}}

In the previous example, the matrix has the eigenvalue 5, which is a double root of $\det(A - \lambda I_n)$, and eigenspace $E_5 = \text{Span}\left\{ \begin{bmatrix} 1 \\ 0 \end{bmatrix} \right\}$.

If you're interested in computing a basis for the eigenspace of a particular eigenvalue, you can certainly use `Eigensystem` and then pick out the nonzero vectors corresponding to your eigenvalue of interest. However, if your matrix is large enough, it may be easier to use the command `NullSpace[A − λI_n]` to find a basis for the null space of $A − \lambda I_n$ since E_λ is the null space of $A − \lambda I_n$.

- To find the dot product of two vectors, simply use the same · symbol that we used for matrix multiplication. When used between two vectors it computes the dot product.

- To compute the norm of a vector, use the command `Norm[ ]`. This will give you the exact value of the norm, which will typically involve a square root. If you'd prefer a decimal approximation, an easy fix is to put a decimal point into one of your matrix entries. This will cause Mathematica to give you a numeric approximation of your answer. (In fact, you can use this trick inside most commands to get a decimal version of your answer.)

- Mathematica will also normalize a given vector, i.e., find a vector in the same direction of your input whose length is 1. To do this, use the command `Normalize[ ]`. As with the `Norm` command, the default output will be exact and so usually involve square roots in the denominators of your matrix entries. Use the "add a decimal point" trick to get decimal entries in your normalized vector.

- Mathematica computes the orthogonal projections of a vector onto another vector using the command `Projection`. If you want to compute the orthogonal projection of $\vec{x}$ onto $\vec{y}$, you'd use the Mathematica input `Projection[$\vec{x}$,$\vec{y}$]`. Remember, it matters for our computation of orthogonal projections which vector is being projected and which is having the other vector projected onto it. This means it also matters which order you enter your two vectors into the `Projection` command. I remember the order by inserting the word onto between my two vectors, i.e., `Projection[$\vec{x}$, $\vec{y}$]` would be read "projection of $\vec{x}$ onto $\vec{y}$."

- Mathematica runs the Gram-Schmidt process we used to create an orthonormal basis, via the command `Orthogonalize[ ]` and its input is the list of vectors you want to use as inputs to the Gram-Schmidt process. (In our discussion of this algorithm in 5.4, these were the vectors $\vec{v}_i$.) If these input vectors are a basis for $\mathbb{R}^n$, then the output is an orthonormal basis for $\mathbb{R}^n$. However, Mathematica will accept any list of vectors as the input for `Orthogonalize`, and its output in this more general case will be an orthonormal basis for the span of the input vectors.

Of course, Mathematica has many capabilities not discussed in this appendix. For those uses, I encourage you to take advantage of both Mathematica's own help section and the universal solution method of googling what you want to understand.

A.3   Solutions to Odd Exercises

1.1:

1. 4

3. $\begin{bmatrix} -2 \\ -4 \\ -6 \end{bmatrix}$

5. $\begin{bmatrix} -12 \\ -4 \end{bmatrix}$

7. $\begin{bmatrix} 12 \\ -8 \end{bmatrix}$

9. $\begin{bmatrix} -3 \\ 0 \\ 30 \\ -9 \end{bmatrix}$

11. (a) impossible, (b) $\begin{bmatrix} 9 \\ 1 \end{bmatrix}$, (c) impossible

13. (a) impossible, (b) $\begin{bmatrix} -2 \\ 2 \\ 9 \end{bmatrix}$

15. (a) impossible, (b) impossible, (c) $\begin{bmatrix} 14 \\ 3 \end{bmatrix}$

17. $\vec{v}_2 + \vec{v}_3 = \begin{bmatrix} 6 \\ -3 \\ -4 \end{bmatrix}$.

19.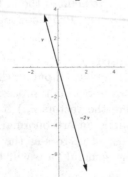

21. (a) $\begin{bmatrix} C \\ H \\ O \end{bmatrix} = \begin{bmatrix} 6 \\ 12 \\ 6 \end{bmatrix}$, (b) $\begin{bmatrix} South \\ West \end{bmatrix} = \begin{bmatrix} 2 \\ 7 \end{bmatrix}$, (c) $\begin{bmatrix} height \\ weight \\ temp. \\ age \end{bmatrix} = \begin{bmatrix} 68.5 \\ 158 \\ 98.7 \\ 38 \end{bmatrix}$

23. $\begin{bmatrix} 0.75 \\ 1 \\ 1.625 \\ 0.125 \end{bmatrix}$ where the entries are in gallons and the vector has the

format $\begin{bmatrix} red \\ blue \\ white \\ yellow \end{bmatrix}$

25. Rotates 180° and doubles distance from the origin.

1.2:

1. $\begin{bmatrix} -7 \\ -11 \end{bmatrix}$

3. $\begin{bmatrix} 2x_1 + 3x_3 + 5x_4 \\ x_1 - 4x_2 - x_3 \end{bmatrix}$

5. 2

7. $x_1 \begin{bmatrix} 1 \\ 0 \\ 0 \end{bmatrix} + x_2 \begin{bmatrix} 0 \\ 2 \\ 0 \end{bmatrix} + x_3 \begin{bmatrix} 0 \\ 2 \\ 1 \end{bmatrix} = x_4 \begin{bmatrix} 1 \\ 3 \\ 3 \end{bmatrix}$

9. No

11. No

13. The line $y = -\frac{1}{2}x$.

15. The xy plane in $\mathbb{R}^3$.

17. No

19. No

21. No

23. Smallest is $\{\vec{0}\}$, largest is $\mathbb{R}^n$.

1.3:

1. linearly dependent

3. linearly independent

5. linearly dependent

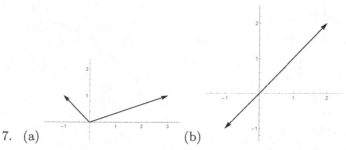

7. (a)   (b)

9. (a) linearly independent, (b) dim $= 3$

11. dim $= 1$

13. Their span has dim ≤ 2 while $\dim(\mathbb{R}^3) = 3$.

15. They could span

17. Need four linearly independent 4-vectors to get dim $= 4$.

2.1:

1. $\begin{bmatrix} 5 \\ 2 \end{bmatrix}$

3. $\begin{bmatrix} 0 \\ 8 \\ 2 \\ 6 \end{bmatrix}$

5. Domain is $\mathbb{R}^2$, codomain is $\mathbb{R}^3$

7. Domain is $\mathbb{R}^5$, codomain is $\mathbb{R}^4$

9. $f\left(\begin{bmatrix} x_1 \\ x_2 \end{bmatrix}\right) = \begin{bmatrix} x_2 \\ x_1 \end{bmatrix}$

11. $f\left(2\begin{bmatrix} x_1 \\ x_2 \end{bmatrix}\right) = \begin{bmatrix} 4x_1 - 2x_2 \\ 2x_1 + 2 \end{bmatrix} \neq \begin{bmatrix} 4x_1 - 2x_2 \\ 2x_1 + 4 \end{bmatrix} = 2f\left(\begin{bmatrix} x_1 \\ x_2 \end{bmatrix}\right)$

13. $f\left(\begin{bmatrix} x_1 \\ x_2 \end{bmatrix} + \begin{bmatrix} y_1 \\ y_2 \end{bmatrix}\right) = \begin{bmatrix} x_2 + y_2 \\ x_1 + x_2 + y_1 + y_2 \end{bmatrix} = f\left(\begin{bmatrix} x_1 \\ x_2 \end{bmatrix}\right) + f\left(\begin{bmatrix} y_1 \\ y_2 \end{bmatrix}\right)$

    and $f\left(r\begin{bmatrix} x_1 \\ x_2 \end{bmatrix}\right) = \begin{bmatrix} rx_2 \\ rx_1 + rx_2 \end{bmatrix} = rf\left(\begin{bmatrix} x_1 \\ x_2 \end{bmatrix}\right)$

15. $f(\vec{0}_n) = f(\vec{v} - \vec{v}) = f(\vec{v}) - f(\vec{v}) = \vec{0}_m$

17. Show $f(2\vec{v}) \neq 2f(\vec{v})$.

2.2:

1. (a) 6-vector, (b) 4-vector

3. (a) 4-vector, (b) 5-vector

5. (a) $\mathbb{R}^4$, (b) $\mathbb{R}^2$

7. (a) $\mathbb{R}^2$, (b) $\mathbb{R}^3$

9. (a) $\mathbb{R}^3$, (b) $\mathbb{R}^2$, (c) 2×3

11. (a) $\mathbb{R}^4$, (b) $\mathbb{R}^3$, (c) 3×4

13. $\begin{bmatrix} 2 & -1 & 1 \\ -1 & 0 & 10 \end{bmatrix}$

15. $\begin{bmatrix} 1 & 1 & 0 \\ 2 & 1 & 0 \\ 0 & 2 & -1 \end{bmatrix}$

17. $\begin{bmatrix} 3 & 0 & -7 & 1 \\ 1 & 0 & 1 & 1 \\ -1 & 0 & 8 & 0 \end{bmatrix}$

19. $\begin{bmatrix} 0 & 1 \\ -1 & 0 \end{bmatrix}$

21. $\begin{bmatrix} 0 & -1 \\ -2 & 0 \end{bmatrix}$

23. $\begin{bmatrix} 16 \\ 5 \end{bmatrix}$

25. (a) $\begin{bmatrix} 11 \\ -9 \end{bmatrix}$, (b) impossible

27. $A\vec{v}_1 = \begin{bmatrix} 11 \\ 5 \end{bmatrix}$

29. $A\vec{v}_1 = \begin{bmatrix} -7 \\ 20 \end{bmatrix}$

31. $\begin{bmatrix} 2 & 4 & -2 & 4 \\ 1 & 2 & -1 & 0 \end{bmatrix} \vec{x} = \begin{bmatrix} -1 \\ 9 \end{bmatrix}$

33. (a) $x_1 \begin{bmatrix} -10 \\ 1 \\ 7 \end{bmatrix} + x_2 \begin{bmatrix} 3 \\ 0 \\ -2 \end{bmatrix} = \begin{bmatrix} 8 \\ -2 \\ 3 \end{bmatrix}$

35. $f_\theta \left(\begin{bmatrix} 1 \\ 0 \end{bmatrix} \right) = \begin{bmatrix} \cos(\theta) \\ \sin(\theta) \end{bmatrix}$ and $f_\theta \left(\begin{bmatrix} 0 \\ 1 \end{bmatrix} \right) = \begin{bmatrix} -\sin(\theta) \\ \cos(\theta) \end{bmatrix}$, (b) $\begin{bmatrix} 10 \\ 0 \end{bmatrix}$, $\begin{bmatrix} 5\sqrt{3} \\ 5 \end{bmatrix}$, $\begin{bmatrix} 5 \\ 5\sqrt{3} \end{bmatrix}$, $\begin{bmatrix} 0 \\ 10 \end{bmatrix}$, $\begin{bmatrix} -5 \\ 5\sqrt{3} \end{bmatrix}$, $\begin{bmatrix} -5\sqrt{3} \\ 5 \end{bmatrix}$, $\begin{bmatrix} -10 \\ 0 \end{bmatrix}$, $\begin{bmatrix} -5\sqrt{3} \\ -5 \end{bmatrix}$, $\begin{bmatrix} -5 \\ -5\sqrt{3} \end{bmatrix}$, $\begin{bmatrix} 0 \\ -10 \end{bmatrix}$, $\begin{bmatrix} 5 \\ -5\sqrt{3} \end{bmatrix}$, $\begin{bmatrix} 5\sqrt{3} \\ -5 \end{bmatrix}$

2.3:

1. $\begin{bmatrix} 3 & -3 \\ 6 & 0 \\ 9 & -6 \end{bmatrix}$

3. $\begin{bmatrix} 1 & 2 \\ -3 & 0 \\ \frac{1}{2} & 4 \end{bmatrix}$

5. $\begin{bmatrix} -6 & -14 \\ -2 & 2 \end{bmatrix}$

7. $\begin{bmatrix} 8 & 8 \\ 4 & -16 \end{bmatrix}$

9. $\begin{bmatrix} 1 & 2 \\ 2 & 4 \end{bmatrix}$

11. $A + C = \begin{bmatrix} 2 & 4 \\ -2 & 0 \\ 4 & 8 \end{bmatrix}$

13. $A + C = \begin{bmatrix} 4 & 3 & 7 \\ -4 & -2 & 8 \\ 0 & 2 & 8 \end{bmatrix}$

15. $A + C = \begin{bmatrix} 2 & -2 & 7 \\ -1 & 0 & 5 \end{bmatrix}$

17. (a) $9 \times n$, (b) $m \times 5$

19. $\begin{bmatrix} 1 & 0 \\ 0 & 1 \end{bmatrix} \cdot \begin{bmatrix} a & b \\ c & d \end{bmatrix} = \begin{bmatrix} a & b \\ c & d \end{bmatrix}$, $\begin{bmatrix} a & b \\ c & d \end{bmatrix} \cdot \begin{bmatrix} 1 & 0 \\ 0 & 1 \end{bmatrix} = \begin{bmatrix} a & b \\ c & d \end{bmatrix}$

21. $\begin{bmatrix} 4 & -6 & -8 \\ 1 & 9 & 19 \end{bmatrix}$

23. $AB = \begin{bmatrix} 35 & -13 \\ -10 & 14 \\ -5 & -2 \end{bmatrix}$

25. $AB = \begin{bmatrix} 3 & 11 \\ 18 & 6 \\ -22 & -4 \end{bmatrix}$

27. $BC = \begin{bmatrix} 8 & -4 & 14 \\ -11 & 8 & 7 \end{bmatrix}$

29. $AB = \begin{bmatrix} 14 & -1 \\ 23 & 8 \end{bmatrix} \neq \begin{bmatrix} 18 & -7 \\ 9 & 4 \end{bmatrix} = BA$

31. $B = \begin{bmatrix} 1 & 1 \\ 1 & 1 \end{bmatrix}$

2.4:

1. $\begin{bmatrix} -3 & -17 & 0 \\ 5 & -2 & 8 \end{bmatrix}$

3. Use the subspace test.

5. Yes

7. Not closed under multiplication by $r < 0$.

9. 0 doesn't have an "additive" inverse

11. Show W is a subspace of M_{33}.

13. $\left\{ \begin{bmatrix} -a_1 & 4a_2 & a_1 \\ 0 & a_1 - a_2 & a_2 \end{bmatrix} \right\}$

15. Yes

15. linearly independent

17. linearly independent

19. (a) $\mathbb{R}^2$, (b) M_{22}

21. (a) M_{22}, (b) M_{23}

23. $\begin{bmatrix} 0 & 0 \\ 0 & 7 \end{bmatrix}$

25. $\begin{bmatrix} -1 & -2 & -3 \\ -1 & 0 & 4 \end{bmatrix}$

27. $f\left(r\begin{bmatrix} a & b \\ c & d \end{bmatrix}\right) \neq rf\left(\begin{bmatrix} a & b \\ c & d \end{bmatrix}\right)$

29. $f\left(\begin{bmatrix} x_1 \\ x_2 \end{bmatrix} + \begin{bmatrix} y_1 \\ y_2 \end{bmatrix}\right) = f\left(\begin{bmatrix} x_1 \\ x_2 \end{bmatrix}\right) + f\left(\begin{bmatrix} y_1 \\ y_2 \end{bmatrix}\right)$ and
$f\left(r\begin{bmatrix} x_1 \\ x_2 \end{bmatrix}\right) = rf\left(\begin{bmatrix} x_1 \\ x_2 \end{bmatrix}\right)$

2.5:

1. No

3. $\left\{ \begin{bmatrix} -2x_3 \\ x_3 - 2x_4 \\ x_3 \\ x_4 \end{bmatrix} \right\}$

5. $\left\{ \begin{bmatrix} a & 2a \\ 0 & a \end{bmatrix} \right\}$

7. $\left\{ \begin{bmatrix} a & a \\ -a & 0 \end{bmatrix} \right\}$

9. $\left\{ \begin{bmatrix} -3x_3 \\ x_3 \\ x_3 \end{bmatrix} \right\}$

11. $\left\{ \begin{bmatrix} x_2 - 2x_4 \\ x_2 \\ 3x_4 \\ x_4 \\ 0 \end{bmatrix} \right\}$

13. A has more columns than rows.

15. Yes

17. No

19. $\left\{ \begin{bmatrix} x_1 \\ x_2 \\ -12x_1 + 3x_2 \end{bmatrix} \right\}$

21. $\left\{ \begin{bmatrix} a & b \\ b+d & d \end{bmatrix} \right\}$

23. M_{22}

25. No

27. $\left\{ \begin{bmatrix} x \\ 0 \\ z \end{bmatrix} \right\}$

29. A has more rows than columns.

31. Yes

33. (a) $\begin{bmatrix} 1 & 0 & 0 \\ 0 & 0 & 1 \end{bmatrix}$, (b) $\begin{bmatrix} 1 & 0 & 0 \\ 2 & 0 & 0 \end{bmatrix}$, (c) A has more columns than rows

2.6:

1. $\left[\begin{array}{ccc|c} 3 & -4 & 0 & -5 \\ 4 & 1 & 17 & 8 \end{array} \right]$

3. $\left[\begin{array}{ccc|c} -1 & 3 & 5 & 0 \\ 2 & 4 & 6 & -6 \\ 0 & 9 & -2 & 10 \end{array} \right]$

5. $\left[\begin{array}{ccc|c} 2 & 1 & -11 & 5 \\ -5 & 0 & 3 & -1/2 \\ 10 & -9 & 2 & 0 \end{array} \right]$

7. $\left[\begin{array}{cccc|c} -4 & 0 & -1 & 0 & 7 \\ 0 & 1 & 1 & 2 & 9 \\ 1 & 0 & 0 & 5 & 13 \end{array} \right]$

9. $\left[\begin{array}{cccc|c} 2 & 4 & -2 & 4 & -1 \\ 1 & 2 & -1 & 0 & 9 \end{array} \right]$

11. $\left[\begin{array}{cc|c} 3 & -2 & 0 \\ 0 & 1 & -5 \\ -6 & 8 & 1 \\ 7 & 4 & 2 \end{array} \right]$

13. (a) $\begin{bmatrix} 0 & -1 & 4 \\ 6 & 0 & -5 \end{bmatrix} \vec{x} = \begin{bmatrix} 2 \\ 11 \end{bmatrix}$, (b) $-x_2 + 4x_3 = 2, \ 6x_1 - 5x_3 = 11$, (c)

$x_1 \begin{bmatrix} 0 \\ 6 \end{bmatrix} + x_2 \begin{bmatrix} -1 \\ 0 \end{bmatrix} + x_3 \begin{bmatrix} 4 \\ -5 \end{bmatrix} = \begin{bmatrix} 2 \\ 11 \end{bmatrix}$

15. Yes, $\begin{bmatrix} ① & -3 & 0 & 0 \\ 0 & 0 & ① & 0 \\ 0 & 0 & 0 & ① \end{bmatrix}$

17. Yes, $\begin{bmatrix} ① & 0 & 1 & 4 \\ 0 & ① & 0 & -3 \end{bmatrix}$

19. $\begin{bmatrix} 1 & -2 & 0 & 0 & 1 \\ 0 & 0 & 1 & 0 & 1 \\ 0 & 0 & 0 & 1 & 1 \\ 0 & 0 & 0 & 0 & 0 \end{bmatrix}$

21. $\begin{bmatrix} 1 & 0 & 0 & -3 \\ 0 & 1 & 0 & -\frac{5}{2} \\ 0 & 0 & 1 & -\frac{1}{2} \end{bmatrix}$

23. $\left[\begin{array}{cc|c} 1 & 0 & 0 \\ 0 & 1 & 0 \\ 0 & 0 & 1 \end{array}\right]$

25. $\begin{bmatrix} -7 \\ 2 \\ 6 \end{bmatrix}$

27. $\begin{bmatrix} 2 \\ -1 \\ 5 \end{bmatrix}$

29. $\begin{bmatrix} 5 \\ 3 \\ -1 \end{bmatrix}$

31. $\begin{bmatrix} 0 \\ 1 \\ -1 \end{bmatrix}$

2.7:

1. $\left\{ \begin{bmatrix} 5 + 3x_3 \\ -3 - 2x_3 \\ x_3 \\ 0 \end{bmatrix} \right\}$

3. $\begin{bmatrix} 11 \\ -4 \\ 3 \end{bmatrix}$

5. $\left\{ \begin{bmatrix} x_3 \\ x_3 \\ x_3 \end{bmatrix} \right\}$

7. $\left\{ \begin{bmatrix} 3x_3 \\ -2x_3 \\ x_3 \\ 0 \end{bmatrix} \right\}$

9. No

11. Yes

13. linearly independent

15. linearly independent

17. (a) linearly dependent, (b) 2

19. (a) Their reduced echelon form is $\begin{bmatrix} 1 & 0 & -4 \\ 0 & 1 & 2 \\ 0 & 0 & 0 \\ 0 & 0 & 0 \end{bmatrix}$. (b) $\vec{v}_3 = -4\vec{v}_1 + 2\vec{v}_2$,

(c) dim $= 2$

21. $\left\{ \begin{bmatrix} x_1 \\ x_2 \\ -8x_1 + 4x_2 \end{bmatrix} \right\}$

23. $\mathbb{R}^2$

25. No

27. Yes

29. $\begin{bmatrix} 1 & 0 & 0 \\ 0 & 1 & 0 \\ 0 & 0 & 1 \\ 0 & 0 & 0 \end{bmatrix}$

31. No

33. Yes

35. Yes

37. See if their reduced echelon form had a leading 1 in every row.

39. $284,639 of agriculture, $492,470 of manufacturing, $893,072 of service

2.8:

1. One unique solution

3. No solutions

5. $\begin{bmatrix} 5 \\ 3 \\ -1 \end{bmatrix}$

7. $x_1 = 4k$, $x_2 = 3k$, $x_3 = 6k$, $x_4 = 4k$ for some positive integer k

9. (a) No leading 1 in rightmost column and leading 1's in all variable's columns, (b) $\begin{bmatrix} 1 & 0 & 4 \\ 0 & 1 & -1 \\ 0 & 0 & 0 \end{bmatrix}$

11. Span $\left\{ \begin{bmatrix} 3 \\ 1 \\ 0 \\ 0 \end{bmatrix}, \begin{bmatrix} -4 \\ 0 \\ 5 \\ 1 \end{bmatrix} \right\}$

13. Span $\left\{ \begin{bmatrix} 2 \\ 1 \\ 0 \\ 0 \\ 0 \end{bmatrix}, \begin{bmatrix} -3 \\ 0 \\ 4 \\ -1 \\ 1 \end{bmatrix} \right\}$

15. Span $\left\{ \begin{bmatrix} -2 \\ 1 \\ 0 \\ 0 \end{bmatrix}, \begin{bmatrix} 3 \\ 0 \\ 1 \\ 1 \end{bmatrix} \right\}$

17. (a) Span $\left\{ \begin{bmatrix} -2 \\ 0 \\ 1 \\ 0 \end{bmatrix}, \begin{bmatrix} 1 \\ -3 \\ 0 \\ 1 \end{bmatrix} \right\}$, (b) It is the span from (a) plus any solution to $A\vec{x} = \vec{b}$.

19. The solution set of $A\vec{x} = \vec{b}$ is a shifted version of the solution set to $A\vec{x} = \vec{0}$.

21. No, because the matrix will have more columns than rows.

23. 3

25. 2

27. $n = 3$, $\mathrm{rk}(A) = 2$, $\dim(\mathrm{Nul}(A)) - 1$, and $2 + 1 = 3$

29. 2

2.9:

1. Horizontal cut between rows 1 and 2, vertical cut between columns 1 and 2

3. Horizontal cuts between rows 1 and 2 and rows 3 and 4, vertical cuts between columns 1 and 2 and columns 3 and 4

5. $\begin{bmatrix} 15 & -10 \end{bmatrix}$

7. $\begin{bmatrix} -4 & 16 \\ 8 & -12 \end{bmatrix}$

9. $\begin{bmatrix} -3 & 13 \end{bmatrix}$

11. $\begin{bmatrix} 7 & -1 \\ -4 & 2 \end{bmatrix}$

13. A: Horizontal cut between rows 2 and 3, B: vertical cut between columns 3 and 4

15. A: Horizontal cuts between rows 1 and 2 and rows 2 and 3, B: vertical cuts between columns 1 and 2 and columns 3 and 4

17. $\begin{bmatrix} 6 \\ 4 \end{bmatrix}$

19. $\begin{bmatrix} 20 & 17 \\ 25 & 9 \end{bmatrix}$

21. $\begin{bmatrix} 4 \\ -1 \\ 2 \end{bmatrix}$

23. $\begin{bmatrix} 1 \\ -2 \\ -1 \end{bmatrix}$

25. $\begin{bmatrix} 1 & 0 \\ 0 & -2 \end{bmatrix} \begin{bmatrix} a & b \\ c & d \end{bmatrix} = \begin{bmatrix} a & b \\ -2c & -2d \end{bmatrix}$

27. $\begin{bmatrix} 1 & 0 \\ 6 & 1 \end{bmatrix}$

29. $\begin{bmatrix} 0 & 0 & 1 \\ 0 & 1 & 0 \\ 1 & 0 & 0 \end{bmatrix}$

31. $\begin{bmatrix} \frac{1}{4} & 0 \\ 0 & 1 \end{bmatrix}$

33. $\begin{bmatrix} 1 & 0 & 0 \\ 0 & 0 & 1 \\ 0 & 1 & 0 \end{bmatrix}$

35. $\begin{bmatrix} -1 & 0 & 0 \\ -5 & 3 & 0 \\ 1 & 2 & -10 \end{bmatrix} \begin{bmatrix} 1 & -1 & -3 \\ 0 & 1 & 8 \\ 0 & 0 & 1 \end{bmatrix}$

37. $\begin{bmatrix} 1 & 0 & 0 \\ 0 & 2 & 0 \\ 1 & 3 & 4 \end{bmatrix} \begin{bmatrix} 1 & -2 & 4 \\ 0 & 1 & -1 \\ 0 & 0 & 1 \end{bmatrix}$

39. (a) $\begin{bmatrix} a & 0 & 0 \\ b & c & 0 \\ d & e & f \end{bmatrix} \begin{bmatrix} x & 0 & 0 \\ y & z & 0 \\ w & u & v \end{bmatrix} = \begin{bmatrix} ax & 0 & 0 \\ bx + cw & cz & 0 \\ dx + ey + fw & ez + fu & fv \end{bmatrix}$, (b)
Use the associative property of matrix multiplication.

2.10:

1. $AA^{-1} = A^{-1}A = I_2$

3. $A^{-1} = \begin{bmatrix} -18 & -3 & 5 \\ -12 & -2 & 3 \\ -5 & -1 & 1 \end{bmatrix}$

5. Not invertible

7. $A^{-1} = \begin{bmatrix} -\frac{3}{2} & -12 & -2 \\ 0 & -1 & 0 \\ 1 & 6 & 1 \end{bmatrix}$

9. Not invertible

11. $A^{-1} = \begin{bmatrix} 5 & -2 & -1 \\ 3 & -1 & -1 \\ -10 & 4 & 3 \end{bmatrix}$

13. $f^{-1}\left(\begin{bmatrix} x_1 \\ x_2 \\ x_3 \end{bmatrix}\right) = \begin{bmatrix} -x_1 + x_2 \\ x_1 - x_2 + x_3 \\ -2x_1 + 3x_2 - 2x_3 \end{bmatrix}$

15. Not invertible

17. $(A^{-1})^{-1} = A$

19. $\vec{x} = (I_3 - A)^{-1}\vec{b}$

21. $y = 22.124 - 0.274x_1 + 0.034x_2$ where x_1 is bikers and x_2 is smokers

2.11:

1. If A is invertible then $A\vec{x} = \vec{b}$ has the unique solution $\vec{x} = A^{-1}\vec{b}$.

3. If f is onto then A's reduced echelon form has a leading 1 in every row.

5. $8 \rightarrow 5 \rightarrow 6 \rightarrow 7 \rightarrow 3 \rightarrow 4$

7. 5, A has a column of zeros

11. Use a nonzero $\vec{x}$ in B's kernel to show AB's kernel isn't $\{\vec{0}\}$.

3.1:

1. No

3. Yes

5. $\begin{bmatrix} 1 & 1 \\ 1 & -1 \end{bmatrix}$ has reduced echelon form I_2

7. There are $\dim(M_{22} = 4$ matrices, and they are linearly independent.

9. Yes

11. $\begin{bmatrix} 1 \\ 0 \\ 0 \\ 0 \end{bmatrix}, \begin{bmatrix} 0 \\ 2 \\ 0 \\ 1 \end{bmatrix}, \begin{bmatrix} 4 \\ 1 \\ -1 \\ 2 \end{bmatrix}$

13. $\begin{bmatrix} 2 \\ 0 \end{bmatrix}, \begin{bmatrix} 0 \\ 1 \end{bmatrix}$

15. $\begin{bmatrix} 4 \\ 1 \\ 0 \\ 0 \\ 0 \\ 0 \end{bmatrix}, \begin{bmatrix} -1 \\ 0 \\ 2 \\ 1 \\ 0 \\ 0 \end{bmatrix}, \begin{bmatrix} 0 \\ 0 \\ 0 \\ 1 \\ -5 \\ 1 \end{bmatrix}$

17. $\begin{bmatrix} -2 \\ 3 \\ 1 \\ 0 \\ 0 \end{bmatrix}, \begin{bmatrix} -2 \\ -5 \\ 0 \\ -2 \\ 1 \end{bmatrix}$

19. Yes

21. 36

23. 2

25. $\mathbb{R}^{12}$

27. (a) $\begin{bmatrix} 5 \\ 28 \\ -2 \end{bmatrix}$, (b) $\begin{bmatrix} \frac{1}{5} \\ -2 \\ 1 \end{bmatrix}$

29. (a) $\begin{bmatrix} 5 \\ 10 \\ -14 \\ 4 \end{bmatrix}$, (b) $\begin{bmatrix} -1 \\ 1 \\ 1 \\ -2 \end{bmatrix}$

31. (a) $\begin{bmatrix} 6 & 5 \\ 4 & 8 \end{bmatrix}$, (b) $\begin{bmatrix} \frac{1}{2} \\ 3 \\ 1 \\ 3 \end{bmatrix}$

33. (a) $\begin{bmatrix} 5 & 0 \\ 5 & -5 \end{bmatrix}$, (b) $\begin{bmatrix} 4 \\ 3 \\ -1 \\ 1 \end{bmatrix}$

35. linearly dependent

37. do not span M_{22}

3.2:

1. The sum of two even numbers is even, the sum of two positive numbers is positive, but a negative scalar times a positive number is negative.

3. W is a subspace of P_2

5. There are $\dim(P_2) = 3$ polynomials, and they are linearly independent.

7. No

9. (a) $x^2 + 3x + 3$, (b) $\begin{bmatrix} 2 \\ -1 \\ 3 \end{bmatrix}$

11. (a) $8x^2 + x + 5$, (b) $\begin{bmatrix} -4 \\ 5 \\ -5 \end{bmatrix}$

13. It is in their span.

15. It is in their span.

17. linearly independent

19. linearly dependent

21. (a) M_{22}, (b) P_2 (or any P_n with $n \geq 2$)

23. (a) $\left\{ \begin{bmatrix} 0 & b & -b \\ d & e & d \end{bmatrix} \right\}$, (b) $\{ax^4 + bx^2 + c\}$

25. (a) $\{ax^2 + ax - a\}$, (b) $\left\{ \begin{bmatrix} a & b \\ a+2b & 0 \end{bmatrix} \right\}$

3.3:

1. Use the subspace test.

3. There are $\dim(\mathbb{C}) = 2$ complex numbers, and they are linearly independent.

5. No

7. (a) $-3 + 2i$, (b) $\begin{bmatrix} 2 \\ -8 \end{bmatrix}$

9. (a) $-13 + 10i$, (b) $\begin{bmatrix} \frac{26}{7} \\ -\frac{20}{7} \end{bmatrix}$

11. linearly dependent

13. linearly dependent

15. they span $\mathbb{C}$

17. they span $\mathbb{C}$

19. (a) $\mathbb{C}$, (b) M_{22}

21. (a) 0, (b) $\left\{ \begin{bmatrix} a & -a \\ 0 & b \end{bmatrix} \right\}$

23. (a) $\left\{ \begin{bmatrix} 2b & b \\ 0 & d \end{bmatrix} \right\}$, (b) $\mathbb{C}$

25. Use the subspace test.

27. Show $\mathscr{D}$ is a subspace of $\mathscr{C}$.

29. f is a linear map.

4.1:

1. $\begin{bmatrix} 81 & 0 & 0 \\ 0 & 16 & 0 \\ 0 & 0 & 49 \end{bmatrix}$

3. $\begin{bmatrix} 1 & 0 & 0 \\ 0 & 0 & 0 \\ 0 & 0 & 64 \end{bmatrix}$

5. (a) $\begin{bmatrix} 6 & 3 & 30 \\ 4 & -3 & -20 \\ 0 & -21 & 10 \end{bmatrix}$, (b) Multiply each column of A by the corresponding diagonal entry of B.

7. Use the subspace test.

9. Not an eigenvector

11. Yes, $\lambda = 3$

13. $\begin{bmatrix} 9 \\ 153 \\ 36 \end{bmatrix}$

15. $\begin{bmatrix} -128 \\ 64 \\ 64 \end{bmatrix}$

17. $\begin{bmatrix} -117 \\ -315 \\ 45 \end{bmatrix}$

19. $\begin{bmatrix} -108 \\ 54 \\ 27 \end{bmatrix}$

21. $\lambda > 1$ gives population growth, $\lambda < 1$ means population dies out

4.2:

1. -33

3. -14

5. Scales area by 3, doesn't flip the plane

7. Doesn't change area, flips the plane

9. -40

11. 16

13. $\begin{bmatrix} 2 & -1 \\ 1 & 0 \end{bmatrix}$

15. $\begin{bmatrix} -5 & 2 \\ 3 & 1 \end{bmatrix}$

17. Multiplies it by -1

19. Multiplies it by -4

21. 22

23. -160

25. Invertible

27. Invertible

29. (a) -18, (b) $\frac{1}{3}$

31. (a) -24, (b) $\frac{1}{12}$

33. 48

35. $a(-1)^2(d) + b(-1)^3(c) = ad - bc$

4.3:

1. $\lambda = 4$ one time, $\lambda = -1$ two times, $\lambda = \frac{1}{2}$ one time

3. $\lambda = -2$ one time, $\lambda = 0$ one time, $\lambda = 3$ one time

5. $\lambda = -4$, $\lambda = -1$, $\lambda = 0$

7. $\lambda = -9$, $\lambda = 7$, $\lambda = 0$

9. $E_{-2} = \mathrm{Span}\left\{ \begin{bmatrix} 0 \\ 1 \\ 0 \end{bmatrix} \right\}$

11. $E_3 = \mathrm{Span}\left\{ \begin{bmatrix} -3 \\ 2 \\ 1 \end{bmatrix} \right\}$

13. $\lambda = -9$, $E_{-9} = \mathrm{Span}\left\{ \begin{bmatrix} 89 \\ -82 \\ 66 \end{bmatrix} \right\}$, $\lambda = -3$, $E_{-3} = \mathrm{Span}\left\{ \begin{bmatrix} -1 \\ 2 \\ 0 \end{bmatrix} \right\}$,

   $\lambda = 2$, $E_2 = \mathrm{Span}\left\{ \begin{bmatrix} 2 \\ 1 \\ 0 \end{bmatrix} \right\}$

15. $\lambda = 9$, $E_9 = \mathrm{Span}\left\{ \begin{bmatrix} -13 \\ 35 \\ 5 \end{bmatrix} \right\}$, $\lambda = -6$, $E_{-6} = \mathrm{Span}\left\{ \begin{bmatrix} 1 \\ 0 \\ 0 \end{bmatrix} \right\}$, $\lambda = 3$,

   $E_2 = \mathrm{Span}\left\{ \begin{bmatrix} -1 \\ -1 \\ 1 \end{bmatrix} \right\}$

17. $A = \begin{bmatrix} x & 1 \\ 1 & x \end{bmatrix}$, $x = \pm 1$

19. 2

4.4:

1. 3

3. 2

5. No

7. Yes

9. Yes

11. Yes

13. Yes

15. No

4.5:

1. $\begin{bmatrix} -18 \\ 28 \end{bmatrix}$

3. $\begin{bmatrix} 9 \\ 2 \\ -1 \end{bmatrix}$

5. $\begin{bmatrix} 3 \\ -7 \end{bmatrix}$

7. $\begin{bmatrix} -7 \\ -9 \\ -14 \end{bmatrix}$

9. $\begin{bmatrix} -1 & 5 & 2 \\ 3 & -1 & 1 \\ 4 & 0 & -3 \end{bmatrix}$

11. $\begin{bmatrix} -1 & 5 & 3 & 2 \\ 1 & 0 & 0 & 3 \\ 0 & -2 & -1 & -1 \\ 2 & 4 & 2 & -1 \end{bmatrix}$

13. $\begin{bmatrix} 1 & -4 & 3 \\ -1 & 0 & 1 \\ 1 & 2 & -1 \end{bmatrix}$

15. $\begin{bmatrix} 3 & 1 & 5 \\ 7 & 0 & 9 \\ -1 & -3 & 1 \end{bmatrix}$

17. $\begin{bmatrix} 8 & -11 & 4 & -2 \\ 0 & 6 & -5 & 3 \\ 9 & 3 & -1 & 4 \\ 1 & -3 & 2 & 11 \end{bmatrix}$

19. $\begin{bmatrix} -\frac{1}{4} & -\frac{3}{4} & -\frac{3}{4} \\ -\frac{1}{2} & -\frac{1}{2} & -\frac{1}{2} \\ \frac{1}{2} & \frac{3}{2} & \frac{3}{2} \end{bmatrix}$

21. $\begin{bmatrix} -\frac{1}{6} & \frac{1}{3} & -\frac{1}{2} & \frac{1}{2} \\ \frac{1}{2} & 0 & \frac{1}{2} & -\frac{1}{2} \\ -\frac{1}{2} & 1 & -\frac{1}{2} & 1 \\ \frac{1}{6} & -\frac{1}{3} & \frac{1}{2} & \frac{1}{2} \end{bmatrix}$

23. $P = \begin{bmatrix} 4 & -1 & 3 \\ 1 & 0 & -4 \\ 0 & 6 & 2 \end{bmatrix}$, $D = \begin{bmatrix} -7 & 0 & 0 \\ 0 & 2 & 0 \\ 0 & 0 & -1 \end{bmatrix}$

25. $P = \begin{bmatrix} -13 & 4 & 0 & -1 \\ 10 & -1 & 3 & 1 \\ 0 & 0 & 2 & 5 \\ 1 & 2 & 0 & -3 \end{bmatrix}$, $D = \begin{bmatrix} -1 & 0 & 0 & 0 \\ 0 & -1 & 0 & 0 \\ 0 & 0 & \frac{1}{2} & 0 \\ 0 & 0 & 0 & 6 \end{bmatrix}$

27. Repeated multiplication by the same matrix of many vectors.

5.1:

1. -9

3. 0

5. 32

7. $\sqrt{21}$

9. $\sqrt{74}$

11. $\sqrt{17}$

13. $\sqrt{34}$

15. 9

17. $\begin{bmatrix} \frac{1}{\sqrt{2}} \\ 0 \\ -\frac{1}{\sqrt{2}} \end{bmatrix}$

19. $\begin{bmatrix} -\frac{2}{\sqrt{5}} \\ \frac{1}{\sqrt{5}} \\ 0 \end{bmatrix}$

5.2:

1. Yes

3. No

5. Yes

7. $\begin{bmatrix} 1 \\ 1 \\ 0 \end{bmatrix}$

9. No

11. No

13. Yes

15. No

17. $\left\{ \begin{bmatrix} -\frac{1}{5}x_3 \\ -\frac{3}{5}x_3 \\ x_3 \end{bmatrix} \right\}$

19. $\left\{ \begin{bmatrix} \frac{1}{3}x_3 - \frac{2}{3}x_4 \\ \frac{1}{2}x_3 - \frac{3}{2}x_4 \\ x_3 \\ x_4 \end{bmatrix} \right\}$

21. $\left\{ \begin{bmatrix} x_4 \\ 4x_4 \\ -3x_4 \\ x_4 \end{bmatrix} \right\}$

5.3:

1. $\vec{x} = \begin{bmatrix} -3 \\ 4 \\ -4 \end{bmatrix}, \vec{y} = \begin{bmatrix} 0 \\ 2 \\ -2 \end{bmatrix}$

3. orthogonal projection: $\begin{bmatrix} 6 \\ 1 \end{bmatrix}$, component orthogonal: $\begin{bmatrix} 2 \\ -2 \end{bmatrix}$

5. orthogonal projection: $\begin{bmatrix} 2 \\ -2 \\ 2 \end{bmatrix}$, component orthogonal: $\begin{bmatrix} 1 \\ 1 \\ 0 \end{bmatrix}$

7. Yes

9. Yes

11. orthogonal projection: $\begin{bmatrix} -1 \\ 8 \\ 1 \end{bmatrix}$, component orthogonal: $\begin{bmatrix} 4 \\ 0 \\ 4 \end{bmatrix}$

13. orthogonal projection: $\begin{bmatrix} 7 \\ -3 \\ 4 \\ 2 \end{bmatrix}$, component orthogonal: $\begin{bmatrix} 3 \\ 6 \\ 0 \\ -8 \end{bmatrix}$

15. $\sqrt{5}$

17. (a) $\begin{bmatrix} \frac{225}{\sqrt{10}} \\ \frac{675}{\sqrt{10}} \end{bmatrix}$, (b) Span $\left\{ \begin{bmatrix} 1 \\ 1 \end{bmatrix} \right\}$, (c) $90\sqrt{5}$

5.4:

1. No

3. Yes

5. $-\frac{3}{5}$

7. $\frac{11}{5}$

9. $\begin{bmatrix} -\frac{5}{2} \\ -1 \\ \frac{3}{2} \end{bmatrix}$

11. $\begin{bmatrix} \frac{1}{5} \\ \frac{1}{3} \\ \frac{1}{2} \end{bmatrix}$

13. $\left\{ \begin{bmatrix} 1 \\ 3 \end{bmatrix}, \begin{bmatrix} 6 \\ -2 \end{bmatrix} \right\}$

15. $\left\{ \begin{bmatrix} -\frac{1}{\sqrt{5}} \\ \frac{2}{\sqrt{5}} \end{bmatrix}, \begin{bmatrix} -\frac{2}{\sqrt{5}} \\ -\frac{1}{\sqrt{5}} \end{bmatrix} \right\}$

17. $\dim(W) = 2$, $\dim(W^{\perp}) = 1$, $n = 3$

19. 3

21. 6

23. The dot product of the ith and jth standard basis vectors is always 0 because their 1s occur in different entries.

Bibliography

[1] D. Hill B. Kolman. *Elementary Linear Algebra with Applications, Ninth Edition.* Pearson, 2008.

[2] H. Caswell. *Matrix Population Models, Second Edition.* Sinauer Associates, 2000.

[3] *Commutative Diagrams with TikZ.* CTAN Archive, http://ctan.math. washington.edu/tex-archive/graphics/pgf/contrib/tikz-cd/ tikz-cd-doc.pdf.

[4] J. McDonald D. Lay S. Lay. *Linear Algebra and Its Applications, Fifth Edition.* Pearson, 2016.

[5] G. Grätzer. *Math into LaTeX, Third Edition.* Brikhäuser Springer, 2000.

[6] *Introductory Mathematical Economics, Second Edition.* Oxford University Press, 2004.

[7] D. Lay. *Linear Algebra and Its Applications, Third Edition.* Pearson, 2006.

[8] A. Hadi S. Chatterjee. *Regression Analysis by Example.* 4th. Wiley, 2006.

[9] L. Spence S. Friedberg A. Insel. *Linear Algebra, Third Edition.* Prentice Hall, 1979.

[10] D. Taylor. *The Mathematics of Games.* CRC Press, 2015.

[11] *TeX-LaTeX Stack Exchange.* https://tex.stackexchange.com.

Index

Printed in the United States
By Bookmasters